# WOMEN IN THE MINES

*Stories of Life and Work*

TWAYNE'S

# ORAL HISTORY SERIES

*Donald A. Ritchie, Series Editor*

PREVIOUSLY PUBLISHED

*Rosie the Riveter Revisited: Women, the War, and Social Change*
Sherna Berger Gluck
*Witnesses to the Holocaust: An Oral History*
Rhoda Lewin
*Hill Country Teacher: Oral Histories from the One-Room School and Beyond*
Diane Manning
*The Unknown Internment: An Oral History of the Relocation of Italian Americans during World War II*
Stephen Fox
*Peacework: Oral Histories of Women Peace Activists*
Judith Porter Adams
*Grandmothers, Mothers, and Daughters: Oral Histories of Three Generations of Ethnic American Women*
Corinne Azen Krause
*Homesteading Women: An Oral History of Colorado, 1890–1950*
Julie Jones-Eddy
*The Hispanic-American Entrepreneur: An Oral History of the American Dream*
Beatrice Rodriguez Owsley
*Infantry: An Oral History of a World War II American Infantry Battalion*
Richard M. Stannard
*Between Management and Labor: Oral Histories of Arbitration*
Clara H. Friedman
*Building Hoover Dam: An Oral History of the Great Depression*
Andrew J. Dunar and Dennis McBride
*From the Old Country: An Oral History of European Migration to America*
Bruce M. Stave and John F. Sutherland with Aldo Salerno
*Married to the Foreign Service: An Oral History of the American Diplomatic Spouse*
Jewell Fenzi with Carl L. Nelson
*Her Excellency: An Oral History of American Women Ambassadors*
Ann Miller Morin
*Doing Oral History*
Donald A. Ritchie
*Head of the Class: An Oral History of African-American Achievement in Higher Education and Beyond*
Gabrielle Morris
*Children of Los Alamos: An Oral History of the Town Where the Atomic Age Began*
Katrina R. Mason
*Crossing Over: An Oral History of Refugees from Hitler's Reich*
Ruth E. Wolman
*A Stranger's Supper: An Oral History of Centenarian Women from Montenegro*
Zorka Milich

MARAT MOORE

# WOMEN IN THE MINES

## *Stories of Life and Work*



TWAYNE PUBLISHERS
*An Imprint of Simon & Schuster Macmillan*
*New York*

PRENTICE HALL INTERNATIONAL
*London Mexico City New Delhi Singapore Sydney Toronto*

Oral testimonies in *Women in the Mines* have been edited for reasons of privacy and purposes of space. Complete audiotapes and transcripts are available at Archives of Appalachia, East Tennessee State University, Johnson City, Tennessee.

*Twayne's Oral History Series No. 20.*



Twayne Publishers
An Imprint of Simon & Schuster Macmillan
1633 Broadway
New York, New York 10019

**Library of Congress Cataloging-in-Publication Data**

Moore, Marat.
Women in the mines : stories of life and work / Marat Moore.
p. cm.—(Twayne's oral history series; no. 20)
Includes index.
ISBN 0-8057-7834-9 (cloth)
1. Women coal miners—United States—History—20th century. 2. Women coal miners—United States—History—20th century—Sources.
3. Women coal miners—United States—Interviews. I. Title. II. Series.
HD6073.M6152U65 1996
331.4'822334'0973—dc20 95–45620
CIP

The paper used in this publication meets the minimum requirements of American National Standard for Information Sciences—Permanence of Paper for Printed Library Materials. ANSI Z39.48–1984.∞™

10 9 8 7 6 5 4 3 2 1 (hc)

Printed in the United States of America

Women miners in the bathhouse, Mary Lee mine, Alabama.
Photo by the author, Marat Moore.
In the text, all photos not otherwise credited are by Marat Moore.

*"Coal Mining Woman":*
*A song written by Hazel Dickens, an activist for miners' rights.*

I've got the woman coal miner blues
Just like you, I've got the right to choose
A job with decent pay, a better chance to make my way
And if you can't stand by me, don't stand in my way.

Well we had the babies, kept the home fires burning bright.
Walked the picket lines in the thickest of the fight.
Yes we helped you open doors, and we can help you open more
And if you can't stand by me, don't stand in my way.

Well I'm entitled to work a job that is free
From intimidations that are forced on me
From men who are out of line, out of step with time.
And if you can't stand by me, don't stand in my way.

Now union brothers, don't you think the time is right?
That we all stick together and unite?
Some better seeds to sow, and we'll help this union grow.
You stand by me, I'll surely stand by you.

We must work together to change the things that's wrong.
For better conditions, we've waited much too long.
Health and safety have to be a first priority,
And the change can only come through you and me.
Yes the change can only come through you and me.

*For my mother,*
*Geraldine Atkinson Moore,*
*and to the memory of my father,*
*Herbert Leslie Moore*

*And to the memory of*
*my union sister who inspired this book,*
*Linda Lee Fields Thompson*

# Contents

# Foreword

Within the past generation, women have entered many untraditional lines of work, but none that was as previously identified with masculinity as coal mining. The social prejudices and superstitions against women in the mines were so strong that in 1936 Eleanor Roosevelt had to cancel a planned tour of a working mine in Illinois when she realized the strength of local opposition; she visited an idle mine instead. While a few women worked the mines during the early decades of this century, generally the mines did not open to women until the 1970s. At that time, the first women miners encountered harassment and community opposition, in addition to the dangers that all miners risk to life, limb, and lung. Their oral histories record their aspirations, personal tragedies, and family histories and recount the relations between men and women miners. No one could tell this story better than those who went into the mines themselves.

Oral history may well be the twentieth century's substitute for the written memoir. In exchange for the immediacy of diaries or correspondence, the retrospective interview offers a dialogue between the participant and the informed interviewer. Having prepared sufficient preliminary research, interviewers can direct the discussion into areas long since "fogotten," or no longer considered of consequence. "I haven't thought about that in years" is a common response, uttered just before an interviewee begins a surprisingly detailed description of some past incident. The quality of the interview, its candidness and depth, generally will depend as much on the interviewer as the interviewee, and the confidence and rapport between the two add a special dimension to the spoken memoir.

Interviewers represent a variety of disciplines and work either as part of a collective effort or individually. Regardless of their different interests or the variety of their subjects, all interviewers have a common imperative: to collect memories while they are still available. Most oral historians feel an additional responsibility to make their interviews accessible for use beyond their own research needs. Still, important collections of vital, vibrant interviews lie scattered in archives throughout every state, undiscovered or simply not used.

Twayne's Oral History Series seeks to identify those resources and to publish selections of the best materials. The series lets people speak for themselves, from their own unique perspectives on persons, places, and events. But to be more than a babble of voices, each volume organizes its interviews around particular essays that place individuals in the larger historical context. The styles and format of individual volumes vary with the material from which they are drawn, demonstrating again the diversity of oral history and its methodology.

Whenever oral historians gather in conference, they enjoy retelling experiences about inspiring individuals they met, unexpected information they elicited, and unforgettable reminiscences that would otherwise never have been recorded. The result invariably reminds listeners of others who deserve to be interviewed, provides them with models of interviewing techniques, and inspires them to make their own contribution to the field. I trust that the oral historians in this series, as interviewers, editors, and interpreters, will have a similar effect on their readers.

DONALD A. RITCHIE
*Series Editor, Senate Historical Office*

# Acknowledgments

Oral history is collaborative by nature, and my first debt is to the women miners who have shared the struggles and victories that make up a working life. My deep respect and gratitude go to the members of the Coal Employment Project, a remarkable organization nearing its twentieth year, and in particular to Cosby Totten for more than a decade of valuable insights and friendship, and to Kipp Dawson, whose perspective on social and political issues relating to women miners has enriched this work. Other women miners who provided concrete assistance include Libby Lindsay, Linda Lester, Lisa Parnell, Charlene Griggs, Andrea Thomas, Barbara Angle, Rose Woodlee, and others too numerous to name.

At the UMWA, my thanks go to my former colleagues: the safety specialist Linda Raisovich-Parsons, Holly Syrrakos, Maier Fox, Tom Johnson, Donna Pearce Watson, and the staff of the *UMW Journal*.

The UMWA is blessed with a membership that has long struggled to improve conditions in the coalfields. For their example, my gratitude goes to the families of the Pittston coal strike and the Massey strike, and to veterans of other strikes with whom I worked in the 1980s.

The poet and writer Elisavietta Ritchie has nurtured many writers in Washington, D.C., and her writing group provided crucial support and encouragement for this project. Support also came from the Kentucky Oral History Commission, the West Virginia Humanities Council, Blue Mountain Center, and the District of Columbia Commission on the Arts and Humanities.

As with any long project, many unseen hands guided this one toward birth; special thanks go to the creative midwives Martha Loyd, Joyce Lichtenstein, Deborah Smith, L-R Berger, Dee Naquin, and Pat Arnow. My thanks go to Mary McCoy, who introduced me to my editor at Twayne; and to the oral historian Andrea Hammer, the late feminist writer Eve Merriam, Norma Myers and the staff at the Archives of Appalachia at East Tennessee

State University, and Betty Jean Hall, who founded the Coal Employment Project and has been generous in her assistance with this book. The lawyer Tony Oppegard, who has long fought for miners' safety, offered insightful feedback. Don Ritchie and Mark Zadrozny have been patient and encouraging editors, and Chester Kulesa, curator at the Anthracite Museum in Scranton, Pennsylvania, unearthed early twentieth-century fiction about women in the mines.

My mother, Geraldine Atkinson Moore, an athlete and musician in her youth, broke gender barriers by joining the WAVES in World War II. She has had the courage to pursue her dreams, an important example for any daughter. My brother Scott Moore has been an invaluable support with his lifelong wise counsel. My husband, Steve Lindner, is a former miner and elected union leader who has never wavered in his commitment to democratic trade unionism. Without his knowledge of UMWA contracts and political structure, his patience, and his love, this book might still be buried in my computer.

# Preface

On my window ledge sits a slab of gray slate, the diameter of a dinner plate and several inches thick, covered with the dark lace of fossil ferns. In an imprint of coal, the fronds overlap, sealing the moment more than 250 million years ago when a swamp forest sank and plant became stone.

Ferns, an old miner I worked with told me, are the flowers of darkness. They are what blooms in the underground. In the mine, I foraged through roof falls like a child searching out shells on a beach, hunting for signs of towering horsehair ferns or the elusive curve of a fish. Fields of ferns are embedded in the flanks of the Appalachians, while dinosaur footprints stalk the old workings of Utah mines.

In the city, my finger traces the serrated edge of a fern leaf caught forever in the act of dying. This piece of earth-memory, unlike our own, is exact in its detail. But I wonder: What kind of power would crush massive life and sustain the fragile curl of a fern?

Digging out human memory is no less mysterious a process. Memory, the great shape-shifter, can soothe, torment, blur, distort, tease, play hide-and-seek, reverberate beneath the surface, and disappear only to leap back out with startling clarity.

*In the mountain the darkness waits, watchful as a close friend, patient as death. Seamless, impenetrable, the blackness is absolute, the aura of the coal. With time, it becomes a sheltering silence, a buffer between the miner and the cares of the outside world. With time, the darkness soaks down into the blood, becoming a warm dark river calling you home.*

In this book, women coal miners describe their working lives in a hazardous industry that remained virtually all-male until 1973. Women miners fought the odds for a measure of economic freedom with a determination and blood-and-guts courage normally attributed to men. They built a women's support network and made their voices heard in the United Mine Workers of America (UMWA). Their story is part of the larger history of social upheaval created by the broad entry of women into the workforce of the United States in the last several decades.

These narratives reflect the impact of shifting sex roles on the coal mine workplace and on domestic life. As male territory, underground mines were a frontier in gender integration. While friction and male resentment triggered numerous cases of gender harassment, the underground mine also proved to be a workplace that forged new understanding between women and men. This step forward was partly due to the occupation itself. The bond of solidarity among coal miners can transcend divisions of race, culture, and gender. Mining traditions include pride of work, razor-sharp humor in the face of danger, and a legacy of militant action against repression.

Like the tradition of diaries and journal-keeping, the oral history of women can be intensely personal, illuminating the truth beneath a bloodless chronology of events. In this kind of storytelling, private life and public life are reconnected in a way that shows how personal choices shape the values of the society in which we live.

The voices of the women included here are proud, strong, angry, exhausted, bitter, impassioned, jubilant, determined, and at times buoyant with humor. But there is also silence, and the desire to forget painful realities. Several women said they could not bring themselves to listen to the audiotaped interviews. "Maybe sometime," one woman said, "but not now, and not any time soon. I'd just as soon forget."

*Can a miner ever forget? A choking breath of black dust; the prickly nervousness of watching dust trickling from the mine roof as the roof prepares to collapse; the heart-bumping blast of explosives set off in a confined space; the damp smell of earth and rock and water; the roar of machines underground; the tension at midnight on a picket line; a voice rasping, "Yeah buddy," over the mine phone; quick-snapping, rapid-fire jokes; the exhaustion of juggling a strenuous job with the work of raising children alone; the tears that come on during a workshop with other women who have fought sexual harassment; and the sudden anger and debilitating grief over the loss of a coworker or loved one in an unsafe mine. Against the hostility of the environment, other feelings emerge: the warmth of a longtime buddy; the confidence that accompanies physical strength, the satisfaction of workplace victories, and the awe of uncovering earth never before seen by human eyes.*

The work of gathering stories can weave its own tale. My journey began with a rise of floodwaters in Appalachia in the spring of 1977. The Tug River, which divides a portion of eastern Kentucky from southern West Virginia, began to rise during spring rains. Choked with strip-mine sediment, the muddy Tug flooded trailer parks, covered low-lying cemeteries, erased the railroad tracks, and filled up the black mouths of mines near the river.

As the floodwaters crept up, families whose homes were threatened packed quilts and photo albums and family Bibles into boxes and stored them on the highest closet shelves, which would have been out of reach of any floodwaters in recent memory. As swiftly as the river was coming up, it was impossible to pack and move everything of personal value. The water rose higher, shattering windows, swallowing children's beds and dressers, and knocking family photographs into its brown tide. Precious artifacts of family memory were lost in one muddy gulp. Finally the Tug claimed the West Virginia towns of Matewan and Williamson (the Mingo County seat), filling the main streets and rising against the vaulted store windows until they exploded like cannon fire.

A month later, the river had retreated to its banks, leaving a layer of oily mud that backed and billowed up in a dusty haze. In Williamson, everyone was coughing. Just graduated from college, I drove an old station wagon stuffed with clothes from east Tennessee up to Mingo County, where I joined a flood relief team shoveling muck out of houses under a broiling sun.

On the grimy porch of a flood recovery center, Linda Thompson appeared soon after the flood, cracking jokes and walking her dog Tramper, a flood refugee. She and her husband Harold, a disabled and decorated World War II veteran, had lost their home in the flood and been forced to live in their car.

Linda was a small, compactly built woman with auburn hair and an irrepressible sense of humor. She struck up conversations with people on the street and admired their pets. Almost everything about her was quick: her gestures, her speech, her jokes that spilled out as she waved a cigarette through the air. She sped down Route 49 every night and up the graveled road to pull the midnight shift at a U.S. Steel mine. The front door of her rented home announced "Day Sleeper" in peel-off letters.

As the months passed, I spent a lot of time with Linda and Harold. They had no children but adopted stray cats and dogs the way Linda had befriended me, with an open heart and quick wit. She took the role of an older sister: advising, inquiring, laughing. Harold had blue eyes and a stern, kind gaze; he moved slowly and very erectly as he walked Tramper along the railroad tracks. His military medals were displayed on the living room wall next to a small oil portrait of him in uniform. He hand-fed Tramper M&Ms and canned some of the best sauerkraut in Mingo County.

For mining families like the Thompsons in Mingo County, the flood was not the last blow of 1977. Flood victims were still struggling to find housing when wildcat strikes broke out during the summer over the issue of losing the union health card. A bitter winter followed; in early December the UMWA contract expired and a national coal strike began that dragged on until March. One night after the strike ended, Linda had offered me a tour of a low-coal mine, where the mine roof was only a yard high. Mining equipment was built to fit the cramped dimensions of the coal seam, not the workers. Miners with kneepads crawled to mine the coal face, shovel, and set timbers.

At the U.S. Steel mine where Linda worked, the coal seam was nearly six feet thick, high enough for a miner to walk upright.

The coal industry, however, enjoyed a historic boom in the mid- to late 1970s. With Linda's encouragement, I took the training course for new miners. Going around to mine offices, I was informed that no hiring was being done, although most of the young men in my class had boasted of jobs waiting for them as soon as they passed the test.

When I submitted my application to U.S. Steel, where Linda worked, she vowed to help. Soon after, the superintendent called me in for an interview. He took me outside and offered me a job, then threatened to run me out of the state if I wrote anything about the mine.

One Sunday afternoon Linda drove me down Route 49 to North Matewan to join the United Mine Workers of America, Local 8840. In a small frame building next to the railroad tracks I raised my hand and repeated the oath. "Now you're a man," Linda said, laughing.

As a general inside laborer, I shoveled coal under conveyor belts, built concrete-block stoppings, and timbered and hung fireproof canvas curtain near mine ribs or walls. Other jobs included laying cast-iron water pipe and working at the coal face. As the months passed, I developed a deep respect for older miners and the rigors of mining coal.

*On our way into the mountain, we dozed, lulled by the communal warmth and by the drummed, atonal sound of the rails. Caplights were shut off in the mantrip dark. The rail car dipped, then climbed, as it followed the coal seam the mine was built to accommodate. The air cooled as tunnel intersections, called breaks, blinked by. The coal walls, or ribs, were powdered with gray limestone dust, a safety measure that reduced the risk of a deadly methane fireball.*

*Grinding, rining steel upon steel, the mantrip moved through the darkness toward the coal face, the dead end of the mountain where crews prepared every shift for the ripping of the seam.*

*We screeched to a stop halfway to First South section, just over a mile inside. It could have been ten. We lost daylight, it seemed, hours ago.*

*"Your stop, Hung Low," the driver called to Pete Lau, a South Vietnamese refugee who had grown up in French Saigon. "Watch out for your redhat. You stick with her* all *the time. Got that?"*

*"Yeah boss," shrilled Pete with a high-pitched giggle. My face burned. With no toilets underground, miners were left to their own resources. A law enforcing the buddy system for new miners complicated the situation for women.*

*With a clatter, the mantrip moved off, gathering speed. The headlights disappeared around a bend. It was quiet and, except for the twin beams of our caplights, completely dark. From the next entry came the deep thrumming of the conveyor belt. Our job was to shovel out the coal spillage beneath it.*

*Pete lifted the metal door that led into the belt entry. I stumbled through and grabbed my hardhat as the door slammed with a massive, echoing boom. Before us stretched the beltline, the aorta of the mine that carried coal to the preparation plant near the portal, where it was loaded onto rail cars to travel to the Virginia docks and then floated to auto factories in Japan.*

*Pete picked up a wide-mouthed shovel. With a twist, he slid it under a layer of muck, pulled it up, and, grunting, heaved it onto the belt. I took the other shovel and moved on down the beltline. With each step, mud sucked at my boots and I yanked them free.*

*The air wrapped around us, heavy and warm. To reduce the risk of fire, ventilation was restricted in the belt entry, which made breathing difficult. I hoisted a shovelful of cementlike muck and heaved it onto the belt, passing at shoulder height.*

*Gradually my muscles relaxed as the rhythm settled in. My hands were slippery with black mud, and I smiled, thinking of my childhood, which was clean and orderly. We lived in a midsized town in east Tennessee where my father cared for many of the town's teeth. On his worktable lay crowns and bridges, gold inlay and porcelain, all perfect and shining and white. I don't remember the needles and drills and blood. He had a talent for not inflicting pain in the dental chair.*

*Coal dust floated around us, glittering in the path of our caplights. Bending, lifting, sweating in the dark, we worked our way down the beltline. After two hours of steady labor, my glasses were fogged, but the mine floor was scraped clean. I wiped my forehead with a dusty bandanna.*

*Pete made a circling motion with his caplight, a miner's signal.*

*"Take a break," he said. His pale face was mottled with grime. "You work too hard, you get hurt. You get hurt, I lose my helper." He grinned and his teeth flashed white against coal-smeared skin. I glanced upward.*

*"What, you scared of the roof?" His caplight played over my face, then up to the roof, seamed with cracks of stress.*

*"Just respectful, Pete."*

*He nodded. "It's good to watch. Me, I'm a great chicken. A chicken in the war. Now a chicken digging in a hole." He laughed softly and pulled out a battered thermos. Steam hung in the air as he poured. His fingers, like the features of his face, were delicate as a child's.*

*He stood up, looking at the belt's dark burden. "It's good here. No bosses. Quiet. You can sing if you want. Like in Saigon before the war. We always sang in Mass." He closed the thermos, and we moved further down the beltline.*

I left the mine in 1980 with the idea of returning to journalism, but I grew homesick for the coalfields. Oral history became a way to remain involved with mining issues. Questions persisted: How many women had worked as underground miners in the United States before their official entry in late 1973? How did their experiences compare with those of the later generation? And what would be the legacy of the 1970s women in the industry and the UMWA?

Since accounts of early women miners were sparse, I wrote to coalfield newspapers and to the cities where millions had migrated during the coal bust of the 1950s and 1960s—Columbus, Cleveland, Detroit, Chicago, Salt Lake City—and to retirement meccas in Florida and Arizona. Gradually, responses arrived in spidery handwriting from women who recalled their experiences in coal mines during the handloading era, which led to oral history interviews. Through the 1980s, as time permitted, I continued interviewing and photographing both generations of women miners.

The published interviews were selected from oral history sessions with 68 narrators (see appendix 1) and reflect an attempt to provide diversity in experience, age, race, ethnicity, class, job classification, geography, and union membership. But the selection and the interviews themselves are doubtless affected by my perspective as a white, southern Appalachian feminist who spent nearly a decade working within the structure of the UMWA.

The interviews were loosely structured: I followed a list of topics to be explored rather than a specific set of questions. The interviews took place at kitchen tables, on picket lines, in bathhouses, at a senior citizens' center, and in moving cars; and in one case in a hospital room after the striker-narrator was injured by gunfire. The sessions observed their own bounds of place and content, lasting from a single hour to multiple sessions over several days. Interviews with one person or group would lead to others. Family and friends would move close to share their perceptions. The camera, too, became an important tool, although it falls impossibly short in the effort to capture the darkness, the roar, the luking dangers, the jokes, the sweat, and the solidarity among workers in an underground mine. Some women I photographed at work and at home, while with others taking pictures was not possible.

Transcribing interviews myself was a financial necessity that yielded unexpected rewards. Some interviews required days to transcribe and edit, but I discovered that long hours of transcription force a deeper level of listening. So although I often dreaded confronting a stack of taped interviews, when the voice emerged from the tape recorder I would sit mesmerized for hours at the computer, my hands moving across the keys while I listened to the storyteller as an audience of one.

This book has taken more than ten years to complete and represents many miles and many hours of conversation. Although it was not my original intention, the length of the project allowed the narrators to reflect on the 1980s, a decade marked by mass layoffs, weakened enforcement of occupational health and safety regulations, speedups, corporate aggression against unions, and a political backlash against affirmative action.

The period was marked by personal change as well. Children grew up and left home. Marriages were celebrated or dissolved. Illness and unemployment struck. Particularly in the narratives gleaned from interviews that span the decade, some goals and beliefs remain, despite the changes.

In oral history the passage of time alters the perspective of both the narrator and the interviewer. To borrow an image from photography, some documentary work is like a snapshot, with every element focused in the visual field. This book more closely resembles a time-lapse photograph, which reveals the blur of change set against the sharp outlines of what has stayed constant in the narrators' lives.

*By the end of 1983 Linda Thompson had survived a stressful year. A layoff at the mine had hit all the women especially hard because they had less seniority than men hired earlier. Harold's health had declined further, and he died in December. His death coincided with her recall to work underground, after many years working on a tipple—the preparation plant where coal is sorted and sized before being transported by rail.*

*Linda returned to work underground with two other women, Betty Jo Snodgrass and Linda Staten. In February 1984 Staten narrowly escaped death as she operated a roof-bolting machine to install steel rods overhead. A piece of rock the size of a car hood caught her in its fall. Eight hours of surgery saved her life but left her with a steel plate in her skull and a blind eye.*

*Two months later, on Easter morning, I dialed Linda's number in West Virginia. Long phone conversations had become our occasional habit. I'm not sure why I called that day—perhaps because Easter is a family day and we were both apart from ours. As always, we talked about love, mine gossip, and the union contract. She said she was facing a new problem: sexual harassment. "I never knew how hard it was to be single," she admitted. She said her job had "saved" her emotionally after Harold died, but that the harassment and stress had taxed her.*

*"I just want to get in my ten years and get out of this place," she said. Her goal was only two years away.*

*Two days later a ringing phone interrupted a hectic morning at the* United Mine Workers Journal. *It was a friend from Mingo County. "Sit down," he cautioned. "I've got some bad news." He waited, then faltered. "There was an accident. It was Linda." He paused. "She was killed last night."*

*I went numb, then fled to a basement bathroom in the union headquarters that smelled like our mine bathhouse. With the union's support, I drove to West Virginia, stopping for shelter and comfort from my friend Cosby Totten, a woman miner who had lost relatives in mining accidents. Driving into Mingo County, I saw redbud blossoms spread like a pink mist over the cold, gray hills.*

*At the mine I picked up the form about death benefits for Linda's sister. The superintendent looked flustered. Near the portal, as the day shift was preparing to go in, a boss gave a safety talk. "We don't know what happened, boys," he said, "but it looks like human error." The investigation had barely begun, but already management was assigning blame.*

*The accident had occurred during midnight shift while Linda was using a putty knife to clean the pump compartment of a five-ton shuttle car. Linda was standing between the shuttle car and the tunnel wall. The shuttle car was energized. The automatic brake released, and the car rolled toward her, catching her pants leg and knocking her down before rolling over her. She died instantly. Afterwards, investigators cited the company for failure to maintain the shuttle car in a safe condition.*

*After the crews went in, I rode in with the union's inspector down the old main entries and turned off onto newer unfamiliar sections. Finally, we reached the accident site, where inspectors were busy measuring the distance between the shuttle car and the mine wall.*

*After leaving the portal, I drove to the funeral home, a rambling Victorian house near the town's main street. A tiny, wizened undertaker greeted me soberly, and I followed him and his young assistant down the stairs into the cool basement, where Linda's body lay. Her hands were small and delicate. I had never seen them so quiet. Faint rims of coal outlined her oval nails.*

*Williamson being a small town, the funeral director had known Linda all her life. "Her mother knew everybody. Worked for the county," he recalled. His memories of young Linda warmed that cold space. The next day, she was buried next to Harold during a driving thunderstorm.*

*At the mine, Betty Jo was the only woman still working, and she was feeling vulnerable. She called one night, saying that the men were telling her that, given the run of "bad luck," she could be the next accident victim. I drove up to the mine and picked up Linda's dusty work clothes, still hanging in the locker, and a dogtag stamped with her name, her check number, and her nickname, "Red." Miners are issued two check tags: one is attached to their mining belts to identify them in case of an accident, and the other is left outside at shift change for proof of who is underground.*

*At a conference for women miners the following month, Linda Staten walked up to me with a black patch on her eye. She asked me to join her for a jog, adding that she was trying to get her mining job back. She looked like a rakish pirate, and her face shone with energy, despite the invisible metal plate imbedded in her skull.*

Through the years, as my own life became more entwined with the oral history project, the line between being a participant and an observer began to dissolve. I wasn't mining coal, but I was on the UMWA staff and stayed active in the women miners' organization, the Coal Employment Project (CEP).

In 1991 women miners expressed interest in expanding the project during a history workshop at the Thirteenth National Conference of Women Miners in Colorado. With the support of my colleague Kipp Dawson, a veteran coal miner, researcher, and activist, we worked to build a collective, grassroots effort to document a coal mining community that extended beyond the boundaries of place.

As layoffs increased, we intensified our efforts, feeling compelled by the rapid decline of the industry, which disproportionately affects women because of the last-hired, first-fired rule of union seniority. Women fortunate enough to still be employed were nearing retirement in the 1990s. And with the economic recession, deunionization, the impact of clean-air legislation, and the growth of remote-control mining, few inexperienced miners—male or female—were being hired.

By 1992 the Coal Employment Project had laid off its staff and moved its offices to Morgantown, West Virginia. The organization's funds had declined, and a mood of uncertainty and depression had set in. No newsletters were produced; a conference was planned, but registrations were down. Some local unions that had always sent women to participate refused, citing lack of funds. Questions buzzed through the CEP network: Would this be the last national conference of women miners? Would CEP close down? No, vowed a group of activists who called other women miners to salvage the conference. For the next few years CEP struggled to survive.

As this history—our history—unfolded, I tried to keep the tape recorder whirring and the camera loaded with film. Traveling on coalfield roads, I have felt like a shuttle on an unseen loom, moving back and forth in a connection of threads that have grown steadily into the larger, more intricate weave. The history of women, unlike the traditional histories of men, cannot be told in isolation or apart from the context of relationship, which is the burden and the blessing of our tradition.

Women have always, in a sense, labored underground, beneath the surface of the world of men, providing fuel for the continuance of life. If that darkness is to be dispelled, it must be understood in symbol as well as in fact. It is in that spirit that this work is offered.

# Introduction

In late 1973 a West Virginia woman crossed a coal mine portal to become the first beneficiary of the federal affirmative action mandates that opened high-wage jobs in the coal industry to American women.[1] The entrance of more than 4,000 female workers into coal-mining jobs over the next two decades challenged deep-seated prejudice in the industry.

The 1970s generation is the largest and longest-lasting group of women to work in American mines, but it is not the first. Women toiled in mines during the nineteenth and twentieth centuries as family laborers, independent miners, operators of small mines, and wartime workers. But women's entry into the mines during the 1970s and 1980s as permanent production workers marked a significant change in policy and practice in the coal industry and in the United Mine Workers of America (UMWA). It was not a smooth transition for the virtually all-male industry or for a union historically opposed to opening its membership to women. The first women hired by unionized companies earned the same paycheck as male new hires, but they routinely faced obstacles men did not: entrenched misogynist taboos, community opposition, discrimination in training opportunities, sexual harassment, and lack of support from union officials. Often added to these stresses were the family responsibilities of raising children and maintaining a household.

Women miners developed individual and collective responses to these pressures. Individually, they chose humor, rapid-fire repartee, and sometimes protective personas to mask underlying pain. Collectively, women miners participated in the Coal Employment Project (CEP), a grassroots organization through which they built leadership, educated themselves on political and union issues, and developed a network of emotional and legal support.

Libby Lindsay, a member of Local 633 who has mined coal in West Virginia since 1976, reflected on two decades of struggle and achievement in a CEP newsletter published in December 1993:

> Has it really been 20 years? Sometimes I think, already? And sometimes I think: Is that all? Sisters, we fought some big battles to get jobs and keep them, to make the mines safer for all miners, for parental and family leave, for workplace justice and human dignity. . . .
>
> We fought sexual harassment from innuendo to peepholes, fought for bathhouses and for training on equipment. We fought for the union, and sometimes we fought the union itself. Some women were jailed for strike activity. Some women moved up and on. Some women were forced out through injury, and many through layoff. Some were killed. Some have died. . . .
>
> Those who came later . . . owe a tremendous debt to those first women. I'll bet they didn't think what they were doing was remarkable. I wonder if they realize it now.[2]

## Female Labor and Taboo: The Early Years

Written records indicate that women worked in European mines as early as the thirteenth century, and artwork depicting women miners dates to the fifteenth century. In *De Re Metallica* [*On Mining*] (1556) woodcuts by Georgius Agricola showed women employed in a variety of mining trades.[3] Women workers hauled coal, operated windlasses, and performed other heavy labor at coal mines for five centuries in a region that included Germany, Belgium, France, and Great Britain. Coal mining during the medieval and early modern periods was preindustrial labor and could be viewed as an extension of agriculture.

In 1842 an investigatory commission issued a report on children and women working underground that prompted Parliament to pass a law immediately banning them from British mines. The ban stopped women's underground labor, which involved pulling sledges or tubs along the mine floor, with the worker attached to the tub by a harness and chain. Women were still allowed to work on the surface as "pit brow lasses" who operated windlasses and hauled, unloaded, and sorted coal outside the pit mouth. The upper classes condemned the pit brow lasses as an example of degraded womanhood, the opposite of the refined Victorian "lady." In the 1880s, the pit brow lasses took center stage in a national debate on the right of women to become paid manual laborers. Most of the 5,000 pit brow lasses were forced from their jobs; only a few continued working past the turn of the century.[4]

The battle in Britain over women's exclusion from mines rippled throughout the European coalfields. In Belgium, however, the debate focused on charges by legislators and industrialists of alleged sexual license and political dissent arising from women's employment in mining. The president of Belgium's Royal Academy of Medicine spouted venom at the women workers, declaring that they were physically deformed and blaming them for bitter strikes in 1868:

> It is they who provoke the most in popular agitations, in revolts in strikes. . . . See those mannish women with their flabby breasts and wide-set haunches, . . . their insolent looks and bold gestures. It is they who . . . seize the riot banner and carry it along, singing at the head of a wild crowd, breathing forth dissension, urging men on to pillage.[5]

Belgian women miners rejected those charges and staged strikes protesting proposed bans against their employment. By the end of the 1880s, however, their numbers had declined, largely because of increasing technology and the introduction of pit ponies to transport coal.

In Asia, efforts to ban women's work in mines took longer to unfold. Japan's coal industry exploded in the first decades of the twentieth century, and in 1909 about 38,000 women worked in mines, 60 percent of them underground. A new term was coined for women miners: "atoyama" meant "in back of the mountain," the place from which women hauled coal to the surface. Pregnancy rules were enacted, nurseries were established, and some babies were even born underground.[6]

In the 1920s, Japan took the initiative on the world stage of lobbying for a ban on women's mining employment before the International Labour Office (ILO). The ILO passed a series of resolutions through the early 1930s that effectively forced women out of mining labor.[7]

As in other countries, the forced exclusion of women from mining labor was accompanied by myths and taboos. In Japan, a common modern tale holds women in the mines to be bad luck because the goddess of mining is female, and she will become jealous if another woman enters the mine—a story that doesn't acknowledge the thousands of women in past decades who were productive workers.

The historian Christina Vanja suggests that public opposition to women miners coincided with nineteenth-century industrialization, when higher wages made mining jobs more desirable for men. Cultural ideas of femininity accordingly changed, and women were no longer identified with their capacity to work. The status of working-class women fell when they proved unable to conform to the emerging feminine ideal of submissiveness and idleness.[8]

Vanja argues that economic pressures forcing women out of mining spawned folktales and taboos about women and mining, which were later enforced through prohibitory laws. The most common superstition equates the presence of a woman around a mine with bad luck and disaster. Though its origins are unclear, the belief is widespread and has been documented in Europe, Japan, the former Soviet Union, and North and South America.[9] Some folktales, however, suggested that the luck women brought to mines

Martha Netchel and her son, near Pottsville, Pennsylvania.
Drawing courtesy of Martha Netchel

was positive. A European tale told of a husband whose wife bumped her head on a rock; when he removed the rock, he round a rich vein of ore.[10]

In the United States, the origins of women in mining are unknown, but early American women, like their European and Asian sisters, labored in a preindustrial setting, most frequently as part of a family unit. Questions persist about the role of female slaves in coal production. In his 1821 will, one Virginia slave owner listed three women among slaves working at a mine he operated, but he did not list the tasks they performed.[11]

Coal development began in the late eighteenth century in the United States,

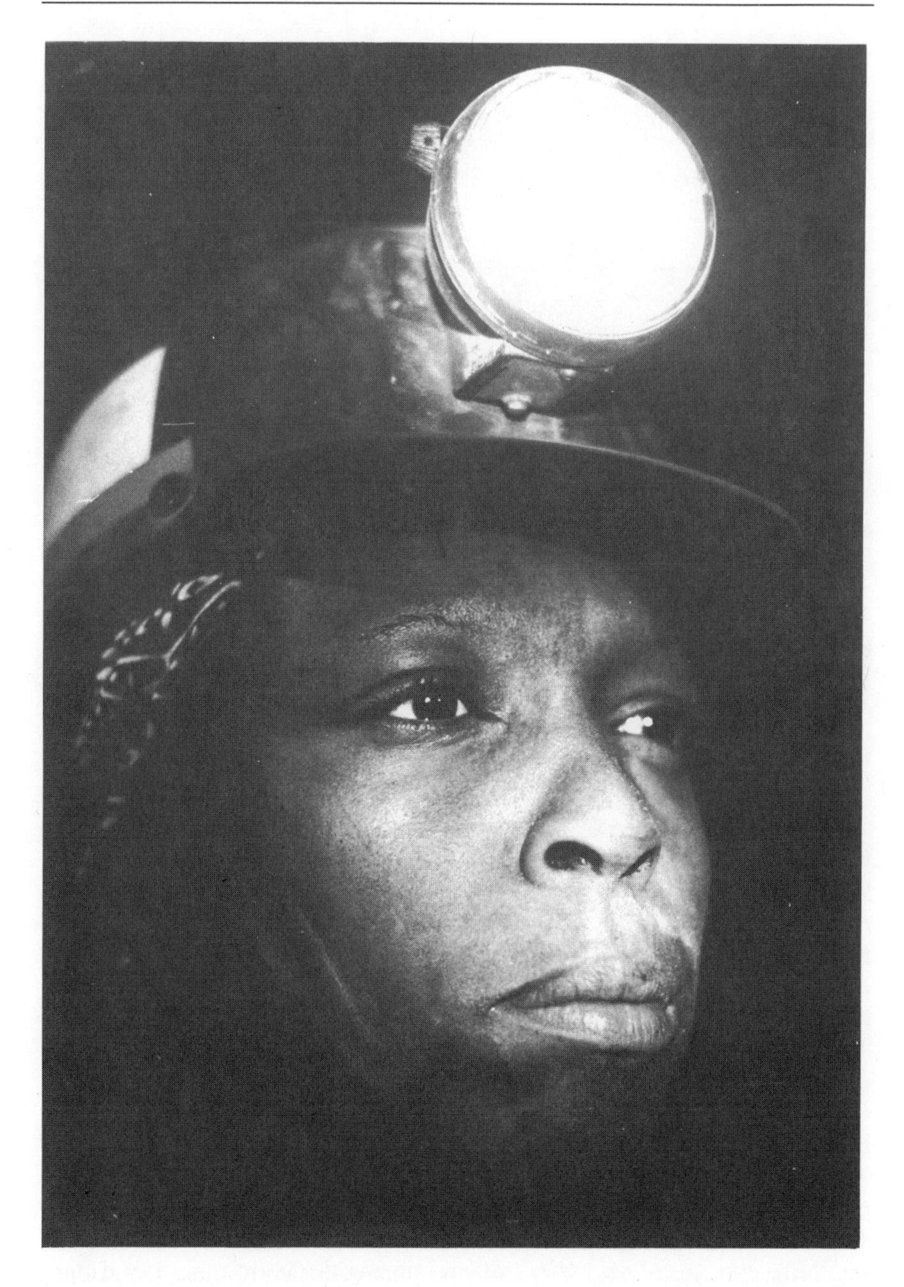

Shirley Hyche, Jim Walters Resources mine, Tuscaloosa, Alabama, 1983

Doris Magan in "low coal," 1980

and industrialization took hold less than a century later. American women miners continued to work in small mines where hand-loading techniques were used well past the turn of the twentieth century. With the wave of European immigrants to American coalfields in the nineteenth and early twentieth centuries came misogynist taboos. George Korson recorded the following folktale among Cornish miners in Pennsylvania: Long ago a race of beautiful women existed whose siren calls made men betray their families. The gods burned the sirens' forest homes into carbonized rocks and imprisoned their spirits in coal seams. A mine explosion was "a sign that more of these mythological sirens were escaping from the wall of coal, accompanied by the poisonous gases which carried death to every miner in their path."[12]

Along with taboos against women's mining labor, earlier traditions of women's ability traveled with European immigrants to the United States. In 1895 a German miner in Pennsylvania told a local reporter that he "introduced the customs of the fatherland" in utilizing the labor of his four daughters to operate the family mine. This proved profitable, enabling him to buy the mine "and a large amount of timber land besides."[13]

States began passing laws banning women's employment in mining in

UMWA black lung rally, Washington, D.C., 1981

the nineteenth century; in Pennsylvania, three laws enacted in 1885 carried penalties of $100 to $500 or a six-month jail term. By 1932, the U.S. Department of Labor Statistics showed that 17 states barred women from the mines; among them, seven had no other exclusion.[14]

Legal bans were tested but not overturned in the United States during World War I, when economic pressures in Europe reopened mines in that region to women. In the United States coal operators lobbied for women's employment in the pages of *Coal Age* after thousands of miners enlisted in the armed services. Faced with a drastic drop in anthracite production, Lehigh Coal and Navigation Company in 1918 took the bold step of hiring 42 women and girls in Nesquehoning, Pennsylvania, to work as timekeepers, weigh-scale tenders, and switch tenders, and in coal breakers.[15] The new workers were "practically all foreigners," and included recent graduates from

Nesquehoning High School.[16] The industry magazine claimed that "hundreds of girls" applied for the jobs, and the company took steps to set up rest rooms for the women workers.[17]

The hiring of the "bloomer girls"—so named because the company provided bloomer uniforms—stirred up immediate protest from the male workforce and an indignant response from the UMWA. As the news spread, miners throughout the Panther Valley region threatened to strike. The *UMW Journal* charged that the company planned to pay the women lower wages and attacked the operators' "sham patriotism," which risked "sowing the seeds of industrial discord."[18]

The president of UMWA District 7, Thomas Kennedy, cited the state law banning women's employment and asked the attorney general's office for a legal ruling. The matter was referred to the state's chief mine inspector, Seward E. Button. On 27 April 1918—four days after they were hired—Button suspended the young women and instructed the state's 25 anthracite inspectors to order an immediate halt to women's employment.[19] According to a local newspaper, the bloomer girls faced harassment even after it became clear that their presence would be short-lived. A reporter wrote that "the girls quit work on Thursday following a hard afternoon when the boys jeered and poked fun at [them]," and noted that several of the boys were arrested.[20]

Both the union and the industry laid the issues of patriotism and social morality like a thin veneer over the deeper economic struggle. The *UMW Journal* declared that the employment of women sorting coal in the windowless breakers, or coal preparation plants at the mine mouth, would subject them "to an atmosphere of immorality,"[21] but later revealed its deeper worry: that the industry was trying "to disrupt the union at the conclusion of the war." The *UMW Journal* charged that women's employment was "an insult . . . while thousands of men are idle in this country and seeking a job." The union argued on class lines that "it was the daughters of workingmen who were sought as an entering wedge to beat down the wages of labor."[22] After the dismissal of the female workers, *Coal Age* decried the "exultation of a certain element of labor over this outcome" in a front-page editorial. But the industry journal also admitted the concern underlying its interest in the fate of women workers: that the union's opposition represented "unnecessary interference with the attempts of companies to maintain their working forces."[23]

## Coal Mining and Family Labor

Despite continuing UMWA opposition, cultural taboos, and a host of state laws barring female employment in the late nineteenth and early twentieth centuries, American girls and women did work underground. They toiled almost invisibly as undocumented laborers in small family contract mines on the periphery of the burgeoning coal industry. Women worked with their fathers, brothers, and husbands, mining coal to ease their family's economic hardship. Oral history evidence suggests that the practice increased during the Great Depression, especially in Appalachia, where hillside coal seams were more easily discovered and mined.

One Ohio woman who leased her own mine challenged a state ban and won. In 1934 Ida Mae Stull defied state mine inspectors who evicted her from her mine near Cadiz and cited her as "a danger of immediate and extraordinary character."[24] Coincidentally, Stull challenged the state with the greatest legal obstacles to women seeking nontraditional jobs. Department of Labor statistics show that in 1932, Ohio restricted women's employment in 23 occupations, more than any other state.[25] She fought the eviction in court, and the ruling was overturned. Newspapers hailed her as "America's first woman coal miner," and her story appeared in the *New York Times*, the *Chicago Tribune*, and newsreels. Stull's jubilant reentry to the mine in 1935 spawned more publicity, and she declared to reporters, "I prefer coal dust to a powder puff. . . . It may sound unladylike, but every woman to her own desires, and mine is digging coal."[26] The victory was short-lived, however; inspectors later shut down Stull's mine for safety violations, and she struggled against poverty until her death in 1980.[27]

Migrating to find work was common among miners during the Depression years. Irene Adkins Dolin recalled her West Virginia childhood, when she hitchhiked and walked with her family on job-hunting treks between eastern Kentucky and southern West Virginia. When the family settled in West Virginia, Irene's father developed tuberculosis. Ten-year-old Irene and her seven-year-old sister ventured into a hillside mine to dig coal for family fuel. From 1937 until 1942, the girls dug coal with no adult supervision, in near-total darkness.[28]

## "The Female on the Tipple": World War II

The entry of the United States into World War II in 1941 opened the door for millions of women to obtain wartime employment in defense industries. The *UMW Journal* reported on employment of coalfield women in defense plants and published advertisements for women workers in shipping and other industries. But the union's tone changed abruptly when coal operators offered mining jobs to women. In late 1942 the *UMW Journal* blasted "two recent grandstand publicity plays" in western Canada and West Virginia, where women were hired as outside workers.[29]

In November 1942 Union Pacific Coal Company hired 25 women to work in mine shops and tipples in Wyoming, and Utah Fuel Company hired a group of female "boney pickers" in Carbon County, Utah. The women received equal pay with the men, and joined the UMWA.

As it had during World War I, the UMWA condemned women's wartime employment in union mines. George Titler, president of UMWA District 29 in southern West Virginia, demanded that the Algoma general manager, William Beury, immediately stop the practice of hiring women, invoking a moral standard and citing the grievous example set by Great Britain. After the UMWA threatened a strike, Beury dismissed the women.[30]

At a 1944 meeting of the UMWA's governing International Executive Board (IEB), President John L. Lewis noted with satisfaction that the situation was "promptly adjusted" in West Virginia after an IEB policy committee had voted to oppose women's employment. But he raised the specter of women's employment in Wyoming, where the number working on Union Pacific tipples had grown from a handful to nearly 100. Lewis expressed concern that a precedent on women's employment in mines could be established; he appointed an IEB commission to investigate the situation and to stop the spread of hiring of women.[31]

Despite the union's opposition, women worked until the war's end in the western coalfields. Nearly half a century later, Madge Kelly described her work on a tipple near Superior, Wyoming, as "one of the happiest times of my life"; her job brought her both responsibility and a decent wage.[32]

Esther Snow worked at U.S. Fuel Company's King mine tipple in 1944 and 1945. Her poem, "The Female on the Tipple," appeared in the company newsletter under the byline "Written by Esther Snow, Boney Picker." This excerpt ends on a note of nostalgia:

> It was in September in forty-three
> When Dan and Duwayne decided to see,
> If women couldn't replace the men,
> That "Uncle Sam" had taken from them.
> So for a week with hammer and saw
> A rest room was built in one sidewall.
> To care for the women who'd be there soon,
> To sweep the stairs and the boney room. . . .
>
> The men all smiled and even laughed
> To think that Females might stand the gaff.
> The boom, the clamor, the roar and clang
> Of machinery moving and going bang!
>
> "Wait til the snow piles white and deep,
> They'll stay at home," says little Pete.
> But we proved to them we could stick it out
> Even the boney we soon learned about. . . .
>
> When the boys come back from the war and strife

We'll quit and go back to being a wife.
To sweep our floors and dust and groom
And think as we sweep of the boney room.[33]

Despite Snow's poetic suggestion, not all the female tipple workers returned to keeping house. Many stayed in the workforce in lower-wage jobs when returning veterans flooded the job market. The former World War II miner Madge Kelly's later jobs included making "spudnuts"—potato-based doughnuts—tending bar, cooking, and working in a gift shop. She said none of the jobs offered wages comparable to her wartime work on a mine tipple.[34]

Women also worked with family members during World War II, and some were openly acknowledged by mine bosses. In Harlan County, Kentucky, Ethel Dixon (McCuiston) began working full-time in 1941 as her husband's helper in a unionized mine. Other miners complained about her presence, but the mine boss defended her right to work.

> Some of the men would . . . be a-cursin' and going on. But the boss said, "If Ethel didn't help Arthur cut coal last night, there wouldn't be no work today. What would you think if your payday came up small?" Some threatened to quit. The boss told them to go ahead, and that I could work any time I wanted to. He told them it was a great honor that a woman would come into the mines to work to help them make bread for their families.

Besides mining coal, she raised her children, kept six boarders, and ran the family farm. "I didn't need much sleep," she wryly observed.[35] McCuiston worked at the mine for more than a decade after the war. She became pregnant with twins in the 1950s and finally left the mine after 14 years.

The postwar decades of the 1950s and 1960s were marked by layoffs created largely by mechanization in the coal industry. Unemployment triggered a massive out-migration of mining families from Appalachia and other coalfield regions to the nation's industrial centers. The oral histories show little evidence of women working in the mines during this period. Department of Labor statistics in 1968 show that mining employed less than 6 percent women, including clerical workers.[36]

By the 1970s, however, the coalfields were poised for dramatic social, political, and economic change. The civil rights movement secured important gains under the Johnson administration, particularly the Civil Rights Act of 1964. The feminist movement followed, and women poured into the work force. The end of the Vietnam War brought young veterans back to the coalfields seeking jobs.

By the early 1970s, male-dominated industries were feeling the impact of the Equal Pay Act, passed in 1963, and Title VII of the Civil Rights Act, which included sex as a class protected from job discrimination.

At the same time, the 1973 oil crisis fueled a coal boom of historic proportions. Coal operators hired 150,031 new miners from 1973 through the end of the decade, with 45,572 workers added in 1974 alone.[37] Migration patterns shifted into reverse as mining families left the cities with the hope of well-paid mining jobs back home. In the west, Peabody Coal Company opened mines on Navajo lands and Navajos, women included, joined the mining workforce.

## Crossing the Portal: The New Miners

The first wave of women hired by unionized companies in the mid-1970s joined a workforce—and a union—that had undergone significant changes. The new hires included more young workers, and college degrees were more common. Despite their youth, the new generation of mine workers—Vietnam veterans, black miners affected by the civil rights movement, Navajos, and women—brought into the mines a breadth of life experience.

Change was also brewing at the UMWA after a revolt by members fed up with the union's international leadership. The Miners For Democracy (MFD) fielded a successful slate of reform candidates headed by Arnold Miller to take the reins of power from Tony Boyle. In 1974 the rank and file regained their right to vote on contracts and to operate union districts autonomously.

In their attempt to break into the nearly all-male coal industry, women miners benefited from a battle by black steelworkers to end employment discrimination. Consent decrees signed in 1970 forced nine major steel companies to implement affirmative action hiring programs in the steel industry. The consent decrees provided the legal impetus for steel-owned "captive" coal mines to begin hiring women in significant numbers.[38]

The first women who sought mining jobs grappled with deeply entrenched gender prejudice about their right to work. In 1972 the *New York Times* chronicled the efforts of four textile workers in southwestern Virginia—including a mother and daughter—to obtain coal-mining jobs. Mrs. Katie Alderson told the *Times* reporter she was earning $2.10 per hour at a garment factory, and suffered a pay cut after taking sick leave. She and other textile workers visited the offices of Clinchfield Coal Company. While Alderson's husband, a disabled miner, supported his wife's efforts to become a miner, her daughter's husband said he would allow his wife to work as a miner "over my dead body."[39] The *United Mine Workers Journal* later carried the story, accompanied by reactions from three male miners, all of whom opposed the idea on the grounds that the women wanted the jobs for "cosmetics money" and that women would have to shower with men.[40]

Legal pressure on steel-owned coal mines proved stronger than social taboos. The National Institute for Occupational Safety and Health (NIOSH) tracked coal employment through X-rays required of new miners; the agency's records show that a West Virginia woman was first X-rayed for a mining job in 1973. Her identity, however, is unknown. In December 1973, Beth-Elkhorn Corporation, a subsidiary of Bethlehem Steel, hired two women at its No. 29 mine in eastern Kentucky.

The hiring of women miners sparked a debate in the *UMW Journal*, which published a series of letters to the editor from miners' wives on both sides of the issue. A West Virginia miner's wife wrote to remind readers that women were still subordinate to men and that "God never wanted them to be equal," adding that women miners "are very ugly or don't care" about their looks and "have no respect for themselves as a woman."[41] The letter prompted three responses from coalfield women in the following issue: two supported women's employment, including the mother of one of the first women hired. A month later, the Indiana miner Pam Schuble wrote an eloquent defense of her right to work and her commitment to the UMWA.[42]

A group of miners' wives in Logan, West Virginia, picketed a local mine in 1974 over the hiring of women. Beulah Whitman told a reporter from the *Wall Street Journal*, "No decent woman would want to work there; . . . there aren't even toilet facilities. . . . Distance is decency," she declared, attesting to wives' discomfort in having their husbands work in physical proximity to women underground.[43]

Shirley Boone went to work in 1976 as the first woman at a mine near her home in Boone County, West Virginia. On her first day she was abruptly reminded of the strength of the local taboos:

> I'd heard all those old superstitions, about how it was bad luck for a woman to go in. . . . The first day I rode into the mine on the mantrip to the mantrip station, a man stepped out of the mantrip car and had a heart attack. He just keeled over, and he never did come back to the mines to work. . . . The boss said, "That's what you get when you start having women in the mines—trouble."

Having grown up in the local community, she found it easier to adapt in some ways, and harder in others.

> Nobody around there had ever done anything like that before. I grew up there, too. That made it easier to get on at work, but harder to deal with people in the community. . . . I was threatened with being shot when I first went in.[44]

In one community surveyed by researchers in northern West Virginia, male miners voiced greater support than homemaker-wives for a woman's right to a mining job. Two-thirds of the male miners favored affirmative action quotas, compared with less than half of the homemaker-wives, and the miners' support extended to passage of the Equal Rights Amendment.[45]

Some women used humor to challenge ingrained attitudes. Cosby Totten of Local 6025 turned the tables on her male coworkers at a large Consolidation Coal mine, who tried to scare her with a tale of a female "haint," or ghost:

> There was a story about a redheaded woman and a couple of kids who went into that mine and got lost and supposedly starved to death. If you saw this redheaded "haint," you didn't have to work that day. It went way back and was part of the company's past practice. The men tried to get me scared on the hoot owl shift, telling me somebody had seen her coming out from under the rocks. I said, "If y'all ever see that redheaded woman in here, it's a vision of things to come. We're in the mines, and we're gonna stay."[46]

Many women miners expressed gratitude and pride about belonging to what they perceived as a strong union and felt they had more job protection than nonunion women, who could easily be fired as soon as companies met their affirmative action quotas. Sandra Bailey Barber, an Ohio native who moved to eastern Kentucky and took a mining job, felt proud to join the UMWA.

> I had always heard about the United Mine Workers being a strong union. So when I knew that I was going to get to join this famous United Mine Workers of America, I was really happy about it, because I knew the benefits they had in the coal industry, and I felt an obligation to at least attend the meetings.

She was not prepared, however, for the community's opposition, which was preached openly at a local church.

> Not being from here, I didn't realize the social stigmas attached to women miners. I didn't know what a disgrace it was going to be. My kids didn't realize it either. They learned afterwards when people made fun of them at school. I'd hear comments from them about people saying I was just there to get a man. . . .
>
> Then there was the sermon preached against me in the church. They didn't use my name, but I was the only woman coal miner in the community. People came up and told me that the preacher said it was sinful, women stepping out of their place, wearing pants, and trying to take men's jobs. I was the issue, and they pointed at me to tell these other women not to go astray.[47]

Women who grew up in UMWA families sometimes faced a special problem: a rich mining tradition that specifically excluded them from the mining workplace. What would be a natural occupation for a son was unacceptable for a daughter. Eighteen-year-old Linda Raisovich-Parsons followed a family tradition when she was hired by U.S. Steel Corporation.

> My grandfather came from Yugoslavia to this country to work in the mines. . . . [He] was one of the first from this area to join the union. When anyone joined, they were thrown out of their company house. I think it makes you a more dedicated union member when your whole family has struggled and suffered to organize and make the union what it is today. You feel like you have to support it and do the most that you can to preserve it. . . . My dad and I were real close when I was growing up. That had something to do with the fact I went underground. He used to sit and tell me all these stories about the mines.

When she took a mining job, her mother was more supportive of her choice:

> My mom encouraged me. She felt that if I wanted to do it, she was all for it. But my dad didn't really want me in the mine. I was sent to work right beside him. He made it kind of rough on me, to make me quit. We were on a backup crew, and he was the crew leader. . . . The shift foreman told him, "Well, we have to have a woman here, and I'm not gonna let you run her off." . . . My dad finally realized that I was there to stay. He never gave me a hard time after that.

Eventually, her relationship with her father was strengthened by the shared time and interest in mining.

> After I started working, we became closer, because we worked the same shift. We rode together. When he worked and I was in school, he was on the evening shift, and by the time I got home he was gone to work. I never saw him, except on weekends. But when you work right beside him, and ride with him, you get the chance to really know each other.

Although she was unaware of it at the time, her father's presence at the mine helped shield her from sexual harassment.

> My dad working at the mine made a difference in how I was treated. He was respected. And he was a big man. I heard later that my dad overheard somebody at the mine talking about me and he nearly got into a fight over it. But my dad never told me about it.[48]

Women from mining families who sought coal production jobs represented an entirely new type of miner, who combined male and female traditions. From their fathers, coalfield women learned mining lore, union history, and manual skills. From their mothers, they absorbed domestic arts and skills of building family and community life. Together, these customs forged workers strongly grounded in union history and mining skills whose lives were indelibly shaped outside the mine by family and community values.

Like most working women, UMWA members struggled to meet family and job responsibilities, and they shouldered the extra burdens of life-threatening job hazards and commutes that could add hours to a workday. African-Americans faced the additional obstacles of racial prejudice in their work lives. As new hires, UMWA women lacked the seniority to bid on day-shift jobs. Many working mothers preferred midnight shift, choosing to lose sleep rather than the evening hours with their children.

Child care was a serious problem for mothers, who often drove long distances to the mine site and then were not accessible for eight hours or more. Women miners relied on family members, worked midnight shifts, and skimped on sleep to minimize time lost with their children. One Kentucky miner found a creative solution to her maternal task of breast-feeding; she brought a breast pump underground and pumped her milk at her solitary outpost on the conveyor belthead.[49]

Despite the job's physical demands, many women reported that mining was easier than working two or three low-wage jobs just to make ends meet. Elizabeth Laird of Cordova, Alabama, went into the mines at 54 after working nearly 25 years in textile plants. After her divorce, she worked five years on the day shift in a textile mill and then worked until two A.M. in a diner. She said that if the coal industry had let women in earlier, "I wouldn't have had to work two jobs and gone five years without a decent night's sleep."[50]

## The Coal Employment Project

In 1977, Appalachian activists accidentally stumbled upon evidence of blatant gender discrimination in the coal industry when a Tennessee coal operator refused to allow a woman to join a mine tour, saying, "Can't have no woman going underground. The men would walk out."[51]

An all-male tour proceeded, but the activists later dug through federal affirmative action policies until they found Executive Order 11246, a 1965 Johnson administration directive that barred sex discrimination by companies holding federal contracts. After confirming that many Tennessee and Kentucky coal operators held contracts with the federal Tennessee Valley Authority (TVA), the group hired a Kentucky lawyer, Betty Jean Hall, to work full-time on the issue.

Hall founded the Coal Employment Project (CEP), a nonprofit women's organization that has proved an essential support to women miners in their efforts to gain jobs, combat discrimination, build a support network, and educate themselves and the public. Through the 1980s and early 1990s, CEP addressed a broad range of issues facing the labor movement and expanded into an international network of coalfield women.

In 1978 CEP filed a landmark sex discrimination complaint with the U.S. Labor Department's Office of Federal Contract Compliance Programs (OFCCP). The administrative complaint charged that the coal industry was "one of the most blatantly discriminatory employers" in the United States, and it targeted 153 coal companies and mines, representing about half of the nation's coal production. Among the companies named were Peabody Coal Company and most of the steel-owned coal firms. The complaint pressed for a hiring plan that would force companies to hire one woman for every three inexperienced men until women made up 20 percent of the workforce.[52] At the same time the Kentucky Human Rights Commission aggressively pursued coal operators who discriminated against women.[53]

CEP's broad-based legal strategy worked. By December 1978 Consolidation Coal Company had agreed to pay $360,000 in back wages and benefits to 78 women who had been refused jobs between 1972 and 1976, and it was forced to implement an affirmative action plan. Other lawsuits followed. The widespread publicity informed coalfield women about new hiring opportunities. By the end of 1979, 2,940 women had been hired as underground coal miners in the United States, with the highest number of them hired in West Virginia, Pennsylvania, Illinois, and Alabama.[54]

Shortly after the complaint was filed, a delegation of UMWA women traveled to the UMWA's international headquarters in Washington, D.C., to ask President Arnold Miller for an official statement of support. Miller promised to support their efforts but later rebuffed CEP's request to approach the International Executive Board for an official support resolution. Union officials claimed that the request had arrived too late to be placed on the agenda.[55] CEP members protested and published a follow-up story in the group's newsletter entitled "Which Side Is the UMWA On?" Women lobbied their IEB representatives, who unanimously passed a support resolution at the next quarterly meeting.[56]

After the Washington, D.C., meeting, CEP worked on the formation of local support teams of women miners, publication of the monthly *Coalmining Women's Support Team News!*, and the first-ever national conference for women miners in June 1979, which attracted about 200 participants. The inaugural CEP conference at Institute, West Virginia, drew UMWA rank-and-file activists of both sexes, national media, and a scattering of leftist political groups—all new to most of the participants. Guests included the singer Florence Reece, who wrote the labor anthem, "Which Side Are You On?" (about earlier mining struggles); and Bill Worthington, an African-American UMWA activist and leader of the Black Lung Association. Participants made it clear that their goals were not restricted to women's issues. Mary

Zins, a West Virginia miner, told the group that the debate on the maternity benefits needed to be broadened: "What it comes down to is we need benefits for sickness, as do our brothers. . . . We need assurance for our safety. We need a better grievance procedure. That's not a matter of sex. That's a matter of being a miner."[57]

A few months later, in November 1979, the union held its own day-long women's conference, which was never repeated. Shortly afterwards, Vice President Sam Church took over at the UMWA headquarters when Arnold Miller suffered a stroke. Church's attitude toward women miners was inadvertently revealed a month later at the UMWA's constitutional convention in Denver, Colorado, where nine women were among the 1,267 delegates. At a hospitality room hosted by CEP, women delegates pressed Church for an affirmative action hiring clause in the contract and improved sickness and accident benefits; he responded with an off-color joke.[58]

In 1982 Church was unseated by Richard Trumka, a young lawyer and miner who promised reform. Women miners vigorously campaigned for Trumka, especially in his home state of Pennsylvania. In 1983 the UMWA officially endorsed the CEP conference, held that year in Pennsylvania. Trumka was the keynote speaker; subsequently, a UMWA international officer addressed the conference nearly every year. Trumka also sent letters of endorsement annually to local unions, urging them to send women to the conferences and approving excused absences for UMWA women to attend.

## Sexual Harassment in the Mines

Soon after women entered the mines in sizable numbers, sexual harassment emerged as a major problem. At every CEP conference, workshops on harassment drew large crowds and elicited gut-wrenching testimonies about harassing behavior from bosses and coworkers. Harassment from management often was more injurious because it could affect a woman's employment and daily working conditions, but UMWA women often complained that local and district officials did not take sexual harassment seriously when it involved coworkers.[59]

In the oral history interviews, women miners describe a wide range of harassing behaviors: sexual comments and jokes, unwanted advances, an initiatory "greasing" rite, groping, bathroom peepholes, verbal abuse, physical assault, and other problems, such as finding human feces in a dinner bucket. Women reported that some male coworkers were uncomfortable with the harassing behavior but frequently did not feel sufficient support to voice their opposition. When the loss of mining jobs in the 1980s left fewer women working and realignments placed some of them with new crews, the isolation and crew changes sometimes triggered a new outbreak of harassing behavior.

At CEP conferences, women miners found emotional support and developed concrete strategies to combat the problem. CEP surveyed women on their experiences of harassment and successful ways to respond, and published guidelines for legal and contractual remedies.

The oral histories suggest that a woman's vulnerability to harassment was affected by a host of factors: the number of women employed at the mine; the length of time she had worked at the mine; her connection to a support network such as CEP; the strength of the local and district union; local management attitudes; the presence of a supportive male relative at the mine; and her awareness and use of legal and contractual remedies.

Like pornography, certain types of sexual harassment are practiced through voyeurism and secrecy. At Consolidation Coal Company's Shoemaker mine near Benwood, West Virginia, the bosses' secret was a pencil-sized hole drilled through a wall into the women's shower area. On one side, women miners dressed and showered under the illusion of privacy. On the other side, men furtively spied on them in their most private moments. In 1981 eight UMWA women filed a $5.5 million lawsuit against Consolidation, accusing the company of invasion of privacy. The group also filed a grievance through their local union, but the contract included no privacy language. The case was widely publicized, and the CBS television show *60 Minutes* featured the story in October 1982.

When the case went to trial, the courtroom drama "took on many aspects of a rape trial," with the defense attacking the characters of the plaintiffs, according to the CEP newsletter. Company witnesses included foremen, other women miners, and union brothers—even the chairman of the local union mine committee. The two parties reached a settlement during the trial, and a gag order was placed on the plaintiffs.[60]

Sexual harassment usually was directed at individual women, however, and often created feelings of shame, increasing a woman's isolation in the mine. An Alabama miner, Patricia Brown, recalled her early years of employment:

> I have a big bust. I never thought about it until I got down there. They made me feel ashamed. They would tease and comment and write things on the walls. They drew pictures. Some guy on the shift before me would go down and draw pictures of my body, and somebody else would come and tell me, so everybody would get a big laugh. . . .
>
> The first time I saw it, I went home and cried. . . . I did have this one black guy who was very supportive. If he went down before me, he would spray-paint the pictures off so I couldn't see them. . . . When I would get upset it would come out in such a way that I was hostile at home. It put a great strain on my children. . . .[61]

In oral history interviews, some women commented that although more older men held superstitious beliefs about women miners, younger miners expressed more hostility at work; Martha Horner of New Mexico noticed the difference:

> I was surprised, because the old-timers took care of me and I had trouble with the young ones. Everybody told me it would be the old-timers who would give me a rough time, because they didn't want women in the mine. But they put me under their wing and took care of me. I think they were raised to take care of women.[62]

The National Bituminous Coal Wage Agreement (NBCWA), negotiated approximately every three to five years between the UMWA and the Bituminous Coal Operators Association, provides limited protection for victims of sexual harassment. Article 25 prohibits discrimination against an employee by the employer or the union and lists sex, age, race, and other factors as protected categories.[63]

Article 25 contains the NBCWA's strongest language on discrimination, but other supporting provisions are contained in Article 19(d), which bans the employer from making a temporary assignment for the purpose of disciplining of discriminating against a worker; and Article 1, an enabling clause that states that management will not abridge the rights of employees as set forth in the contract. If the harassment creates a safety hazard for a worker, she can file a grievance citing violation of individual safety rights under article 3(i), which gives an employee the right to remove herself from an "immediately hazardous" situation.

In their first years of employment, relatively few women miners used the grievance procedure to address sexual harassment or other problems. As time passed, they grew more knowledgeable and confident about using the contract to protect their job rights. As one UMWA member, Bonnie Boyer, explained: "I didn't start filing grievances until about the seventh year. I think it takes that long to learn their whole system of discrimination. I've talked to other women miners. It's a turning point. It's like, you've taken all you can tolerate, and you start filing grievances, and you start pushing things to the limit. You turn your whole attitude around, you don't care anymore, and you're going to fight the company."[64]

The UMWA grievance procedure includes four steps, outlined in articles 23 and 24 of the NBCWA. The grievant first attempts to resolve differences through the immediate supervisor/foreman. If that is unsuccessful, the grievant files a grievance, and a meeting takes place between the union mine committee and management. If the grievance cannot be resolved at this level, it is referred to a union district representative, who sets up a meeting with a different management official for third-step action. The fourth step is a hearing by a district arbitrator, who rules on the grievance after reviewing relevant evidence and testimony.

The degree of discrimination faced by women seeking union representation is difficult to document because of lack of consistent documentation of grievance procedures by UMWA advocates.

Carol Davis, a CEP member elected as a member of her local union's grievance committee, said that her local tried to ignore sexual harassment when it involved another union member:

> In any industry that has nontraditional jobs for women, you should be able to sit down and talk to your own sex about what's going on. Sexual harassment was really big at that time [the early 1980s], and women needed to know they had a recourse. The UMW don't know how to face a woman being harassed. They turn their back. They say, "That's a union brother you're talking about." So women had to take measures into their own hands. Most times you had to go through an agency outside of the UMW, which we shouldn't have to do after all these years.
>
> When I got into the union I took an oath to uphold my union brothers and sisters. That means something to me, and it will mean something to me until I die. The men take the oath to help a sister in need, but when the sister is in need they won't help you.[65]

## Health and Safety Issues

Women miners have emerged as outspoken activists on mine health and safety issues ranging from local mine conditions to national issues such as federal regulations. Of all the mining specialties, mine safety has proved the strongest draw for women, who have been elected to local union safety committees, have been hired as state and federal inspectors, and have served as company and union safety specialists. Mine safety and health have also been a central focus of the Coal Employment Project.

CEP newsletters, conference resolutions, and coverage in the *UMWJ* reflect the commitment of women miners to mine safety, and numerous examples exist of women taking a stand, sometimes at great risk, to make their mines safe. Patsy Fraley and two female coworkers exposed corruption at a federal Mine Safety and Health Administration office and faced threats to their safety and their families'. Five women in Alabama were suspended for refusing unsafe work lifting 1,800-pound rails after a male coworker collapsed from exhaustion. Later, 70 men joined the women protesting the suspension, and the women were reinstated.[66] CEP filed amicus briefs in lawsuits on mine safety, and women miners testified before Congress and staged protests over proposed cuts in federal ventilation regulations in the 1980s and 1990s.

Since 1973 at least seven women have been killed in mining accidents.[67] The first underground woman miner killed on the job, Marilyn McCusker, drew national media attention in 1979; her death raised the issue of spousal benefits for men. McCusker, 35, was installing roof bolts at the Rushton mine in Pennsylvania on October 2, 1979, when she was crushed by a roof fall. A male coworker escaped injury, and the company's safety director later stated that if McCusker had been a man, "she would have been ten steps away."[68] In 1977, McCusker and three other women had sued Rushton for the right to work underground. In the early 1980s CEP launched a study of lost-time accident rates among women miners. It concluded that 4 percent of all female coal miners were involved in lost-time accidents from 1978 to 1980, as compared with 7 percent of all male miners. Female workers lost approximately 1.1 days apiece, as compared with 2.4 days for males.[69] In oral history interviews, many women credited their roles as family caretakers for a heightened awareness of the safety of themselves and others.

Keeping a mine safe is a pressing concern for single parents who provide their children's primary economic and emotional support. A West Virginia miner whose arm was nearly ripped off in a shuttlecar accident described the effect of her accident on her one-year-old son:

> It was a month before I saw my baby, and it was a bad scene. He wouldn't recognize me. He wouldn't let himself hug me. Near the end of the time I was in the hospital, though, he was really hanging onto me. For a couple of years after that, he didn't want me out of his sight. I think he didn't want to go through losing me again. It was hard for people to understand, but for him Mommy had gotten up and gone to work and disappeared for a month.[70]

Pregnancy has been a major health issue for women miners, since many entered the mines in their childbearing years in the 1970s and 1980s. As early as 1980 CEP conference delegates passed a resolution instructing CEP to conduct a study on coal mining and pregnancy.[71] Little was known about the effects on a pregnant worker of airborne coal dust, vibration, electromagnetic fields, chemical agents, noise, and other hazards common in the mining environment. In 1985, with the support of the UMWA, CEP and the National Institute for Occupational Safety and Health (NIOSH) released a report based on a preliminary health survey of 26 women coal miners who had been pregnant on the job, but the results were inconclusive, pointing to the need for further research.[72]

At mine sites, pregnant miners were often surprised at the strong reactions from male coworkers. Brenda Brock was single and in her late twenties when she worked nearly eight months into her pregnancy at a non-UMWA mine in Harlan County, Kentucky:

> When I told my boss I was pregnant, he jumped up and said, "You can't work! You'll be up here with morning sickness!" And I said, "I've puked for two months and you haven't even noticed." . . .
>
> [A]bout a week later [I] told everybody. The men were shocked. They said I should quit. And they talked bad about me. So I said, "If any one of you wants to talk about me, fine, but I can make your life perfectly miserable because I'll go straight to your wife and tell her it's yours." . . . When I told the men that, it shut them up.[73]

A West Virginia miner noticed that some men felt protective toward her after her pregnancy became known but also felt obliged to disguise their feelings:

> The ones who wouldn't harass me were embarrassed to work with me. Some of them felt sympathetic, but they couldn't really express it because so many of the men were resentful. They did what they could in a very unobtrusive way. Others would help if I had to hang brattice curtain above my head. They didn't say anything, and I didn't say anything, but we both understood. . . .[74]

## Sisterhood and Solidarity

The 1980s proved to be a decade marked by decline and retrenchment for labor unions in the United States, but women miners organized aggressively within CEP and the UMWA on a wide range of social issues. CEP conference participants pressed the UMWA for a more activist stance on the environmental impact of acid rain, solidarity with striking British miners, and race and gender discrimination. CEP conferences became arenas for UMWA members to share their views on the Equal Rights Amendment, reproductive rights, gay rights, and AIDS.[75]

In 1983 CEP earned the union's support and involvement on the issue of family leave, and the successful campaign is a landmark achievement in the labor movement. Women miners first raised the issue of extended maternity benefits at their first national conference in 1979; at subsequent conferences they broadened the concept to maternity/paternity leave. In April 1983 CEP sponsored a three-day workshop in Washington, D.C., for 17 UMWA women representing 14 of the union's districts.[76] The group studied the UMWA constitution and drafted contract language on maternity/paternity leave that also protected the jobs of parents of seriously ill children. They developed a strategy to educate their union brothers and to lobby for the issue through local and district unions.

In choosing to launch a campaign for family leave, women miners deliberately chose an issue that would benefit all union members. "Family leave is about equality—for men," said one CEP member, Cosby Totten of UMWA

Local 6025. By publicizing the stories of men whose jobs were threatened because of a family illness, women miners earned the union's support. The family leave provision was passed unanimously in 1983 by delegates to the UMWA constitutional convention for contract language on family leave. Early the next year a UMWA contract was ratified that included a letter of intent for a joint union-industry committee to study the issue. After lobbying for representation on the committee, CEP members were appointed to two of the five union seats.[77]

Women miners also influenced the drafting of federal legislation. After CEP initiated a meeting in 1984 with ten women's groups, feminist leaders expanded proposed language on "parental leave" to include time needed to care for seriously ill children.[78] After more than a decade of work, women miners finally celebrated President Bill Clinton's signing of the federal Family and Medical Leave Act in February 1993.

In the 1970s and 1980s CEP's activist agenda was frequently promoted and expanded by UMWA women who also belonged to the Socialist Workers Party (SWP), a fact that generated an internal debate within CEP.[79] UMWA women's activism, however, was not confined to a single ideology and became a natural expression for women who had fought to be hired, trained, represented, and heard as a collective voice. Women miners active in SWP made CEP's initial contact with embattled British miners striking to prevent the closing of union pits. In 1985 a delegation of British miners' wives representing the activist Women Against Pit Closures (WAPC) made the first in a series of trips to CEP conferences, and women miners forged strong ties to WAPC. CEP later sponsored fact-finding trips to the coalfields of China, the former Soviet Union, India, Europe, Canada, the *maquiladora* region of Mexico, and the Fourth World Conference on Women in Beijing.

Being among the "last-hired" workers also made UMWA women vulnerable to the mass layoffs of the 1980s. But despite unemployment and a declining labor movement in the United States, UMWA women built solidarity in strikes against A. T. Massey, Canterbury Coal, and Decker Coal Company. Women miners also forged links with striking Hormel meatpackers from Local P-9 in Minnesota, who had challenged their international union, the United Food and Commercial Workers.

Another CEP organizing effort focused on UMWA auxiliaries, the groups of miners' wives associated with union locals. UMWA women's auxiliaries have a rich history of militancy during strikes. From West Virginia to Colorado, miners' wives have confronted strikebreakers and mine guards and faced arrest and physical assault to defend their families and the union.

During the historic Pittston strike in 1989–90 in the Appalachian coalfields, the UMWA supported a large-scale organizing of UMWA auxiliaries, an effort led by CEP members. About 1,000 strikers' wives volunteered for strike duties. In 1989 auxiliary members staged a peaceful occupation of the

Pittston regional office in Virginia, were arrested and went to jail in civil disobedience protests, and staffed Camp Solidarity, which hosted more than 50,000 visitors before the strike ended in early 1990.[80]

As the strike progressed, however, some auxiliary leaders privately complained that the UMWA did not support women's organizing efforts.[81] After the successful early takeover of the Pittston building, auxiliary members were relegated to less activist roles. During the later men's occupation of the Moss 3 preparation plant, women reportedly were deliberately excluded from the strategy discussions.[82] Cosby Totten worked for the union to organize the Pittston auxiliaries and found herself disillusioned with the UMWA:

> After [the women's occupation] was all over, [the strike leader] said that he had wanted the occupation carried out so that the women could show up the men. I didn't like that, because a wife does not want to be used to humiliate her husband. . . . It seemed like most of the men in the decision-making part of the union would use the women when they needed them, and when they didn't they wanted them to get back out of the way. . . .
>
> It was frustrating, because the women found a voice in the strike and then were kept from using it.[83]

Four years later women miners were visible during the 1993 strike against targeted member companies of the Bituminous Coal Operators Association, which affected miners in seven states. UMWA women served as strike captains, outreach coordinators, and key organizers in strike relief efforts. While the union's efforts to organize auxiliaries in 1993 appeared perfunctory, women miners stepped in to help organize the groups, with at least one woman miner serving as auxiliary president in her dual capacity as UMWA wife and striking miner.[84]

## Women as Union Leaders

By the early 1980s women miners had gained enough union experience to begin running for elected local union offices. The exact number of women who have served as officers and elected committee members is unknown; at the 1988 CEP national conference, however, when the chairperson asked for all women who had ever held local office to stand up, dozens of women rose to their feet.[85]

CEP in particular worked to promote and educate women for positions of leadership in the UMWA. In 1986 the CEP member Joy Huitt added her name as an independent candidate for secretary-treasurer of the UMWA's District 22, which spans four western states and a diversity of cultures—Navajos, Mormons, Mexicans, and migrants from the eastern coalfields. Although she ran against a slate of candidates organized by the incumbents, she won the race by a narrow margin.[86] Huitt credited CEP with helping her build the skills necessary for a successful political campaign.

After serving one term, she lost her bid for reelection in an even closer race with a slate-endorsed candidate who was also female. Afterward, Huitt filed a worker's compensation claim for stress for alleged harassment from District 22 officials.[87] UMWA women found few job opportunities open to them on the professional staff of the international union during the Trumka administration. In the 13 years Trumka presided over the union, he added *no* dues-paying UMWA women to the professional staff, despite claims to the contrary in the union's centennial history.[88]

## Conclusion

How many women miners are still employed in the mid-1990s is uncertain; a 1991 Bureau of Mines Report listed about 1 percent of 97,500 miners as women, down from 3,328 in 1986.[89] Of those hired in the peak years of the late 1970s, the women still working are nearing retirement as the unionized industry continues to shrink, endangering mining wages and the future of the UMWA.

The collected oral histories suggest that mining women have exercised a high level of safety awareness and activism; strengthened solidarity in labor struggles during the antilabor decade of the 1980s; broadened the UMWA's social agenda; and found vital support and educational opportunities through their nonprofit organization, CEP.

Perhaps it is too soon to know whether women miners have had a lasting impact on the UMWA and on the coal industry in the United States. More research and discussion are needed in key areas of women miners' history: the degree of union representation of UMWA women in grievance and arbitration procedures, community and spousal attitudes toward women in mining over the last 20 years, gender relations underground, and the relationship between women coal miners and their primary union, the UMWA.

Although most of the approximately 4,000 women who have been hired as miners since 1973 have lost their jobs, their identity as women miners remains strong years after their departure from mine portals. "Once a woman miner, always a woman miner," noted the laid-off miner Carol Davis in a comment often repeated by CEP members.[90] Women like Cosby Totten continue to pay UMWA dues long after being laid off, because of loyalty and a commitment to working for social justice.

> I was laid off in 1982, and I have stayed a dues-paying member of the United Mine Workers. I intend to stay a member, because I love my union.

> I stay active for my mental health, although sometimes being active makes me think I'm crazy. I stay involved because I want to make life better for my children. . . . An injury to one is an injury to all. Our children need to understand those words.[91]

By crossing mine portals into a high-wage, virtually all-male occupation, women expanded the horizons of nontraditional work in the wake of affirmative action mandates. The legacy of women's labor in American coal mines, like the fate of affirmative action, will be decided in the years to come, by future generations of women workers. In late 1993, Libby Lindsay—who is still working in the mines—concluded an editorial in the CEP newsletter by noting that the twentieth anniversary of the official entry of women into the mines had almost slipped by without her realizing it. Her words reflect the determined optimism of a lifelong activist:

> No wonder I nearly forgot. We've been busy! We're getting older. We have more health problems than we used to. We're burned out. But we're still here. We're still saying, "Si, se puede"—[Yes, we can].[92]

# A Cast of Voices: The Speakers

Sandra Bailey Barber
b. 1947
Federal mine safety supervisor
Mayking, Kentucky

Bonnie Boyer
b. 1957
Underground miner
Shelocta, Pennsylvania

Brenda Brock
b. 1955
Miner, nontraditional worker
Tempe, Arizona

Patricia Brown
b. 1946
Underground miner
Bessemer, Alabama

Elizabeth Stull Crawford
1907–1989
Migrant miner
Dillonvale, Ohio

Carol Davis
b. 1948
Underground miner
Mather, Pennsylvania

Kipp Dawson
b. 1945
Underground miner
Pittsburgh, Pennsylvania

Irene Adkins Dolin
b. 1928
Family mine worker
Julian, West Virginia

Patsy Fraley
b. 1945
Underground miner
East Point, Kentucky

Alice Fulford
1895–1989
World War II mine worker
Columbus, Ohio

Martha Horner
b. 1952
Underground miner
Raton, New Mexico

Joy Huitt
b. 1935
Miner, union official
East Carbon, Utah

Madge Kelly
b. 1921
World War II mine worker
Rock Springs, Wyoming

Ruby Carey Kifer
b. 1924
Retired miner
Chandler, Indiana

Helen Korich Krmpotich
b. 1906
Survivor of the 1914
Ludlow Massacre
Jackson, Wyoming

Elizabeth Laird
b. 1921
Retired miner
Cordova, Alabama

Evelyn Luna (Evie Tsosie)
b. 1955
Surface miner
Kayenta, Arizona

Ethel Dixon McCuiston
1918–1986
Family mine laborer
Cumberland, Kentucky

Rita Miller
b. 1957
Underground miner
White Plains, Kentucky

Janice Molineaux
b. 1953
State mine inspector
Jenkinjones, West Virginia

Sylvia Pye
b. 1937
Pit canteen worker, activist
Lancashire, England

Ethel Day Smith
b. 1913
Family mine laborer
Evarts, Kentucky

Elizabeth Zofchak Leach Stevens
b. 1917
Family mine laborer
Poolesville, Maryland

Cosby Ann Totten
b. 1941
Underground miner, activist
Tazewell, Virginia

# PART I

## *The Women Before Us*

The industrialization of coal forged ahead in the late nineteenth and early twentieth centuries in the United States as companies developed the capacity for large-scale, continuous production. Besides sinking shafts and digging drift mines, operators built coal camps, coal patches, and large company towns in rural areas to house workers and their families. Most small family mines were drift mines, in which tunnels are dug directly into coal outcrops. Drift mines can go straight into the hillside or angle slightly upward, giving natural drainage. The coal seams of slope mines angle downward, and are also developed from outcrops on the surface. Shaft mines require a cage or elevator to transport miners belowground to deeper coal seams. Company towns often became a means of social control by coal owners, who owned the land and financially controlled nearly every aspect of public life: jobs, stores, schools, churches, medical care, and law enforcement. Exploitation was common; miners rebelled against unsafe working conditions, low wages, and unsanitary living conditions during bitter struggles that broke out throughout the coalfields and lasted through the early twentieth century.

The mine wars provided the bloody backdrop of coal development in the United States. Conflict erupted in nearly every coalfield region between the fledgling UMWA and the highly organized forces of capital. For miners, the names of labor's battlegrounds are legend: Bloody Mingo, Bloody Harlan, Bloody Williamson, the Ludlow Massacre, the Lattimer Massacre, the Matewan Massacre, the Battle of Blair Mountain, and others.

Women were involved in at least five capacities during the decades of industrialization: as wives and mothers who settled the coal camps and raised the next generation of workers; as family laborers in small mines outside the control of the large companies; as union activists during the mine wars; as temporary workers during both world wars; and, less commonly, as operators of mines they leased or owned.

Coal mining involved family labor in small mines where handloading practices lingered in some areas until the 1950s. Handloading equipment included picks, shovels, explosive black powder, carbide lights, wooden timbers for roof support, ponies or mules, and sometimes a large breast auger, or drill, to bore holes for blasting. Usually several family members worked to boost the paycheck of the male breadwinner. These oral histories suggest that the practice of family mining spread during the Depression years of the 1930s, especially in Appalachia, where coal seams were easily found and mined. In a cash-poor economy, coal often became an item of barter.

In the interviews in this section, women recall coalfield life in the first half of the twentieth century. Helen Korich Krmpotich offers a harrowing glimpse of the 1914 Ludlow Massacre in Colorado. Elizabeth Stull Crawford became a migrant miner at 13 and worked over a period of three decades. Elizabeth Zofchak Stevens and Ethel Day Smith labored alongside their fathers; Irene

Adkins Dolin led her younger sister into a family mine to dig house coal. World War II brought Ethel Dixon McCuiston into a large underground mine to work with her husband; she stayed there for 14 years. Alice Fulford and Madge Kelly were employed as wartime tipple workers by large companies, then lost their jobs to returning veterans.

Many of the early women miners voiced mixed feelings about the growth of women's presence during the 1970s in the mining workforce. While they had enjoyed their own mining work and believed in women's abilities, some expressed doubts about women entering large all-male mines. To work in a supportive role with men in the family, they felt, was more acceptable than working side by side with unrelated men. Other women expressed protectiveness toward and hope for the later generation and wished aloud that such economic opportunities had been available to them.

# 1

# BARBED WIRE AND EASTER LACE: THE LUDLOW MASSACRE

## Helen Korich Krmpotich

*"We were good union girls. We knew it was right to stand up for our rights."*

*Seven-year-old Helen Korich and her family endured one of the most brutal labor conflicts in American history. In 1913 immigrant miners rebelled against oppressive conditions in the coal camps of southern Colorado and responded to a strike call by the United Mine Workers of America. Thousands of men, women, and children left the desolate camps and drove their wagons through snow and sleet down the canyons to tent colonies set up by the UMWA. The largest was Ludlow, a tent colony of 1,200 people who spoke 24 languages. The colony was strategically placed at the intersection of the railroad lines that served as a point of arrival for strikebreakers.*

*Eighty years later, Helen Korich Krmpotich still remembered Ludlow as a vibrant multiethnic community. The colony's rich blend of cultures found expression in foods, music, and religious observances by Greek, French, Italian, Mexican, Finn, French, British, and Slavic miners, among others. Ludlow's large meeting tent drew famous labor orators, including Mary Harris "Mother" Jones. Mother Jones fearlessly faced down scabs, thugs, and coal barons, but she also distributed clothing to Ludlow's children, including young Helen Korich, who received a prized pair of shoes from the white-haired leader she idolized.*

*Strikebreakers threatened Ludlow and surrounding colonies, and sporadic shooting took place. The governor called in the Colorado National Guard, which strikers welcomed for its supposed neutrality. The Guard set up camps overlooking the strikers' encampment. But tension rose at Ludlow when mine guards gradually filled the ranks of the militia and moved into the armed camps ringing Ludlow. Fearful of a machine-gun assault, families began digging pits beneath the canvas tents.*

*On April 20, 1914, militia troops dominated by mine guards attacked the*

Helen Korich Krmpotich

*Ludlow tent colony with gunfire that lasted 12 hours and shredded the rows of white tents into lace. Afterwards, the militia poured coal oil on the tents and set them afire. Eleven children and two women were burned to death or smothered in hand-dug cellars. A total of 20 people died in the Ludlow Massacre.*

*In 1985 I was interviewing retired miners at a senior citizens' center in Rock Springs, Wyoming, when I felt a gentle tap on my shoulder. I looked up into the kindly face of an elderly woman. She smiled and said, "We fought for the union, too, honey, at Ludlow."*

*Later I visited Helen Korich Krmpotich's small, immaculate home, where she had lived since the death of her husband, a miner. In an oval frame hung a large family portrait. Her father looked out of the photograph from above a thick mustache, with his six children. Her mother was not pictured, having died in*

Tent colony near Ludlow, Colorado, winter 1913—14.
Colorado Historical Society

*the 1918 flu epidemic at the age of 37. In the portrait's stair steps of oldest to youngest, Helen was second in line: a thin, sober child with straight dark hair.*

My father came from Yugoslavia in 1900. He worked for five years in the steel mills of Chicago before he had enough money to send for my mother. She came over with my sister in 1905, and I was born the next year.

We first came to Colorado because my father had gotten hurt in the steel mills in Illinois. Hot ash fell on his head and burned all his hair off, and he was six months in the hospital. There was no compensation at that time. He'd heard that the coal mines in Colorado were good, and since my aunt lived in Pueblo, that's where we went.

We went to the coal mines in Walsenburg, Colorado, and lived there for about two years, and then moved to Hastings to be with friends. We stayed about three years, and my sister was born there. The strikes started around 1912. The men were tired of working for peanuts and not being able to feed their families. Mother had to cook and bake and wash clothes and iron for other people, and we'd taken in boarders to make ends meet. When the strikes started, my father decided to go with the families to fight for the union. He was the first one who volunteered. I believe that was in 1913.

We went to the Ludlow colony. There were already tents there. We had a double tent with a bedroom and a kitchen. The tent had floors, and regular walls with canvas over them. Four of us children slept in a double bed in the kitchen.

There were hundreds of us in the colony, and rows and rows of tents.

People came to Ludlow from all over the world, and we all stuck together. People spoke many different languages, but we all lived together and were close. It was really wonderful. We loved living there. My brother was the first baby born in the colony, and it was a big occasion.

Going to school we sang our union songs, and the kids who lived on farms didn't like it. We sang until our lungs felt like they would burst. We were happy, and we were proud. We were good union girls. We knew it was right to stand up for our rights.

All these scabs would ride in cars down the highway from Hastings into Trinidad. They used to make fun of us and bark at us. There were colored ones, but there were a lot of white scabs, too. After that, we had to go to school in another district because the teachers didn't like union people. They wouldn't even teach us.

The winter was bad. Sometimes we couldn't get out of our tents unless we made snow tunnels. The union gave us each fifty cents a week to live on, so for the seven of us, the family had three dollars and fifty cents a week. We managed the best we could. My mother sewed for the union men, sewing belts with rifle shells on them, which she delivered at night. She never took money for it. The girls would cook and sew. I helped my mother wash for the bachelors. My mother used to tell me that your hands would be beautiful, like a queen's, if you washed on the board. So I'd scrub and scrub and look at my hands to see if they were getting pretty.

*The Ludlow colony became a focus of western labor organizing and drew prominent labor leaders, including the fiery Mary Harris "Mother" Jones.*

Union people would come to Ludlow from back east and give speeches from the platform in the big meeting tent. Mother Jones would come. She really made the speeches, I'm telling you! We loved her. We clapped and clapped until our hands hurt. We wanted to be as strong as she was. She was like an old schoolteacher. When she was there, we felt safe. She brought us clothes and things we didn't have. She gave me a pair of high-heeled shoes to wear to school, and I cried and cried. They were too big, but I loved them. She brought clothes and diapers and food and would sort them out in boxes. My aunt would help. We never went hungry. There were dances and singing and baseball games. We were scared all the time, but we were happy. Everybody helped each other.

*As the strike wore on, the operators appealed to the governor, who sent in state militia troops in the autumn of 1913. The soldiers set up camps surrounding the Ludlow strike colony. In February the state withdrew most of the troops, and tension rose as company guards began joining the militia.*

When the soldiers would come into the colony, we really got scared. There were some beautiful girls in Ludlow, and the soldiers would want to be entertained. They would come to our parties and want to dance. What were we going to do? They had their brown uniforms on and would try to take the girls out. They were mean. They thought they were bigger than we were, and we just had to go along. The union men couldn't run them out, because if they had started something, the soldiers would have finished it. My sister Pearl was about 12 then, and she loved to go out with her girlfriend and dance. My mother would get mad at her, and so my sister made me go with her. I was five years younger.

*Fearful of an attack, some families began digging out pits beneath their tents to hide from machine-gun fire.*

The men in the union knew something was going to happen, because they got to their guns and started handing them out. The Greeks were smart. They buried their guns, and they had them ready when the soldiers started shooting. We never knew when they were going to start shooting, and we lived there a long time waiting and worrying.

Our Italian neighbor said, "Why don't you build a basement for your family?" His wife was ready to have a baby, and they dug a basement. But my dad asked, "What if they burn the tents down? Then what are you going to do?" The others thought they wouldn't burn the tents down, that it would be a safe place away from the shooting.

*After a long winter, a warm sun melted the snow in time for the celebration of the Greek Orthodox Easter.*

My mother had dyed Easter eggs, and we had a big dinner. Our Easter was Orthodox, like the Greeks'. We had meat and barbecued. We celebrated with baseball games. The women were baseball players, and they were good! They sewed their own outfits of white blouses and black pants that looked like skirts. On Easter they had a big baseball game, and the guards came and watched.

The next day the shooting started. I had my white lace dress on, my Easter dress. The weather was beautiful. They started shooting their machine guns at us, and they knew people were in the basements.

My dad got his gun out when they started shooting at us. I wanted to go with my dad. But he said, "No, you can't go with me. Go home, go with Mama." We were out of the tent, and my sister came after me and pulled me back by my hair. I was so mad! I thought they would kill him. The shells were going right by my ear. We knew they were going to murder all of us.

We started running, and our dog Princess was running with us. We also had a little puppy that lived behind the stove. Princess ran with us, and they shot her. They killed the puppy, too. We loved them so. We ran as far as the water tank. There was an Irish woman and her husband there, and they were great union people. They had two little redheaded girls, and my mother took them with us into the water tank, where the railroad got its water. It was real big and under the ground about three stories. We went down the steps and sat on a plank all day. There were ten mothers and their children hiding there. My little brother was only six months old, and my mother was nursing him.

It was quiet, really quiet, except for the shooting. We were so scared, but nobody yelled. The union knew we were down there and sent us one loaf of bread that day. My mother didn't eat any. It was hard on the mothers. One of our men got wounded, and he was brought down into the water tank. He'd been shot, and he cried and cried. Then the union doctor came to take care of him.

*An arriving train provided a temporary shield for the mothers and children hiding in the water tank.*

The Irish lady knew that the train stopped there. She was shooting at the scabs with her husband from a little toilet. That Irish lady was a fighter. Another woman who lived by the depot knew we were there, and she was sympathetic. She told the train to stop and give us a chance to get away. The train stopped, and the Irish lady ran and told my mother that we had to get out. My mother thought we were safe there, but the woman said, "No, you're not. They're gonna drown us." So we all got out. It was late afternoon, and we went by the bridge. They were shooting all over, and the union men were scattered. We walked down under a bridge where two union men were shooting. One got shot and fell, and I stepped on him. I was so scared I ran straight through some barbed wire and it tore my Easter dress, but I just kept running, pulling that barbed wire and lace, for what seemed like miles and miles.

We went to a farm and slept all over the place. The people there were so kind. The woman fixed us each a boiled egg for breakfast. The next day the union brought wagons and took us into Trinidad, to a union hotel.

We didn't see my father for five days. My mother cried and cried. When all the men came back to Trinidad, everybody was so happy that it was like they'd come back from a war. My father didn't tell us much. He said they were just fighting everywhere around there, but they didn't have a chance. There were too many of them against us. Nobody knows who won it. I don't know exactly who got killed and who didn't.

The men told us that the soldiers knew people were in the basements. So they got brooms, and they oiled them and lit them like torches, and they burned them one by one until they burned them all. Our Italian neighbor's wife and their children smothered and burned. She was four days away from having a baby. I used to play with her little boy. After they burned down the tents, the scabs grabbed some of the women and children that were still there and took them back to their tents and assaulted them. I heard about it from friends of our family. That's something that's never come out about the massacre. One woman was a good friend of my mother's.

*The union constructed a new tent city at Ludlow, and families moved back after the tragedy.*

After the massacre the men were even stronger for the union. We came back to Ludlow after the union built the tents back. It was real quiet when we got back. The men would talk in low voices so us kids wouldn't hear. Our tent was in a different place. We had our stove, but our birth certificates and everything else got burned up. My dad used to be in lodges, and all his uniforms were burned, and my mother's trunk with all our keepsakes from the old country.

But we were happy to get back. A lot of our friends came back after the massacre. We took the train back to the tents, and they shot at us on the train. My mother put us under the seat. After that the union men didn't care if they got killed or not protecting their families. The company put searchlights on the railroad tracks. The men would say it was so bright that you could pick up a pin at night. They swung those lights all over the colony, looking for our union men. But they couldn't find them. And some of the men knocked the searchlights down.

We only stayed at Ludlow a few months. We went to the colony at Walsenburg and stayed there for a couple of years and then moved to Pueblo, where my aunt was. Then my dad heard about Superior, Wyoming, where the mines weren't so gassy. He went up, and we joined him six months later.

When I tell people what happened at Ludlow, many of them don't believe me. They say that couldn't happen in America. And I say, "You don't know. Times will come again like that." They never had to go through it. I tell them, "I know. I lived it." But I feel sorry for younger people now because they don't seem to understand what is really going on. They need to learn. We were taught that you've got to fight for your rights and know where you stand. That's what it meant to us to live in this country.

I am still proud of what I went through at Ludlow. You have to fight for your rights, because nobody's going to just hand your freedom over to you like it was worth nothing. It's worth everything. That's what I keep with me about what happened to us at Ludlow.

# 2

# A CHILD AMONG STRANGERS
## Elizabeth Stull Crawford

*"When I first went to the mine, the boss said, "Hello, little girl. What are you doing up here?" And I said, 'I'm gonna work in the mine.' He just stood there and stared at me."*

*In the early twentieth century miners were often forced to migrate from place to place to find work. Thirteen-year-old Elizabeth Stull joined the ranks of migrant laborers, following her cousin from Ohio to Pennsylvania after she was "tossed out among strangers" from a large family with too many children and step-children to feed. She was hired at a nonfamily mine and earned her own paycheck, a rarity for a girl in 1920.*

*She was not the only woman miner in her family. In 1934 her older sister Ida Mae was removed from her own leased mine near Cadiz, Ohio, by a state inspector enforcing an Ohio law that banned women from working in underground mines. Ida Mae Stull's success in overturning her eviction was widely recorded in newspapers, magazines, and newsreels. Elizabeth, however, worked underground far longer than her sister did, laboring intermittently from 1920 until the mid-1950s. Both Elizabeth and Ida Mae battled poverty all their lives.*

*Ida Mae Stull died in 1980. My interview with Elizabeth Stull Crawford took place in 1982 at her home in Dillonvale, Ohio. She pulled out her collection of antique tools—a breast auger, shovel, and cap with a lamp for lard oil—and proudly posed with her handloading tools.*

*Elizabeth Stull Crawford died in 1989.*

My dad was a coal miner, but after he died, I was put out among strangers. I was six years old. There were a bunch of kids in my family, maybe 28 with my father's and my mother's put together, but only three of us were put

Elizabeth
Stull Crawford

with strangers. In my natural family there was a bunch of kids: my sister Bessie, Roy and Willie, then the set of twins, then Roxie and Elsie and Idie and Jim and Charlie and me. Some of them have died.

This one old lady my mother put me out with was mean as an old dog. She would pinch me and beat me and wouldn't let me have nothing to eat. I run off from there when I was about eight years old. I went to another farm, then went back home again, but they made me go back with that old woman. When I was 13 I went with my cousin to Pennsylvania, where I started working in the mines.

I was born in 1907 and went in the mines around 1920. Coal was running pretty good then. There was about 30 or 40 people in the mine I worked in, and they loaded coal on cars. They didn't have no other kids, just me. I wore some of my cousin's clothes. They were big, and I had to pull them up. I wore a cap with a pit lamp that had a wick coming out of the top. We used lard oil in the lamp.

*Elizabeth was treated with kindness at the mine.*

When I first went to the mine, the boss said, "Hello, little girl. What are you doing up here?" And I said, "I'm gonna work in the mine." He just stood there and stared at me. But I went in every day my cousin went in. I worked for about a year at that mine, carrying water and loading coal. It was the best mine I ever worked in, because the boss treated me so good, which was important because I'd been tossed out among strangers so young. The boss would say, "That's my daughter back there." He brought me dinner to eat, and soda and candy. I never did have to bring no dinner.

I had my own paycheck, and me just a kid! My cousin never took it. They paid me five or six dollars a day. I didn't get paid according to tonnage, like the men. I didn't get near as much as they got. I carried water and loaded coal and sometimes led a mule out. I liked it pretty well. You could stand up in the coal. The only thing I didn't like was when the water dripped off the roof and went down my back.

I was never afraid. I was just so damn stubborn, I didn't care what it was like. My brothers and other people said it was bad luck for a woman to be in the mines, but I just went right on in. The men there were good to me. As soon as they'd hear something cracking, they'd get me out of there. I would tell people I was working in the mines, and they would say I was crazy. So I said, "Come on over to the mine and find out." And some of them did.

After about a year, I started cutting timber with my brother, mostly mine posts. I didn't like it. You had to handle all them posts, and they were too heavy for me. I was only 14. So I went to another farm, and then to the glass factory, where we made bottles. I liked that all right until the boss got a little funny and started trying to feel around on me. I cracked him in the head with a paperweight and knocked him out.

Then I went to a pottery for a while, and then I went and got married. My husband died after I had three kids, so I started in the mines again. A notion just took me. I knew neighbors and friends at this second mine, and I started making pit cuts and using the auger. You would make two cuts, top and bottom, and then blow it out. You would drill your holes back in the coal, then put in your powder and squibs and light the fuse and run. After it blew, you would load it and push the cars out by hand. Another fellow would help me with that.

*Elizabeth Stull Crawford left the mines during the Depression but then returned to help her older sister, Ida Mae, who had leased her own small mine.*

*Evicted by a state inspector as a "danger of immediate and extraordinary character," Ida Mae took her case to court and got the ruling overturned.*

I got married for the second time in 1933 and started doing farmwork, plowing for people and hauling coal. My old man wouldn't work, so I had to go into the mines again. I had been out of the mines for about four years during the Depression, except to work with Idie.

Idie had her own mine. She had it leased. She went in about 1930 and worked it by herself. I worked with her before I got married for the second time and then after for a while.

One day I stayed home to do some cleaning and washing, and that was when the inspectors came and shut her down. So they didn't catch me. Some days the man she lived with went in and worked, but he didn't go in that day.

I worked with my sister for two or three years, on and off. I'd be making pit cuts while she would be making auger holes. Then we would blow it out and load it and push it out together. Some days we didn't do too good. Other days we would make four or five loads a day, plus the stone. We had a good time together. She was good to work with, but honey, don't get her mad! She was like me. She had fire in her behind.

Her man treated her mean. He took all her money and wouldn't let her eat. He'd hide and wouldn't go to work, and she'd have to wash clothes for people to make a living before she went into the mines. One time he was supposed to go to work. She heard the turkeys raising the dickens. He was out there hiding under the straw stack, but his feet were sticking out. She got some matches and set fire to the straw, and he got out of there in a hurry!

The inspectors shut the mine down and told her she couldn't work there. She fought the case in court, saying she was going back into the mines because housework was monkey's work to her. She won her job back, but the inspectors came back later and said the mine wasn't fit to work in. So they shut her down for good.

After that, she quit the mine and worked in a restaurant in Cadiz, then quit that and her property was sold to the railroad company. She had to move. Then she moved here, there, and everywhere. I didn't see her much. The guy she was with was nuts. People always made a fool out of Idie because she couldn't read or write.

Idie was several years older than me, but I started in the mines earlier, when she was still at home. She liked it in the mines. I liked it, too. You have to have spunk to work in the mines. You have to have nerves. If you're scared, you won't make it.

*Ida Mae Stull left the mines, but Elizabeth stayed in, finding intermittent mining jobs until she was in her forties. At one mine she worked in low coal only three feet high.*

My old man left me on the farm, and there was no money coming in, so I had to go to another mine. My old man left me when I was on the farm. So this fellow from the mine said, "Put on your mine clothes and come out in the morning." I made five dollars a day and all the coal I could use. There was only a few of us working—the boss, his brother-in-law, and another fellow, a colored fellow, my son and me. They were afraid of me. I weighed 300 pounds, and I was solid as an ox. I could throw a 100-pound sack of wheat on my shoulder, run across and dump it in the corn crib, and go back for more. And that was in my forties.

At Dry Fork the coal was about three foot high. We wrapped gunny sacks around our knees to crawl in. I would outwork some of the men. One of the bosses would say, "How come that woman's doing more than you're doing?" And he would try to make them come and help me, even though I was doing more. They would say, "That's her job." We got along good. We never fought or argued. I stayed there for about five months.

I liked loading coal better than making the pit cut. With the pit cut you were down on your knees. And making auger holes was hard work. But I was always healthy. I'd get up about four-thirty in the morning and do my milking. Then I'd get to the mines and be there by eight o'clock and work until about five. I'd go home and do the milking and feeding and cook a bit to eat and then go back to the mine and make the pit cut for the next morning. I'd get out of the mine about midnight. I needed sleep, but I didn't take much, and it never seemed to bother me.

Later on, when I thought about it, it seemed like something to have worked in the mines for so long. A lot of people say a man is stronger than a woman. But women make themselves weak. I think women used to be much stronger than they are now. Now, there's always something wrong with their kidneys or bladder or something. I figure it's their own fault. There's too much dope and cigarettes around. If you don't have brains enough or strength enough to make yourself work, you just sit around.

My granddaughter worked awhile at the mines, but they gypped her out of some money, so she quit. I feel about them going in the same way I feel about myself. It's their decision to make. But that doesn't mean it's a good place for them, because of the conditions. You lose your health in the coal mines. You get short-winded and have arthritis and rheumatism. The dampness is what gives you the arthritis.

Everybody now tells me to quit, to sit down. If I sat down, I'd die tomorrow. Call me old-fashioned, but I'm still kicking.

# 3

# HAND-DIGGIN' TIMES
## Elizabeth Zofchak Stevens

*"It was Depression time, around 1930, and I was 13 years old. My father took me to work in the mine dressed up like a boy. I wore shoes like the Boy Scouts, with a little knife in the side. . . . I liked going off to the mine in boys' clothes, with my hair cut short. It made me feel good to know I could help the family."*

*Young Elizabeth Zofchak fidgeted impatiently as her mother moved the cold scissors near her ear and snipped. Dark hair feathered the floor. Her heart beat faster as she thought of going with her father to the coal mine. She pulled on trousers and stuffed her short hair under her cap. She felt confident that no inspector would think she was a girl.*

*In 1930 the nation was sinking into the Depression, and the Zofchak children were needed to help dig coal to trade for flour, coffee, and other staples. John Zofchak taught young Elizabeth and John to lie on their bellies to dig out the bottom of the thin coal seam with picks, then shovel it into a pony-drawn car.*

*Now in her seventies, Elizabeth Zofchak Stevens holds a deep respect for her parents' strength and their sacrifices as newly arrived Czechoslovakian immigrants forging a new life in America.*

*Our first interview was conducted in 1982 in Woodsfield, Ohio, not long after her first husband had died. A decade later I talked with her again in Poolesville, Maryland, where she had moved after remarrying. She lives now in a historic home where she and one of her daughters run a business selling handicrafts.*

When I was born, my mom said my dad didn't speak to her from Wednesday to Sunday because he wanted a boy. Men wanted boys to help them. But I hung so close to my father when I was little that I might as well have been born a boy. If I weren't helping him in the mine, I'd work with him in the fields or garden. When our shoes wore out, I helped him cut the

Elizabeth Zofchak Stevens

patterns to repair them. I felt closer to my dad when I was young, but my mother was the one who taught me how to survive. She came from Europe alone to New York City when she was 15. The crossing took 30 days, and she was sick the whole time and slept on the deck on a feather comforter her mother had given her in Hungary.

Life wasn't easy for my mother. I think coming to this country was easier on my father, because he had brothers and sisters here, but my mother came alone. She planned to save her money and send for her older sister and younger brother. She did send for her sister, but the sister didn't work outside the home, and my mother got married and started having babies. Mother had eight children, with two years between each two. She couldn't afford to send for their brother, and they never heard anything of him again. Later she told me she had planned once to go back to Europe to see her parents and got as far as New York. But my father called her and said my brother was crying and she needed to come home. She never tried to go back again.

Besides raising eight children, my mother took care of the boarders and other men looking for work. She cooked and did their laundry, and they paid whatever they could. Later she went to work cleaning offices at Good-

Elizabeth (far right) with her parents and siblings. Courtesy of Elizabeth Stevens

year Tire and Rubber Company and worked for 25 years, until retirement.

Thinking back on it, I realize how strong my mother had to be. My mother told me she was out pitching hay, and she had my brother John in the field, and she just put him in her apron and brought him home and then called the midwife. When I was born, she stayed in bed a few hours, then got up and fixed dinner for the boarders.

Another time one of my sisters burned to death while Mom was at the store. Our neighbor was supposed to be looking out for us. She was burning trash in the street, because we had no sidewalks. We had on long dresses, and my sister got too close to the fire. She was four and I was two. Another sister died in the flu epidemic. Mother always kept a picture of the two girls in the hall.

*Rubber ruled Akron, and jobs at Goodyear and Firestone drew thousands of immigrant workers from Europe.*

We lived on Seventh Avenue in East Akron, Ohio. The neighborhood was Slovaks, Ukrainians, Russians, and Polish. Some black families who worked at Goodyear lived among us there. We weren't allowed to speak English at home. My father said, "When you're in school, you can learn." Akron was split into four sections. The Italians lived in North Hill. The more educated people lived in West Hill. People who worked for Firestone Rubber Company lived in South Akron.

Nobody I knew worked in a coal mine but my father. More people worked at the rubber shop. They were putting in roads and putting in sewer pipes. Working in the rubber shop was easier than digging ditches all day. Before my father leased the mine, he worked 12 hours a day making bricks and building sewer pipes.

Before we had the coal mine, the coal cars would drop coal off on the railroad tracks. My mother would say, "Why don't you take the bucket and see if you can find any coal?" My sister Ann and I picked up coal that fell off. Next to the tracks was a feed mill, and at night we would jump in open boxcars and scoop corn and wheat out of the cracks. We might come back with a bucket of corn for the chickens. Then if you found any wood along the tracks, you brought that home, and that went into the furnace. We would pick berries for jelly and wild apples in the fall. We gathered mushrooms and dried them and strung them so they lasted through the winter.

*Elizabeth's parents maintained their European customs and nurtured their Catholic faith.*

For punishment, my father would make me kneel on corn kernels, which was an old custom. My mother said they dumped me in cold water when I was born so that I would never get a cold, and I almost never did.

We were Catholic, so the Easter holiday was very special. Each woman prepared a basket of food for Easter and embroidered a special scarf to put over it. My mother's was roses all around. The women cooked special food for their baskets. Crosses were put on the bread, the butter was shaped like a cross, and you made sausage and hard-boiled eggs and horseradish. The night before Easter we took the basket to be blessed by the priest and didn't touch it until after church the next morning. At the table my father was always fed first, because he was the worker, and then all the kids would eat. My mother would eat last, while she was serving the food. It was an old custom with her.

We had Christmas with the old calendar, on January 7. Every Christmas they'd spread hay over the linoleum to represent the manger, and they'd put nuts underneath it for us to hunt on Christmas Day. My aunt and uncle did the same thing. Their two boys were picking up the straw and putting it in a big stove, and they caught on fire and burned to death.

We always had enough to eat, but we all worked. Every year my father butchered a pig, and we made lard. We pastured a cow and a pony on land Goodyear owned. My mother had geese, ducks, and chickens, and my dad raised pigeons. She would pluck the goose feathers to make pillows. When each of us got married, she took the feather comforter she brought from Europe and made six little comforters for us. She made almost everything out of feed bags: dresses, underwear, sheets, curtains, towels, and aprons.

In East Akron, neighbors watched out for each other. A Polish woman who was our neighbor started selling bootleg whiskey when things got rough. The police came, and the neighbor hollered for my mother and threw a fifth of whiskey through the window. My mother grabbed that whiskey and hid it in a sewer pipe. One woman had a lot of children, but she lost her milk. My mother always had milk because her children were born so close together. Mom would go three times a day to nurse the baby so it wouldn't go hungry.

*When the Depression hit, John Zofchak leased a small mine near Talmadge, Ohio, and took three of the children to help dig and handload coal. Her parents disguised Elizabeth as a boy.*

It was Depression time, around 1930, and I was 13 years old. My father took me to work in the mine dressed up like a boy. I wore shoes like the Boy Scouts, with a little knife on the side. My mother cut my hair off so nobody would recognize me, and I stuffed the rest of it up under my hat. I wore pants, overalls, and high-topped shoes. I liked going off to the mine in boys' clothes, with my hair cut short. It made me feel good to know I could help the family. Nobody paid much attention, because everybody had to work.

My father took three of us children to the coal mine so my mother could do her work around the house. My brother George was seven, and my brother Johnny was nine. Taking George to the mine, Dad was like a baby-sitter. Early on Saturdays he would yell up the steps, "Which of you kids are going with me?" One had to stay home with mother. The boys wouldn't, so my sister did. I wanted to go with my dad because I always hung so close to him.

It was a small mine under a big hill that my father had on a 99-year lease. I'd go to school during the week and go with my father on Saturdays. We weren't allowed to work on Sundays.

The opening to the mine was so low that we had to walk in stooped over. Before we went in, my father would set his pick upright and lay his leather cap over it. We knelt down on the dirt while he said a prayer in Slovak to ask God to protect us underground. We would repeat the prayer word for word. He'd do the same thing at the end of the day, to give thanks that we had come out safely.

Coal mining was different then. This was hand-diggin' times. We all had a pick and shovel. Dad would put us in a room, and we would undermine the coal. We worked a lot kneeling down or lying on our sides, with a potato sack as a cushion. We'd dig underneath the coal so it would come loose. We'd make a trench at the bottom of the coal seam and knock the coal out with our picks. We didn't have an auger to bore with, and we very seldom shot the coal.

In the mine we could hear the trucks going by overhead on the road. I was always scared the road would cave in and the trucks would fall in on us. Dad always checked the ceilings for cracks. He went first, and we would follow. We built wooden rafters on the side and on the top and laid a wooden track.

We all had hats with carbide lights. Mom would fix them in the back so they would fit. I'd fill up the carbide and make sure there was enough. Even George had a light so he could see to go outside if he had to go to the potty outside.

Mother would pack our lunch, and we'd heat it up on a little coal stove in a shack outside the mine. She would pack meat or chicken for us, and she made us pickles. I'd come out and warm the food, and my father and brother would come out and eat. Then I'd clean out the shed. I would use a broom or tie some weeds together and sweep it out with that. We even had a cot out in the shack so you could lay down if you were ill.

There were six or eight rooms in the mine. My father would be in one room, and we'd be in another one. If I dug a bushel of coal, I would call for the wagon and dump coal until we got half a ton. Then we'd load it onto little coal cars. We had a pony, and he pulled the cars out, half a ton at a time, until he went blind from the darkness. Then my two brothers and I pushed the coal car ourselves, and my brother Joe delivered it in his truck. It sold for a dollar a ton delivered.

All our coal we dug for a purpose. We dug coal to buy flour, salt, sugar, and other essentials. There was a store near us, and we'd trade coal in the winter for groceries all year. One time the bill was several hundred dollars, and we paid all that in coal.

*The arrival of a mine inspector would send Elizabeth scurrying back into a mined-out room.*

Our neighbors had another small mine across the road, and when the mine inspector came, they sent their son to warn us. We'd just go back into one of the little rooms and hide from him. He had a carbide light, so we could always spot him. It was like playing hide-and-seek. I would go in a place we had already dug out. My father could have gotten arrested for having a girl in the mine, but I never got caught.

I put in three winters at the mine. My dad gave up that little mine when the coal seam ran out, and I had to quit. In 1939 he went off to work in a deep mine with my brother Joe. I couldn't go because it was a union mine, and they didn't let women work. But I had to keep working, so I left high school and got a job doing housework. In one job I cleaned house and cooked for three people and a baby and did all their laundry for a week, and earned three dollars and a bus token. I thought they were wealthy, but now we would call them white-collar.

*A few years later Elizabeth Zofchak married, then worked in the defense industry during the Second World War.*

I married when I was 19. We had met at a carnival. John's mother had died when he was eight years old, and he was in the children's home for about four years. He didn't have a real sense of belonging. He lived with his sister for a while, and then he moved in with a man who taught him how to fix shoes. Then an Italian family taught him the cleaning business. He stayed with them till we got married. He got a job through the WPA [Works Progress Administration] in 1938. We thanked God for the WPA. It paid $57.20 a month. He couldn't get in the service because he didn't have a kidney. Then he worked in the rubber shop and the arsenal.

During the war I worked for Goodyear inspecting gas masks and making rubber rafts. I didn't want to leave my children with baby-sitters, but I felt like I had to. Women did a lot of good by doing the war work, but I think the problem of divorces started in World War II. I think it was a bad time for marriages. Maybe I'm wrong. If I'm working side by side with a man, I believe that I have the right to get the same salary he does, but I don't go for some of the things they have pulled on the men. Some of them will do anything to climb up. I guess I'm just more the home type.

I know there are times a woman has to stand up for herself. In my generation, some of us wanted to fight back. I was always trying to be my own person with my first husband, but it was hard. If I was going shopping, he would want me home in an hour. For 44 years he bossed me around. But now that I've grown up—I say I've grown up in my old age—no man is gonna tell me where to go and when to come home and what to spend.

After 40 years of marriage I told him: I don't have any babies, and I don't have to take this anymore, and if you don't quit drinking, I am leaving. I have my own pension, and I'll do what I damn well please. But isn't it a shame that it took me all those years to stand on my own two feet? I'm 75, and I finally feel liberated.

I believe there's a time in your life when you think that fighting back isn't worth it. You just go with the flow and do whatever makes the man's life better. In our generation we spoiled a lot of men. My first husband was good-looking. He was a boxer. Women flocked to Johnny like honey. He was a good man who would help anybody who needed it. I've been married to two different kinds of men in my life. Johnny was a hard man at times, and then I married a gentle, quiet one. In your lifetime you have to know what true love is, and at this point I think I have it.

Even when I was married to Johnny, I did men's work. I used the power saw and the tractor and repaired the windows and repaired the roof. If I was hungry, I could go out and shoot a deer. I could kill a snake. I'm trying to teach my grandchildren about the woods, how to recognize ginseng and bloodroot and wild lettuce. It's a shame that kids in school don't get taken in the woods for a week to teach them survival.

Women who go in the coal mine, they can't be weak, they've gotta be able to take the slurs that these men are going to throw at them. They're going to have to hear that cussing because the men aren't going to stop. So if you can't take it, you better stay out of the coal mines, because these guys will harass you. Don't you think coal-mining women are different? Don't you think they have to be strong to fight for their rights?

In my mother's time, women had to be strong in order to survive. They were a different breed of women than most you see today. After my father died, my mother managed. He didn't have a bit of insurance, because he was a coal miner and had his own business, and she had three children still at home. She still didn't speak good English, but she worked at Goodyear for fifty dollars a week, and she managed to send my brothers through school. She made house payments and got a loan to pay for the funeral, but she survived.

Women were stronger then. They weren't sick as much. We never had aspirin in the house. And we were taught not to waste things. We were careful with water. We put the baby in the bath first, then we bathed in the same water, and then used it to wash rugs and scrub the porches. The garbage was fed to the chickens. The apple peelings went to the animals, and potato peelings were put in the ground to rot. Somebody on TV said, would women be willing to go back to what we were and not waste? Give me a kerosene lamp and a coal stove and I'd be satisfied. I've tried to teach my three daughters how to survive. They are freethinkers who will fight for what's important to them. Some women depend on a man to do their thinking, and they get weak. We need strong women in this country.

So that's my life. I am proud to say I was a woman coal miner. And I give thanks to my mother, a coal miner's wife, who was the backbone of our family. Our life was hard work, but the experience taught us to deal with whatever happened in life and go on.

# 4

# GIRL MINER
## Ethel Day Smith

*"Back then we dug and opened the mines. Dad would find a place where the coal was, and we'd face it up, bore holes, blast it out, and set our timbers to follow the seam."*

*At 15, Ethel Day mined coal with her father and her younger brother to supply the coal trains that crossed Pine Mountain, Kentucky. They dug out the portal, shored it up with timbers, set explosives, and laid track. Finally, they would strap on heavy breast augers to drill the coal by hand. The family worked on contract for an operator who was unaware that the group included a girl. Her mother occasionally joined them in the mine, although she was occupied with raising 13 children.*

*After her marriage Ethel Day continued to work with her father despite her husband's objections. Her first husband was killed in the mines in 1953. She later married a miner who worked with women hired in the 1970s. She expresses reservations about feminism and the new generation of women miners, even as she is aware of the conflict between her belief in women's abilities and their "place" in the world of work. It is a view expressed by several early women miners who did men's work but toiled within the family sphere.*

*Our interview was conducted on a hot summer afternoon in 1980 at Ethel Day Smith's home in Evarts, Kentucky. I found her in the garden hoeing among the broad rows of corn, squash, and tomatoes. She handed me a paper sack of ripe tomatoes, wiped the sweat from her brow, and motioned me toward the porch.*

I was raised on Pine Mountain, Kentucky. My parents were Saul Day and Mary Susan Day. They raised 13 young'uns in the Depression. Times were hard, but we never thought of it that way. We were healthy. My dad lived to be 90, and my mother lived to be 104. When one of us got sick, Mother would know how to cure us with some weed or something, and we were

Ethel Day Smith

never hungry. Dad would take us and grub new ground, and we'd go up in the mountains with our coffee sacks and get pine knots for the cookstove. We cooked a lot with pine torches.

There were no public works back then. My father took little jobs and had to take us children with him to help so we could make it. My dad had worked us children at so many hard things that coal mining just seemed like another family job to me. He worked us hard, but he was good to us. We had done logging. My dad was master of the logs. Me and Bill would saw trees. Then we'd skin one side of the logs and bald-hoot them out of the mountains. That's when you take a log on its skinned side and use a hook to push it down the mountain to a level place where a truck would pick it up. That work was a lot harder and more dangerous than the mines. The mines could have fallen on us, but those trees could have, too. One time we couldn't get a horse to drag the logs, so we dragged them with a chain. I fell off a cliff and broke my arm, but I just went on.

We took a mining job, me and my dad and my little brother Bill. For four years or longer we got out coal for three trains that ran across Pine Mountain. In the family there were six children older and six younger than me, but me and Bill were the only ones Dad could get to work with him.

Back then we dug and opened the mines. Dad would find a place where the coal was, and we'd face it up, bore holes, blast it out, and set our timbers to follow the seam. We put railroad ties down and then put in the rails. The company paid for materials, and Dad got paid for the tonnage. We never knew how much Dad got paid. Me and my brother was just big enough to bore them holes with a breast auger. I was born in 1913, and I was 15, so that was about 1928. Bill was two years younger, around 13.

*Superstitions existed about women in the mines, but the family's economic need was stronger.*

Old folks believed it was bad luck for women to be anywhere around the mines, but my dad had worked us at so much that it just seemed like any other job. He never said anything special to me about being a girl. He'd just say, "Let's go, children." My mother was kind of scary about us going in. Mother said many a time she would rather be in the mine than her children. She helped Daddy in the mines when the water would be above our knees. She would shovel coal into the cars, and if we didn't have a pony to haul it out, she would push the car.

She didn't like it when Dad took us in on Sunday evenings. One night she got a big bottle and gave it to my brother to scare us away from working. He hid behind a big rock and blew on it, like a warning. My dad didn't take us in because he thought the warning was against taking me in because I was a woman. But I was never worried a bit. Me and Bill weren't a bit afraid. We dug out a big hole back in the mine and would go hide from each other in there.

Me and Bill wore a cloth cap with a carbide light that hooked on with a piece of leather. I wore old overall pants and ordinary shoes. We didn't have money to buy gum shoes. That was the only way we had to make it. Seems like I loved it, though. I loved the coal mines.

*Ethel and her brother not only drilled, dug, and loaded coal but set off explosives to loosen the coal seam.*

We'd start in the evening and would clean up any gob that had fallen down. Then we'd bore our holes with that big breast auger. That was hard, honey. Me and Bill was small, and we had to put that auger plate across our belly to hold it up. Then we'd get the dust out of those bore holes and make dummies out of ordering books like Sears catalogs. We'd roll our paper and fix it like a big stick of candy, with a four-inch fuse. We'd put dynamite or black powder in back of the dummy and tuck it in real snug.

Bill would start at one side, and I'd start at the other, and we'd meet in the middle. We'd set off ten or 15 at a time, and you could hear all those fuses just a-spewing, and then we'd take off a-running. If the blast didn't go off, we'd wait until the next day to check them. We would go back in those rooms where the coal was shot down and take a pick and dig out all the loose coal. Then my brother and I would haul it out with a pony and dump the coal in a chute. We would dump up to 25 cars a day or more. We would pull a big old rag over one of the pony's eyes to blind it when we come outside, so it could see with that eye going back in the mine. Bill was younger than me, and I could do a lot more work than he could.

We worked all the time. We'd shoot the coal of an evening and clean it up of a day. Sometimes we would go in the morning and come out after dark. Mother would pack our dinner buckets with roasting ears and fresh milk and cornbread. We didn't worry about the danger. As long as our daddy was there, we felt like everything was all right. He would peck at the roof to see how it sounded. If it sounded hollow, he would pull it down. We didn't have any fans or any equipment. There wasn't much dust, and we used a hand pump for the water.

We worked at another mine. The man who owned it never knew I worked in there. If they had, they would have fired my dad, because women weren't supposed to go in the mines at that time. But everybody else around there knew I was working in the mines. They said they wouldn't go in there for no amount of money. My sister went in and helped some on one of our mining jobs. She helped push out the coal cars because there was too much water. It was hard, because it was uphill. She was about 16 years old. Later the doctor said pushing those coal cars made her womb drop down and caused cancer.

*Later Ethel Day married a miner but continued to help her father dig coal on occasion.*

My first husband didn't want me under there helping my dad. He thought I should only work at the house, but I didn't think so. I had worked all my life, and I love to be busy. And Dad didn't have anybody else helping him, and I couldn't stand the thought of that.

I was always strong. The day one of my babies was born I walked several miles and hoed corn all day and had the baby that night. I always worked when I was pregnant.

There was fighting around here over the union. Me and my husband had walked over Pine Mountain to get his pay at Evarts, and they had shot and killed some people that day. That was in 1933, just after my first child was born. The guards would go shoot at windows or take the men's dinner buckets and bust them up.

Union, honey, is a fine thing if it's used fine. I tell people union and holiness is two of the best things I know of, if they're used right. But it seems like both of them get out of hand.

My husband got killed in the mines up here in 1953. A big rock fell and mashed him against a timber. He was a good man. I had five children. My baby was six months old when he got killed. He left me something to raise the children on, but I got so tore up and lonesome that I went to work at a boardinghouse.

*She had mixed feelings when coal companies started hiring women miners in the 1970s.*

We never did think women would be going into the mines. I couldn't believe it. It made me feel funny, like the end time is coming closer. The Bible is fulfilling itself. It seemed like the women were trying to take over. Now, I believe women have to work, because times are hard. Women have the right to do anything they want to do, but I don't think it's nice. I think women ought to be where they can work decent. I just don't believe in going in them big mines with that blackguarding, bad-talking bunch of men.

I think it's good for women to have a job. When you've just got a little home to clean up and children to keep, it's a pretty dull life. I don't blame no woman for trying to make a few dollars. Men don't give women what they need. And you can't make enough to live [on] in restaurants and little stores. But they shouldn't try to take over. Eve betrayed Adam, and some women go too far. I think that the government will say, if women can work in the coal mines, that they can go fight in the army. I believe that's what will cause it. I just hope they don't.

My husband works with women. There have been divorces over some of them, but I'm not that jealous. I tell him to keep his mouth shut. But men can tell you anything. If a man wants a woman, he's going to get one anyhow, so I don't pay it much mind. And a coal mine is a filthy place! All that would go on is probably just their bad talk. Men gossip more than women. They will pull a woman down.

A woman's got her children on her mind in a dangerous place. She's worried about what would happen if she got killed. I believe a woman would watch out more than a man for safety. Men just get so used to the mines that they forget about the dangers.

# 5

## LIVING ON THE WILD
## Irene Adkins Dolin

*"We dug coal in the summertime, because you had to have fires to cook with. And we dug coal in blue-cold weather, with snow knee-deep. . . . Lots of times it was so cold, we'd crawl in that coal bank just to get warm. We didn't fear no danger, even with the whole mountain over top of us."*

*Family labor in the mines typically was a matter of daughters and wives helping the male breadwinner produce coal in small mines. In the late 1930s, ten-year-old Irene Adkins and her seven-year-old sister worked underground without an adult. Their father, a miner, was ill with tuberculosis and could not accompany them into the coal hole. They worked in darkness, since he considered carbide too dangerous for the girls to handle. The girls mined coal mainly by touch, feeling for the seam after their father sketched out a map of the mine. Irene and her sister crawled in and dug out the seam, filling 100-pound coffee bags with chunks of coal, which they dragged home.*

*Irene Adkins Dolin knew poverty and homelessness when she was young. When the food canned from the garden was exhausted by late winter, the family "lived on the wild," picking berries and hunting small game for food.*

*Our interview took place in 1992, with family members present, in the living room of her home on Camp Creek near Julian, West Virginia. Widowed and in her sixties, she drew herself up with dignity to begin her story.*

We were on the road a lot, my mother and dad and seven kids. We walked and hitchhiked, because we had no home. My dad would work in the mines in Kentucky, and then he'd leave there and get on at a mine in West Virginia. Then he'd leave again. Sometimes we'd go to a place, and he wouldn't get the job, and we'd have to leave again.

Sometimes it would take a day and a night, and sometimes it would take two days and nights. At night we kept walking. We went to Floyd County, and on to Johnson County. One time in the winter we got a ride

Irene Adkins Dolin. Photo, Tom Johnson

with a man in an A-Model Ford. It broke down, and we almost froze to death.

I remember one time we come from out there, and my grandmother in West Virginia didn't have too much room. She had some of her children home, too. She told us that we could clean up the chicken house and live there. So we cleared the chickens out and had the ground for a floor, and we lived in one room for about a year. I was about nine years old.

Both sides of the family were from Kentucky. My Grandma Adkins was a full-blooded Cherokee. They adopted her out amongst white people when she was seven years old. She didn't talk too much about it.

My mother was doing something all the time. You couldn't set her down anywhere. She'd be washing them clothes, boiling that water with lye in it, and you could hear her for miles, just a-singing those religious songs, like "I'm Comin' Home" and "Farewell, Stranger." She sung in the church choir, and she had a brain like a computer.

*Mine wars broke out in southern West Virginia in the early 1900s when the United Mine Workers attempted to organize nonunion coalfields. Following a shootout in the town of Matewan in 1920, 10,000 miners marched toward nonunion Mingo County during the Battle of Blair Mountain, not far from where Irene Adkins Dolin now lives.*

In 1918 my parents moved to West Virginia and my dad got a job in the coal mines. My daddy fought at Blair Mountain. He was on the union side. There was a scab up there who raised trouble with my dad. Daddy catched him not looking, got him by the neck, and put him down behind a log in the water and held him there until he drowned. Dad was strong. We never knew that happened until he passed away.

My daddy drank a lot. He'd come in and cut shines and jump on the family. My mother would take the kids and leave. One time I had a stone bruise and couldn't get out. I was in bed with my foot tied up to the ceiling. I heard him coming in the door—I was six years old, but you don't forget—and I jerked that rope loose and got down under the bed. My daddy came into the house and got down on the floor with a knife, and he was searching under that bed. I never breathed, because the knife blade went across the edge of my blouse. I never moved until he left the house. I was still there when Mother sent a boy to get me and take me to where the others was at. You talk about a hard time? We had it. But I still loved my daddy. I forgave him.

*When her father became ill, Irene and her younger sister took over the task of digging coal for the family. They worked without adult supervision in near-total darkness.*

My father got tuberculosis from the mines. I was ten, and my sister was seven. We had to go in and dig the coal to keep our family warm. My sister and me wouldn't let my mother go in this coal bank. I would rather have gotten killed myself than let her go in. We just said, "We'll go. You stay outside." She was scared, but she knew we had to get the coal. She'd had six boys before I was born, but the older ones left, and our younger brothers were too small to go in. So we went in the coal hole.

We had to get the coal, or the family would have froze to death. The coal hole was only about three foot high. There were probably six or seven posts in the whole place. We were back in there with picks. We'd dig, and it would start popping, and we'd quit, and then we'd start digging again. If it started cracking, we'd go into another place and start digging there. Daddy told us to peck on the rooftop, so we'd peck on it. But we didn't know no danger. All we knew was that we had to get the coal out.

We were lying down, and I'd take a sharp pick and hit it. I'd be digging out that coal and hear it falling. I couldn't see it, but I knew it was falling. We'd lay our picks down, and me and my sister would push the coal into a

100-pound coffee sack until it was almost full. Sometimes we'd fill up two sacks, because it was a long trip crawling back out to that hole.

Then we'd drag the sacks home, one at a time. We'd each grab hold of one corner and drag it out, and then come back and drag another one. We'd make five or six trips back in there to get the coal. It was hard. But my mother and father was freezing, and we had no choice.

Daddy had carbide, but he wouldn't let us use it for a light because it was too dangerous. If you didn't know how to use it, it could blow up in your hands. We was afraid of it. We just went to the face and felt around to dig the coal out and then put it in our sacks. We felt our way. We knew we were hitting coal, and we'd sack it up and drag it out. We dug it in the dark, we sacked it in the dark, and we drug it out in the dark until we got out of that hole.

We could see the opening of the mine, and it would guide us back out. We knew the mines like a book. Daddy told us where all the rooms were. We'd feel around and find them, and we didn't forget nothing. He told us to watch and not get lost. And he said, if it started popping, to get out of the face. Sometimes big boulders would fall down. One fell down right beside us one day. It could have killed us, but we weren't even scared. We knew we had to get the coal out, so we just moved over to the next room.

*Their mother worried about the girls' being alone in the mine and would wait outside the mine mouth.*

Our mother would be out at the coal bank. She'd holler, "Hurry up, girls! That's enough!" And we'd holler back, "Just wait, Mommy. We'll get a little more." She was scared, although she tried not to show it. She would come and stand there a lot of times, thinking we'd been in there too long. She feared it might fall in. But we didn't know no time when we were in there.

It was better for us to be in there than for her. Mommy couldn't have done it anyway, not in that low-down coal. We got all of the wood, too. We done it all, to help. And we felt like we had accomplished something when we got done.

My older brothers would come in to visit, but they wouldn't go in the mine. One time my brother came in from Ohio, and me and my sister were there. He stood out there with two eight-pound lard buckets, while we went in there and filled up those lard buckets with coal. My brother wouldn't go in the coal bank. He had a white shirt on, and dress pants.

We dug coal in the summertime, because you had to have fires to cook with. And we dug coal in blue-cold weather, with snow knee-deep. Sometimes we had to dig out the snow to get into the hole. Lots of times it was so cold, we'd crawl in that coal bank just to get warm. We didn't fear no danger, even with the whole mountain over top of us. We were just glad to get warm.

*Digging coal for heat was a more pressing need than grade school. In the winter the children had a hard time getting to school.*

We went to school, but I'll put it to you this way. We had to walk two miles and half to get to the mouth of that holler to catch the bus to Van Grade School. Wintertime, we had to walk on logs across the creek. They'd be covered with ice, and we'd slide off into the creek. So we didn't go to school very much. I got promoted to the fourth. I'm thankful the little bit I went, but most of our lives was spent in the woods. They just let us run loose and told us what we could eat and what we couldn't, like pokeberries, which were poisonous.

We didn't have a farm or a cow. We would take a can of milk and water it down to make it last four days for six people. My mother said I never took a step until I was three years old, because I had the rickets. My oldest brother carried me around all the time.

We had a hog in the wintertime, and a garden spot, but the garden wouldn't last all year. For part of the year we lived on the wild. We lived on mountain dew, on raspberries, blackberries, and birch bark. We ate all that. We ate it right off the limbs. We'd climb high up in the trees to eat them.

We'd go out and get our own food. We killed rabbits, and we had a dog that would run down ground squirrels. They were good fried. Our dog would run them down and kill them and lay them down for us to pick up. And the dogs would run groundhogs out of the hole. Then we'd beat them over the head with a stick until we killed them.

*When a family member was sick, her parents used medicinal herbs and other traditional remedies.*

My mother boiled up yellowroot all the time and kept it for the croup. When you had tonsillitis, they'd take lamp oil lard and put it on your chest. The next morning it was gone. If we had a cut that was bleeding, Daddy would go to the stove and get soot out of the stovecap and smear it on. They said the kids ate coal for some sickness, but I can't remember that.

My sister was always down in her back, and she always said it was from dragging that coal. She died when she was 51. Sometimes it's hard for me to get my breath. It was dusty in there.

We worked that mine for four or five years, until about 1942. I was 14 years old. After that, I left and got married, and the coal bank just fell in.

My sister left and came to stay with me. Then my two little brothers moved away.

When I was married, I chopped wood and carried tubs of water myself, and I had eight kids—nine, but I lost one. I washed on washboards when my first three children were born. Then I got a wringer-type washer. When my kids left home, I got an automatic washer, and I said I wouldn't go through that again for no amount of money.

I've been thankful for most of my life that my children had a roof over their heads. That's something I never had, and something you don't forget. The problem of homelessness is happening now. I say that if a person's got a roof over their heads and they ain't happy, there's something wrong with them. When they've been down the road I've been down, they'd be happy.

It was a hard life, but we lived it. It's something you don't forget. You try, but you don't. It just comes in your mind every so often. Sometimes I think, how did we ever do that, at our age? I wonder why we weren't afraid. To be girls, we had it rough. We wore boys' hard-toed shoes to school. My sister kicked a girl for making fun of her and almost killed her with those hard toes. The Adkinses have always been a proud people. We don't like to ask for help.

# 6

# A LIFE SWEET AND PRECIOUS
## Ethel Dixon McCuiston

*"Some of the men would come out in the morning and find out I'd been in the mine that night, and they'd be a-cursin' and going on. But the boss said, "If Ethel didn't help Arthur cut coal last night, there wouldn't be no work today. What would you think if your payday came up small?" Some threatened to quit. The boss told them to go ahead, and that I could work any time I wanted to. He told them it was a great honor that a woman would come into the mines to work to help them make bread for their families."*

*Ethel McCuiston brought together the traditions of family labor and wartime work. When the Second World War led to increased demand for coal, she worked with her first husband, Arthur, in a larger mine where management and supported her presence. Unlike that of most wartime women workers, her employment outlasted the war's end. She worked over a period of 14 years, quitting only in her forties after giving birth to twins.*

*Our interview was conducted in 1980 on the front porch of her home in Gilley Hollow, near Cumberland, Kentucky. She was working at the local senior citizens' center and filled her extra hours with quilting and church work.*

*McCuiston proudly recalled the days when she stuffed her hair into her cap and went off to help her "beloved companion." It was a story she told often to the senior citizens, who remembered the days of hardship in the handloading era of the early twentieth century, when the tools of the miner's trade were a pick, a shovel, and a stout-limbed mule.*

*Ethel Dixon McCuiston died in 1986.*

When I was a small child, my grandmother told me about women being bad luck in the mines. "Don't go around those Benham mines, honey," she'd say. "That's bad luck for a woman to be around the mines." They had big gates over the driftmouth to keep the children out.

Ethel Dixon McCuiston

I was born in Benham, Kentucky, in 1918. There were eleven of us children, six brothers and four sisters. My father was a carpenter all his life, and my brothers didn't work in the mines at that time, but they did later. Being raised on a farm, they were afraid of the mines, and naturally Mom was, too. I was the oldest daughter, and I'd try anything once. If somebody broke out a window light, I was forevermore going up to replace it. I was that way all the way down through life.

I got married in 1937, when times were very hard. I kept five or six miners to board besides my family, and I raised a big garden and had cows to feed, along with chickens and hogs. My husband Arthur was a timberman. We had been married about two years when mining started booming around 1939. He got a job at a mine where little ponies would pull the coal out and then they'd put it on the main track with a mule. The men would take the coal out and dump it in the tipple.

*When the United States entered World War II after the attack on Pearl Harbor, many miners enlisted to fight overseas. Arthur stayed on the home front, where the wartime need for coal skyrocketed.*

During wartime the men all had to work overtime. The water would accumulate in the mine, and they had to keep the pumps running. We lived near the mine, and my husband would go out to check the pumps. So I persuaded him to let me go in one Sunday evening, just to look it over, because I'd never been in a mine. I started taking meals in to these miners. When they doubled back on overtime, ours was the closest farmhouse to the mine, and they would send a little boy to tell me to cook and bring food to the mines.

During the war about all the miners were drafted into service, and he lost his helper. I'd go over there and find out no one was helping my husband. Finally I borrowed one of the boarders' belts, and I stuffed my long hair up under the cap. Then I went to the lamphouse and put that big old battery on my hip. The boarders got a kick out of seeing me dress like that. I'd climb on the deck of the motor and go hunt my husband. I just couldn't stand the thought of Arthur working over there by himself. The other wives called me a fool. I told them Arthur's life was just as sweet and precious to me as it was to him. With both of us in there, if one got hurt, the other could go get help. I thought, well, fiddle, there's my husband in there making our living, working for our children. Arthur knew how desperate they were for coal. He would go in there and undertake to do the whole job himself.

I'd go in there and help Arthur shoot coal. I'd make the dynamite dummies, and I'd get down on my knees and shovel coal just like any man. I was really scared sometimes, with that old popping and cracking. Sometimes there wouldn't be anyone in the mine but me and my husband. I would be in the back shoveling dust and watching those big poles holding the top up, and if I saw them a-giving, I'd always holler. He'd cut the machine off and jump. That way, he had a buddy.

I would say a prayer every time we'd cut a place. My husband was a sinner, but he knew how well I loved the Lord. I'd say, "All right, let's have a little talk with Jesus." We'd get down on our knees, and I'd pray while he sat there with his head bowed down. And we never did get hurt. We were in several rockfalls, but it always escaped us. We were married 27 ½ years, and we never once got hurt bad in the mines.

I didn't work regular in the mine until 1941. I was 21. My oldest was two years old, and my other baby was one year. I worked about three nights a week. I would work from about seven in the evening until five in the morning. I would go home and build a fire in the coal cookstove and boil me a big kettle of water and take a bath and clean up. When I washed my things out, they'd be as black as the dust I had shoveled.

Then I'd start cooking breakfast for the day-shift boarders and the school kids. I'd get them off, then I'd make the beds and wash the dishes and feed the cows, then cook breakfast for my husband about nine o'clock. Then I'd go to the field, or whatever I had to do. I didn't need much help. I could fall down and sleep for 15 minutes and get right back up and go on. The good Lord just gave me strength.

Besides that, I had to do my washing on a board. I washed the men's dress clothes and starched and ironed them. That would take one day a week. Another day I went to the grocery store. Two days a week I did cleaning and ironing, and on Sunday I was off. I have been a very stout woman in my lifetime.

Some of the men would come out in the morning and find out I'd been in the mine that night, and they'd be a-cursin' and going on. But the boss said, "If Ethel didn't help Arthur cut coal last night, there wouldn't be no work today. What would you think if your payday came up small?" Some threatened to quit. The boss told them to go ahead, and that I could work any time I wanted to. He told them it was a great honor that a woman would come into the mines to work to help them make bread for their families. Some would quit and then think better of it and come back after two or three months' time.

After a while the men didn't mind my being there a bit. I was even secretary of the union for a while. My husband would take notes, and I would do the bookwork. I wrote all the signs for the men to come to special union meetings. The men didn't care as long as the work got done. I really enjoyed it.

I worked for 14 years, from 1939 to 1953. After the twins came in 1950, I had to slow down. Taking care of them was a two-person job, and I had them when I was in my forties. But I still went back when Arthur needed me. I would not let him work by himself.

*When her husband died, she was left with four sons to raise.*

When I lost my husband, I thought about going back to apply at the big mines. I told one boss, "I've got experience. You name anything in the mines, and I've run it." He laughed at me. This was before women had gone into the mines. I had four sons working up there, and they said, "We don't want our mother up there. If we were to hear somebody cursing in front of you, there would be a fight." They just couldn't stand the thought. So I went to work at the senior citizens' center instead.

After all those years in the mines, I am bothered with smothering a lot. I call it sinuses. Some have suggested it might be black lung. I never put in for black lung benefits, since there's no record of me working. The company would just put in extra money on Arthur's payday.

When women went to working in the mines, it was terrible what people would say. And I'd say, "Girls, back up there. It's not that way at all. If you have to get out and work for a living, I don't blame them a bit for going where they can make the most."

It's bound to be different now for the women. Going in there with my companion, he would tell me what to do. If I was doing wrong, he'd holler and tell me how to do it right. But another man, he might think you were in there for women's lib and not want to show you how to do things.

I tell you one thing, little sister. If all the women would go spend one day, just one day, in the mines and see how their husbands earn their living, they'd squeeze that old greenback dollar two or three times before they'd spend it.

See, I know what it's all about. I've been back in there, and I know. I just say about those girls in the mines, Lord bless 'em! Help 'em! Don't let them get hurt.

# 7

# HELPING THE SOLDIER BOYS
## Alice Fulford

*"If a woman ain't afraid, I think it's a good thing for them to be working in the mines. If I was 18 years old and could get me a good-paying job in the mines, I'd be right in there with them."*

*During World War II a handful of coal companies hired women to work in tipples—the preparation plants where coal was sorted and sized. In 1942 William Beury, the general manager of Algoma Coal and Coke Company in McDowell County, West Virginia, hired five women as "bone pickers." The women picked refuse rock or "bone" from the coal passing through the preparation plant on shaker tables. Alice Fulford worked at Algoma with her niece and three other women. Fulford was in her forties and needed the job to support her children and her disabled husband.*

*The wartime hiring of women stirred a storm of controversy in the pages of the* United Mine Workers Journal, *and the Algoma general manager was the main target of the union's wrath. Under the headline, "Necessity for Women in Coal Mines Does Not Exist," the* Journal *blasted Beury for threatening men's wages by hiring women. In McDowell County, a union official persuaded Fulford to leave her work by promising a job to her husband despite his physical problems.*

*In 1982, when this interview was conducted, Alice Fulford was living with her daughter Beulah in Columbus, Ohio. She was 87 years old, an imposing woman with snow-white hair and a fierce look in her eyes.*

*Alice Fulford died in 1989.*

I grew up in Adair County, Kentucky. We lived on a farm, and I worked in the fields from five in the morning until five in the evening. I've worked hard all my life.

My sister lived in Jenkinjones, West Virginia. When I was young, she asked me to come stay with her and her husband, who was a contractor. I moved there, and there was a passageway in one of the mines in Elkhorn, West Virginia, that we would go through many a time. From the other side it was just a little

Alice Fulford

piece to town. Me and my boyfriend, and a whole lot more, would come through the mines to get ice cream and pop on Sundays. There were maybe eight of us. We rode the motor. The track ran all through the mine, and there were lights on. Any of them boys could run the motor, because they worked in the mine. We had a lot of fun and dressed up nice and didn't get dirty.

When I was older I worked in boardinghouses, cooking for a big gang of men, and I did housework for people. I got married in 1919. My husband was crippled then, but I didn't know it. He worked in the mines as a brakeman on the motor car.

*By the time the United States entered World War II, Alice Fulford was in her forties, with three children. In 1942 William Beury took the bold step of hiring five women to work for Algoma as bone pickers on the coal tipple.*

It was during the time of the war, the Second World War. The boss said we would help the soldier boys if we went to work, because most of them was gone. I think it was 1942 or '43.

Bone pickers working at Algoma Coal and Coke Company tipple, 1942.
Left to right: Viola Vickers, Alice Fulford, Julie Powers, and Minnie Sanders.
Courtesy of Pocahontas Coalfield Centennial Archives, Bluefield, West Virginia

I was married at the time, but my husband was crippled up with a bone disease. He'd had his leg taken off. Since he didn't have no job, I thought I could make more money working at the mines. We were paid as much as the boys was paid. If they hadn't paid us as much, we wouldn't have worked.

Before I got the job at the tipple, I'd been doing housework for a woman who had seven children. Those girls asked if I was aiming to quit that tipple and come back and iron their dresses.

There were five of us women who worked together, four white ones and a colored one. None of us worked inside. We were in the tipple on the outside, where they brought the coal through. In the mine they would shoot the coal down with dynamite and load it in cars and pull it out with an electric motor car. Then it would come down through the tipple, onto tables. The tables would shake back and forth, and we would pick the bone out of the coal. The bone wasn't any good. Then the coal went down on through a chute and into the railroad cars at the bottom of the tipple. We learned fast, and we worked fast.

I was in my forties when I worked at the tipple. It was Algoma Coal Company. One woman was a widow, and she was too old to go to work.

But when she signed up, she told them she was thirty years old, and the doctor asked her where all those barefoot days had gone. We had the best old company doctor there. He was the whole county's doctor.

My niece was the youngest. She wasn't married, but she took care of her mother. So she and I would walk from the coal camp in Powhatan to Algoma. We'd leave about seven and get there about seven-thirty. My sister took care of the children.

I worked about a year. There were five of us women who worked together, and there were women in other places. We had a lot of fun, and the men were nice to us. Most of them were single boys.

*Her daughter, Beulah Staley, adds her own memories. "We had our first new clothes when you went to work at the mine," she said. "I remember you going to work in those big overalls with big boots, and coming home just as black as could be." Her mother nods and continues.*

There was one man who didn't like it much that we women was working there. He looked at me so hard I asked him if I owed him anything. I said, "I am a union member, and I pay union dues the same as you do." We had our picture in the Welch newspaper, and we sent them to the soldier boys. I sent mine to my son in the Pacific.

The union didn't want the women working in the mines. They wanted us to quit. I never did know why. I paid union dues. A head union man came over from Beckley to talk to me. I told him that I needed work to support my children. The union asked me if I would quit if they got my husband a job at the tipple, with his artificial leg. I told them I would.

I thought we were helping the soldier boys. That's why I went to work. I don't know if I could have kept my job after the war. Maybe yes and maybe no. But I would not have fought the union, because they had been nice to me. They'd gotten my husband a job. They went to talk to my niece at the same time, and she quit, too. I don't know what they offered her. The others worked a little longer, but then we heard the union was going to shut the mine down, so the rest of the women got laid off.

After that I worked for the government in a sewing room, making quilts and overalls, and I took in ironing at fifty cents a bushel. I did a lot of housework. The job at the tipple was just another job to me, except that the pay was better.

Women nowadays are doing men's work, and nobody thinks anything about it. If a woman ain't afraid, I think it's a good thing for them to be working in the mines. If I was 18 years old and could get me a good-paying job in the mines, I'd be right in there with them.

# 8

# A GOOD JOB AND A HAPPY TIME
## Madge Kelly

*"We never thought we were making history.*
*We just knew it was a job the country needed us to do."*

*In the western coalfields, coal operators in Utah and Wyoming hired groups of women to work on mine tipples. Union Pacific hired Madge Kelly in 1945, near the end of the war, after she moved to Wyoming from Kansas, where she had worked at a munitions plant. Recently separated from her husband and with four children to feed, she welcomed the paycheck.*

*Although the job lasted less than a year and a half, she recalled it wistfully, saying that it was the best job she had in more than 40 years in the workforce. The rewards were more than financial: she recalled the camaraderie and high spirits of the "boney pickers" as they stood over vibrating tables of coal and rock and unloaded heavy timbers in the rail yard. Madge Kelly soon was promoted to operating the tipple's control panel.*

*At the time of our 1985 interview in Rock Springs, Wyoming, she worked as a cook at the senior citizens' center, which was filled every day with a crowd of retired miners remembering the old days.*

My first real job was early in the war, when I worked in an ammunition plant in Kansas. We filled chemical mortars and 75-millimeter shells with TNT. If we didn't do it right, we'd have to drill it out and start over. It was hard, but it paid good and I liked the work.

I was an only child, and there was only me and my mother. I married at 15, and my husband was on the WPA. He made $17.50 a week, and that's not much to live on when you have four kids. He came to Wyoming to get a job in the mines, and I followed him a few months later. It was 1944. We separated over his drinking, and I eventually divorced him and moved from Superior to Rock Springs. When I came to Rock Springs, the town had around 54 nationalities represented.

Madge Kelly

I needed a job to support the kids, and when Union Pacific took applications from women, I got hired on the tipple at Stansbury in 1945. There were seven or eight women on our shift. There were Slavs and Italians. We were boney pickers and stood over shaker tables that separated the coal from the rock, or boney.

The men brought the coal out of the mine on "mantrips" made up of 32 cars of coal. They took it up and dumped it into a hopper, and then the coal came down to where we worked. It got pretty dusty. We wore bandannas on our foreheads, and sometimes we'd put them over our mouths so we could breathe. Union Pacific built us a bathroom but no showers, so we had to go home to clean up.

It got really cold in the winter. We had a little potbellied stove, but it didn't throw off much heat. We wore long underwear and coveralls and big jackets and gloves. When we had a break, we'd get close to that little stove to get warm.

*Soon she was offered a higher-paying job operating the panel board, which controlled the tipple's production process.*

The company asked one woman who had been there longer than me if she wanted to run the panel that operated the tipple. She refused it, and they gave it to me. I loved that job. I pushed 21 buttons that controlled the coal cars and kept the process moving along. It was one of the most responsible jobs, and it paid more. There was a man who did the same job, and we were paid the same. One time I forgot to push a button, and we had a coal spill and had to shovel it up, and he raised Cain. Then the next week he did the same thing, and he caught it from me.

During breaks at the tipple, when the coal wasn't coming in, we'd have to go out on the yard and unload the railroad ties and other materials. But we were strong women. We all were. When we'd hear the 32-car mantrip with the coal come toward the tipple, we'd run back to pick the boney. We had a lot of fun out there. It was mostly married women. We never thought we were making history. We just knew it was a job the country needed us to do.

I made good money. I made $9.57 a night, $12.46 on Saturdays, and double time on Sundays. We worked seven days a week during the war. The only other jobs around then were working in the cafeteria or cleaning motel rooms. They didn't pay much. I needed the job because I wasn't married and I had four little children.

*The women joined the United Mine Workers of America and attended union meetings.*

It was good to be a union member. We attended the UMWA meetings here in town. The men were good to us. We could go to the local union with problems, and they always helped us. We took part in the celebrations. But people always told the story around here that John L. Lewis never approved of a woman working in a coal mine because his mother had worked underground when he was young. I think Lewis would have liked to have gotten rid of us before we finally did get laid off.

The company wouldn't let us go under to see the mining operation. They said it was bad luck. But then the actress June Haver came to Rock Springs, and they took her down and showed her around. She was going with a Rock Springs man, and we couldn't understand why she wasn't bad luck, too.

One time there was a strike while we were there, and we went out with the men. It was during the war and affected all the coal miners.

*She worked until late 1946, when she was laid off as veterans gradually returned to reclaim mining jobs.*

We knew that when the servicemen came back they'd get their jobs back. That was all right with us. We would have liked to have stayed, but we knew we couldn't. If we'd had the choice, we would have stayed in a minute. I was the next-to-last woman laid off. I worked for 17 months in 1945 and 1946.

After I got laid off I got married to a man who worked at the mine. Then I had several jobs. I made spudnuts—you know, those doughnuts made out of potato flour. I tended bar, worked in a gift shop, worked in a motel, and cooked in the senior citizens' center. I've worked almost all my adult life, around 50 years. The tipple was the best job I ever had. It was a great group of people, and a job with responsibility, and we got paid well and even got a paid vacation. But the opportunities weren't there in those kinds of jobs after the war. The jobs I had later were hard, but they didn't pay much. I hated cleaning motel rooms, especially as I got older.

I got tested for black lung but got turned down. I still have that cough, and I think part of it's due to the coal dust. There isn't a woman I know who has gotten black lung benefits.

I worked at the tipple more than 50 years ago, but it was one of the happiest times of my life. If there was a job open today at a mine and I was younger, I'd be the first in line.

# PART 2

## *The New Miners*

# Crossing the Portal

The coal-mining workforce ballooned in the early and mid-1970s, when coal enjoyed its biggest boom in decades following the oil crisis of 1973. From 1974 through 1978 the coal industry hired more than 138,000 new miners—45,501 in 1974 alone, according to statistics compiled by the National Institute for Occupational Health and Safety (NIOSH).

Accompanying this influx of new workers was the revolt of UMWA miners against their union's leadership. Rank-and-file miners built the Miners For Democracy movement within the United Mine Workers of America after the 1969 murder of the reform presidential candidate Joseph A. "Jock" Yablonski and his family. Rank-and-file miners elected the former MFD leader Arnold Miller as UMWA president and won the right to vote on union contracts and to expand the number of autonomous UMWA districts.

According to NIOSH hiring statistics, the nation's first woman coal miner was hired in December 1973. The first women who gained jobs underground challenged generations of prejudice and superstition. Most were isolated in their communities and on the job. The legal initiatives launched by the Coal Employment Project in 1978 created widespread publicity about gender discrimination in coal mines and a spate of lawsuits that brought women into the mines. By 1986, even after widespread layoffs, the Bureau of Mines reported that 3,328 women were working in mines in the United States.

The interviews in this section reflect the diversity of these women. Ruby Kifer came of age during World War II and served in the armed services as well as on wartime assembly lines before joining the 1970s generation of women miners. Elizabeth Laird entered the coal mines at 54, after toiling nearly 25 years in an Alabama cotton mill.

Younger women who were divorced and raising their children alone sought mining jobs for economic survival. Many had been raised in mining families. Evelyn Luna (Evie Tsosie) challenged the mores of the Navajo reservation by taking a "man's" job after following her father into a uranium mine. In northern New Mexico, Martha Horner followed in her father's footsteps in order to provide for her son. Janice Molineaux, originally from Trinidad, West Indies, worked as a miner and truck driver before becoming West Virginia's first black female employed as a state mine inspector.

# 9

# RUBY THE RIVETER
## Ruby Carey Kifer

*"My work [during the Second World War] was airplane mechanics. It was fantastic! What we did was tear the airplane engines down after so much air time and replace the old parts. . . . I got to fly as the crew chief to test it. The fun part was going into those power dives, when you go up and peel off and go straight down. I loved it!"*

*In her work life, Ruby Kifer crosses the generational divide. She entered the armed services in 1942 and worked on airplanes until she was sent to a military supply house after the war. In the accounting field in the 1950s she felt the bite of sex discrimination before the term was known by most working women. She struggled to raise her children after a divorce and in the 1970s joined the mostly baby-boomer women entering the mines.*

*Her love of adventure and her buoyant energy are still evident. Half a century ago, at an air show, she leaped out of a plane unparachuted and tied to a Congressional Medal of Honor winner. Her deep laugh remains vibrant. Our interview took place in the home that has been her family's for three generations, in the coal town of Chandler, Indiana.*

I live in my grandmother's house. I've lived within six blocks of this house for most of my life. My grandparents raised me because my mother worked all the time. She worked in a tomato ketchup factory, and then she worked in a factory making bullets during the Second World War. Later she did typesetting for Sears Roebuck. And she did odd jobs: cooking at a school, cooking in a country club, and sewing for people. I remember her making dresses for one fat lady who wanted eight dresses all alike, with buttons and buttonholes all the way down the front. She got fifty cents for every dress, and she would give me and my sister a quarter apiece.

My father left us when I was four weeks old, my sister 15 months older. Both my grandfathers were coal miners, and my father was a coal miner.

Ruby Carey Kifer

They worked underground. When I was about five or six they were laid off, and they'd go to the company store for food. We called it the commissary. In this town there was a big concrete-block commissary where people stood in line to get their food, their clothing, everything. Everybody owed the commissary.

There were about six or eight small underground mines around Chandler. My grandfather was an engineer, but everybody was out of work. They might work two days a week and then be off. I remember going with my grandpa to the commissary to get food and blankets. My grandparents never had a dime. Everybody in Chandler worked in the coal mines. My grandfather started when he was 14, and then the small mines started to close. People stuck it out, but everybody was poor.

We had no transients in Chandler. People didn't move in from outside to work like they do now. We were a very close community of about 300 people. Everybody had big gardens. All I can remember is that everybody had a hard time, but it didn't seem that bad to me. I was just a kid. It seems like we always had everything we wanted. Maybe I didn't know there was other things out there to want.

No women were allowed around the coal mines. People were very superstitious. If my grandfather went off and left his lunch at home, we didn't take his lunch to him, although it was only four blocks to the mine. If you forgot your lunch, you did without.

My idea all through high school was to be a nurse. I didn't want to get married. I wanted to support myself. I remember my grandmother and her friends would all gather on wash day or quilting day, and they would talk about what a hard time they were having and how expensive everything was. All my relatives were coal miners—cousins and uncles, everybody. My father's father fell down the mine shaft and broke every bone in his body—and lived to be 96. He was a tough old cuss! He laid in the bed for years it seemed like. Eventually he got up.

*She graduated from high school in 1942, in the midst of World War II.*

When the war started people went off to war, or to war plants, and the mines closed down. I went to join the navy in 1942, when I graduated from high school. They had a recruiting office set up in Evansville, in the window of a big electrical building, which was not unusual. They wanted people to be able to look in and think about signing up. I was already working for Chrysler, but I quit my job to go into the service.

Well, I went in to join up, but the man told me I was too young. I just cried and cried. Men could get in at 17, but the limit on women was 18. So I went next door and got another job. It wasn't a problem, because most companies had defense contracts. I worked that job until I turned 18, and then I joined the air force. I was still mad at the navy.

For a woman to go into the service was considered a bad thing in Chandler. They thought you were going down a bad path. If a woman went to work in a factory, that was almost as bad. In my high school I was the only girl who went into the service. There was a waiting period of maybe two or three weeks between the time you signed and the time you left, and that was the hardest part to get through. I got worried, and I think my parents did, too.

Des Moines was where we had basic training. It was rough, but I loved it. It taught you respect. We went on maneuvers and went through obstacle courses. We got out in the fresh air and snow in the morning and took physical training. I got a lot of awards and commendations, and I could have gone on to officers' training school, but I turned that down.

If I'd been an officer, I never would have gotten to jump out of an airplane without a parachute. I was tied onto the first Congressional Medal of Honor winner in World War II. We were in an air show in Monroe, Louisiana. I was tied to him with a nylon cord, and I jumped out of the plane. I'd never done any parachuting before. They weren't going to let me do it, but the commander wasn't at the base that day, and my cousin was a major, and he okayed it.

I didn't have no sense. You don't have sense at 18. I can't describe what it was like. It was great! It was unbelievable. It was perfect. And every time I see one of those hot-air balloons go floating by the coal mine, I think about being up there and think, man, I wish I was up there instead of down here.

My work was airplane mechanics. It was fantastic! What we did was tear the airplane engines down after so much air time and replace the old parts. We cleaned and repaired the parts that were usable. Then I got to fly as the crew chief to test it. The fun part was going into those power dives, when you go up and peel off and go straight down. I loved it!

*When the war ended, she was transferred to a military supply house in Texas.*

When the bombardiers and navigators started coming back from overseas, they had to get their flying time in, so they bumped us from our mechanics' jobs. They were second lieutenants, and I was nothing but a corporal, so I was transferred to the supply house in Dallas. Naturally I was disappointed. I felt like a misfit with somebody coming and taking the job I was good at.

At the supply house they put the women to going through the personal effects of the guys who had been killed in the war. We had to go through everything and pick out what we didn't think the family should see. It was hard on you emotionally. Sometimes we'd pull out pictures of them with a woman, and we didn't know if it was his wife or a girlfriend. If it was his wife, and he kept her picture with him, she would have been hurt if it wasn't with his things. On the other hand, if it wasn't her, it would have been worse. One day I pulled out a picture of myself and a girlfriend of mine. You'd better believe I took that one out! Mostly we threw all the pictures out, but it's hard on you to have to make decisions like that. And that was a job they didn't have any men doing.

*She married an air force serviceman and gave birth to a daughter in 1946.*

Don't ask me about marriage! You really opened up a can of worms when you said that word! Which one do you want to know about first? The first time I got married was in Texas, when I was in the service. They didn't force you out if you got married, but you could get out, so I did. I was discharged in March 1946. I've still got my papers with me. Ruby Lois Carey, wing structure assembler—not that anybody ever wants to see it.

My oldest daughter was born in October 1946. Along about February I got married. He was a twin from Michigan, in the air force, stationed in Detroit. I decided to come back to Chandler. The marriage just wasn't for me. By the time I got home, I had a kidney infection. It went all through my pregnancy and for about three years more.

When I came home I found that my name was on a plaque where I worked before I went into the air force, for being in the service. That factory had made shells. I was proud of that job. With them, you had the right to get your job back after the war. I'd been a riveter. Ruby the riveter!

After my daughter was born I went to business college under the GI Bill and took accounting. I got a job with the army reserve and kept going to school. All together I guess I went to college nine years, all of it the hard way, at night. But I've never regretted it, because I was able to become a certified public accountant.

*As a CPA, she encountered the obstacles of working in a traditionally male occupation.*

You've never seen discrimination like there was against women in the 1950s in accounting. I didn't know what discrimination was. I would work right alongside of a man, and he'd be making twice as much money as me while I was doing twice as much work. I worked for big companies. At one place I trained 14 people on computers and sold computers along the way and didn't even get a thank-you. That was back when computers were very new. I kept telling them I wanted a raise. Finally I said, "If I don't get a raise, I'm quitting." And I did. They were shocked.

After I got the CPA, I went into business for myself for a while. Then I was elected justice of the peace. They'd never had a woman before, and someone suggested I run. It wasn't a bad job, but I didn't get much help from the sheriff's department or the state police because they were all Democrats and I was a Republican. They kept calling me at two in the morning just to bother me, so I resigned in my second term. But it gives you a lot of pride to be elected to public office. Then I got married again. I didn't have any sense. I guess I got swept away by the moment, but then one day I woke up and said, "What about that dream?" There was no love left.

Are you sure you want all of this true? Next comes my boy. He was born in 1956, ten years after my girl. I wasn't married to the father. That was the biggest thing that ever happened in this town. It was so important they got up and announced it at the Kiwanis Club. The guy was a really big Kiwanian. He's dead now, so I can talk about it. It was big news in Chandler.

Around that time I was working at General Electric as a timekeeper and getting discriminated against. I was on second shift and had 750 people to keep up with on the old punch-type time cards. One day Ronald Reagan came for a tour. He was head of General Electric Theater at that time. We cleaned the place up and put corsages on all our bosoms, you know! He came and sat at my desk and bent down and smelled my corsage, and about that time his contact lens popped out. He spit on it and put it back in and apologized. Maybe that's why I'm a Republican today!

Then I got married again, and this time had a baby. At work the personnel director demanded a copy of my marriage license. I told him that wasn't any of his damn business. He told me to bring in the license or I'd lose my job. When I took a copy in, he told me it wasn't any good, so I got fired. I had never heard of discrimination, or I would have done something about it.

The twins were born after five months. By that time the marriage was on the rocks. The doctors said the babies wouldn't live, because Paula weighed two pounds and Beth weighed one pound. The medical bills were astronomical. My husband had Blue Cross, but he refused to pay for it.

*One of her infant daughters was deaf.*

These notices started coming from the hospitals, saying they were going to sue me if the bills weren't paid. The bills were for $26,000, and that was around 1960. I had to file for bankruptcy, or I'd lose my job as an accountant at a tool and dye company. That wasn't fair. My husband should have been forced to pay those bills. I was discriminated against then. I thought about taking him to court, but what chance would I have had? The judges were all men. I didn't waste my time.

In 1963 the tool and dye place shut down, and I was left with four kids and no help. All this time I had been doing taxes for my friends and everybody else for practically nothing. All this time I was trying to keep the kids fed, and then I had to file for bankruptcy, and then the plant closed and I lost my job. My oldest girl was graduating from high school, and she wanted nurse's training. She went to the Methodist church, and they raised some money for a scholarship, so finally I got a break.

When I had a job I was always looking for another one. I was running around everywhere trying to find a job. I went to a mine site that was under construction. The company kept telling me they weren't hiring, but I could see them grading and putting in culverts. I wanted any kind of job. But with my background, I never dreamed of anything but office work.

I was working for Alcoa for minimum wage when I got called to the mine for an interview. I was exactly 50 years old. I was shocked to get called. They asked me how I was at greasing and cleaning grease. I said, "If you've cleaned the rear ends of four kids, you can sure clean grease."

*She was hired in 1975, the second woman at the mine.*

The first woman started in April of 1974. That woman was refused a job at first, and then she went back with an attorney and a tape recorder, so they hired her. I was hired in January 1975. In October I bought me a brand-new pickup truck with air conditioning and paid cash for it. Very few people can say that.

My family and friends were really happy for me, because they knew how hard I'd had it trying to raise four kids by myself. I couldn't believe it. I pinched myself black and blue because I didn't believe it. When I went for my physical, I was scared to death I wouldn't pass.

If you do your job and keep your mouth shut, you don't have any trouble. I have filed grievances. I won 17 Sundays in a row because they worked somebody out of turn. I've turned in a lot of grievances for other people. And I'm not afraid of turning in a boss for working, or a grievance for anything else.

Any way you look at it, there is discrimination against women in such a male-oriented job. But I've also seen women bid on jobs they couldn't handle.

I didn't really hear the superstition about women being bad luck until I went to work at the mine. It's hard for women to fit in. There are very few underground miners at our mine, but the few that were there respected us. But the young guys had no respect. They said, "Why don't you quit and let a man have that job?" You don't get that attitude from underground miners. They're a different breed. They're a lot closer than strip miners. They look out for each other.

I'd never seen so much money. I went to work on cleanup in the welding shop. Most of the men treated me like a queen. Word got out that I was an accountant, and the local union was looking for a financial secretary, so I took on the job.

I'd never been in a union, since I'd always worked in office jobs after leaving the service. But most of the places I'd worked in had a union that the office people weren't in—even at my first job at Chrysler. At the mine I started going to all the union meetings right away.

The men had a great big bathroom in the shop, but the women didn't. We did have a rest room, and I would take the union check-off sheets and lay them on the floor and mark them off. Things went pretty good until about 1979, when one of the mines closed and I got 64 people transferring into the local in one day.

I started to realize that unions are what protect you—or should protect you—from discrimination. Since then I've stressed to my children, especially to my daughters, look for a job with a union. Even the people in the warehouse who weren't union got a big raise because of the union talk. I think a person's got to have a union. You've got to get stronger, not weaker.

Holding a union office is a thankless job. You put in more hours than people realize, and there are retired miners and widows who call you at all hours to read a paper or fill one out. It got to where it was taking 24 hours a day. Some of them couldn't read or write, and they needed help. After a while I felt like it was time for somebody else to step in.

*At the same time she was grappling with serious health problems.*

I got down to 79 pounds, and I was choking on everything. The doctor found a lump on my neck and found out I had an overactive thyroid. Then they found two big lumps that were malignant. There was 44 things wrong with my thyroid. Even the fluid in my eyes dried out. And the surgeon said I was five days from Alzheimer's disease. I had become a bleeder, and I had a bad heart. They gave me medication, and I gained 39 pounds and stabilized. They took all my thyroid out except for one centimeter.

*In 1977–78 she participated in a national coal strike that lasted 111 days.*

We had a lot of trouble. One evening we had 198 people arrested. I was very active in the strike. I was on the picket line every day. I felt like that was part of my job. Except for one day when there was an ice storm, I went every day during that four-month strike. I'd go out to other mines. Sometimes there was just me and one other guy, and we'd leave at six in the morning to go to these places that were way out from anywhere. No one was around, and sometimes all you could see was shotgun barrels sticking out. A lot of scab coal got dumped in the middle of the highway.

The men were very supportive of me being out there. When you look behind you and see 800 men backing you up, you feel pretty big. At Rockport we had it all worked out that so many cars and pickups were coming from the east and so many from the west. They were bringing nonunion coal to the coal loading dock. We were the first car coming from the west. We pulled up and parked. Being there on that day was the most wonderful thing you can imagine. It was unbelievable. Everybody knew something was going to happen, but nobody knew just what.

They called out the National Guard and the SWAT teams with their big bubble hats on and their billy clubs. They came to our truck, and [one of them] told me I was under arrest. I asked the boy if he had a policewoman out there. He said no, and I said, "I'm not getting out until you get me a woman." So I wasn't arrested. In court they fined us $350,000, but everybody contributed, and we paid all of it off.

*She hoped to work until the Equal Rights Amendment was passed.*

The [social security] agency was set up in 1936 by Roosevelt, and it was strictly a man's world. If I retired tomorrow and had worked with a man all along and we were paid the same, I draw less than what he draws because I'm a woman. If you've been married to someone for ten years, you can draw off them. The miner's pension that comes through the union is fair, but the government isn't.

I paid into social security long before I ever went into a coal mine. What happened to all that money? Somebody else is getting my benefits. The only way I can ever get full social security is if the Equal Rights Amendment goes through. It's not fair. A lot of women don't realize it. Take a teacher who's never married, or a woman doctor. Some are going to have a rude awakening. My mother, who lives next door, is 81 years old and living on $284 a month after working for years at Sears Roebuck. They didn't pay a pension.

*Ruby Kifer retired in 1992 after working more than 15 years at the mine.*

# 10

# WHAT I HAVE TO DO
## Martha Horner

*"Because of the mining paycheck, I could support my child. I had money to help my parents. It gave me the freedom that I didn't have to depend on anybody but me."*

*The mines of northern New Mexico reflect the rich Hispanic and Native American heritage of the southwestern coalfields. Martha Horner was born to those traditions and enjoys digging into history when she has the time, which hasn't been often since she was hired at Pittsburg & Midway's York Canyon mine in 1975. She has worked there on both underground and surface operations. Having spent nearly two decades in mining, Horner describes the growth in her awareness of mine safety and the resolution of conflicts with local union members.*

My grandfather came from Taos to work in the mines near Raton, New Mexico. My mom's side is Taos Indian and Spanish, and my dad is half-Spanish and half-Irish. In Raton, all the coal miners and the Spanish people lived on the east side of town. The whites lived on the other side of town. They always got the better jobs, you know.

My grandfather started with a railroad company that eventually became Kaiser Steel. He was half-Indian, and he retired from the mines and died of black lung. My father was a coal miner, and he broke his neck and his back in the mines. He's disabled now. I also had a few uncles on my mother's side that worked in the mines near Raton. One of the mines was called the Swastika mine, but that was changed during World War II. The last coal camp was Koehler, and it was closed in 1957.

Growing up, I remember going to pick my dad up, and him coming out dirty and making $30 a week. During layoffs they worked at nonunion mines that used mules. I'd hear them talk about conditions and about how they

Martha Horner

didn't get paid for the coal they loaded that was mostly rock. There would be three or four days they would work for nothing until they got their place set up. My dad would blast coal for my grandfather and some other people who didn't know how to. They put their checks together to make ends meet.

When the mines were on strike, they had families to support, so they went and worked the nonunion mines. There were no relief programs. If you were laid off, you went and found work. Everybody was doing it, even union presidents. They were trying to just feed their families.

*She grew up hearing about the Ludlow Massacre, which had taken place in 1914 just over the Colorado border.*

Ludlow has always meant a lot to the miners around our area. I heard about it growing up, about people who had been there and got shot at. A friend of mine who retired from the mine, her mother was one of the survivors. You learn that Ludlow really brought people together and showed them what the union was about.

My grandfather used to say he was against the union when it was first brought in. Like most old-timers at that time, he felt that taking out any money from their check for any reason was taking money from their families.

Gradually he saw that the union couldn't just push him back to doing the laborer's job because he wasn't educated. He could learn the other jobs. Years later, when he got real sick and the union was picking up all his bills, then his attitude changed. Toward the end of his life he told me, "Always respect the union, because that's your life."

My grandmother was one of those who took in every kid that didn't have anybody. She had 16 of her own and wound up raising another 16 from relations that couldn't make it. They would drop the kids off. My grandfather worked in the mines and had a better job than them. My mom used to laugh and say that my grandmother finally got a front room, because before it was always just beds lined up.

I could never do what my grandmother did. She did all the laundry, and they had outside toilets. She had an old-fashioned oven to bake her bread. My mom used to say it was like Ma and Pa Kettle on TV. That was us.

*Catholicism was central to their family life.*

My grandparents and my mom and dad have all been very religious people. I think my mom spends more time in church than she does anywhere else.

We were mostly Catholic. My mom's a bad diabetic, but she is active in the funeral ministry and cooks for them. She also cleans the church. My grandmother and grandfather believed in it. That was their whole life.

My grandfather was one of the *penitentes* who were outlawed. They were a group of men who were very religious, and they whipped themselves around Easter and the Crucifixion. Back in those days, that was their way of proving their faith in God. It was outlawed, but they continued doing it until finally most of them died off. My grandfather still believes in it. He had scars on his back. My grandfather was one of the last ones I heard of. But it was a very secret organization. My grandmother belonged to the Carmelitas, a women's organization. They prayed to St. Carmel and always wore brown. She believed in that until the day she died.

I'm not as good a Christian as the rest of my family. I don't go to church as often as they do. But I still believe, and I'm the type of a miner who goes in to work with a little rosary in my pocket.

*When her father broke his back in a mining accident, her mother sought work, since there was no workers' compensation.*

My dad's a very quiet man. He doesn't really express his feelings, but he's always there with anything that you need. When he got hurt, it was bad. The top came down on him. We were all small when he broke his back. My mom had never worked. She never finished high school. But she went out and found jobs that kept us fed and clothed while he couldn't. There was never anything like workmen's comp. She supported us for many years.

The worst accident was 11 years ago, when he worked for CF&I and the company didn't even send him to the hospital. They sent him home with a broken neck. We drove him to the hospital. My dad was never a big man. He stood maybe five feet, seven inches. He never weighed more than 140 pounds, but now he weighs around 105 pounds. The company went out of business, and it took more than 11 years for him to get a settlement, which was nothing really.

My mom, you did what she said. She was a good mother. She was a good Spanish Indian. She showed you right and wrong. You learned to respect your parents and take care of them. You respected your job and other people and their property. You treated everybody like you want to be treated. If people treated you bad, in return you tried to treat them better. My mom has always been very firm. She believes that if you treat people right, they'll treat you the same. She's always been a hard worker. Growing up, we always knew what it was like not to have money. But we always had food. My mom would always save. We got our school clothes from a store where they would let her have them, and she would make her payments every month. We'd each have three or four dresses, and my brother would get his clothes, and that would be our clothes for the year. I saw other kids that were worse off.

As we were growing up, we would take my grandparents with us to the store. They had never seen a department store because all they had known was the company store. They couldn't believe it. It was a different world. They had gone to these stores that had charged them a fortune. My grandfather was still paying off what his family had owed to these stores. So we would take them to other stores, and they would buy there but would still have to make payments to the company store.

*The family found ways to enjoy the beauty of their natural surroundings.*

My dad was an avid fisherman, so we always went fishing on weekends. That was his way of relaxing. So we'd go fishing, we'd go piñon-picking, or just go out and sit at the lakes. I think that was the best time, because we were always together. When you're poor, you don't have much for the other things. Occasionally they'd have dollar nights at the drive-in, and that would be our treat. But when you're poor, you learn to love the outdoors. That's your pleasure.

My parents always wanted me to be a person that didn't have to depend on other people. That's the way we were raised. I guess they figured I would just get married and be a housewife, or I would become a teacher or go into the nursing profession. I thought about both of those. I took some fashion courses, some beauty classes. They told me I'd make a good secretary. I really wasn't interested in school. I didn't know what I wanted to do with my life. I got mixed up with the wrong guy and got pregnant and thought, well, I've got to support my child, and I knew I couldn't do it on the wages I was making. That's one reason I went to the mine. At the time I was working two jobs. I was working at an electronics place during the day, making radios and stuff for the Vietnam War. I worked at the sweetshop as a waitress and barmaid. Then there was a little print shop that stayed open late, and we would go down there and do work for him. I would go home for a couple hours of sleep and get up and do it all again. It was the only way I could make ends meet.

I never really thought a job at the mine would come through. I had applied to the secretarial department. The manager said he had enough women working. So I asked if I could put in an application for underground, and he told me it would be a rainy day in hell before I'd get a job. One of our congressmen was in town that day, and I went in and told him what I wanted. About six months later he called and asked if I was still interested.

My hiring date was January 25, 1975. I was 22 years old. I was the first woman underground at the York Canyon mine in Raton, New Mexico. The mine was noted as one of the best and safest in the country. We even had lights going down the beltway.

My mother thought I was totally crazy. My dad didn't want me to because he knew what the mines were about. But I told them I had to support my son. They finally accepted it. It was hard. My uncle was working underground at that time, and I had a couple of cousins that were working there. They said to some of the men, watch out for my niece, she'll be working there. In our mine you were always related to somebody somewhere.

I was surprised, because the old-timers took care of me and I had trouble with the young ones. Everybody told me it would be the old-timers who would give me a rough time, because they didn't want women in the mine. But they put me under their wing and took care of me. I think they were raised to take care of women. My dad told me, "If you want to go to work at the mine, go. But if you're going to go to sleep around, don't bother. Don't embarrass me." I told him, "Don't worry."

I worked hard. They treated me bad at that mine. My own union walked out on me. I started on the longwall because the company thought if they put the women there, they'd run us off right away. We had a radical union official that hated the company and hated the idea of women being in the mine. He convinced the guys to walk out on me because I couldn't lift a bale of wire that weighed 150 pounds. There was very few men who could do that. The union was out to get my job. The Labor Commission came to a meeting, and the company came, and the union. The company said, "She does a good job. We have no complaints." And here was my own local union trying to hang me.

At that time I wanted nothing to do with them. The company put me on probation with old-timers, and I learned a lot. Finally I went into a section, and I got to know the guys. I had a good crew. It took quite a few years before I finally got faith back in the union.

After they walked out on me, the company sent me to day shift to work as a timberman and pumpman and mason's helper, and they finally put me on the belts. I worked with the old-timers on the belts, and then finally I worked as material car operator. I've worked as a roof bolter and a fireboss [checking for methane gas in abandoned sections of the mine]. I think I was the first woman in the state of New Mexico to get my mine foreman papers. That was around 1980. Outside I worked as a driller's helper. Then I operated a highwall miner, but the company decided not to keep that machine.

We used the highwall miner on the strip jobs. It's 90 or 100 feet long, and it's self-contained. The miner is on a platform or launch, and from there you launch into the coal seam. My job was operator. I sat in a little room and looked at four cameras and mined coal. It's supposed to be the cheapest new method of mining. You're supposed to get a ton of coal out for five dollars. You can average from 500 to over 1,000 feet a shift, but my company didn't seem to think it was the right time for it. There are still a lot of bugs. It's a new machine. It only takes five people to run this machine.

After this layoff I bid the first job that came up, as a beltman underground. It's a big pay cut. They're still talking about another possible layoff in September. So you see where the safest place might be [to keep a job].

I'm also a CONSPEC operator. CONSPEC is a system originally designed for monitoring gases in two-entry mines. Now we've gone to three- and four-entry systems, so all we do now is monitor the fans, water pumps. You sit at a computer and watch it. But that's a job they tried to keep for the men. They said you had to be an electrician to sit in front of this computer.

The old-timers would let you get on equipment, but nowadays if I go outside or on a piece of equipment, they'll say, "Well, she's got more time than I do, and if I teach her, come layoff she'll have my job." When I worked on the highwall miner machine and it wasn't running, they would pay me eight hours to sit and do nothing. I had to threaten the bosses with a discrimination case before they taught me how to run a truck. They had a compactor I wanted to learn. It's what you learn before you learn to run a dozer. They would send me to clean the bathhouse when I wanted to learn how to be a driller's helper. I always had to keep pushing.

You learn to take care of number one. When I started, I had to get an unlisted number because wives would call and cuss me out. Half these people I didn't even know. To this day it's still unlisted, because to this day you don't know if some wife will want to jump on somebody because their husband's doing something. I work with their husbands, and that's as far as it goes.

I had problems with one of the guys. When I went to the bathroom, he would always be watching. Once when I dropped my drawers, he came right at me. I let out a scream, and the guy I worked with came and took care of him.

I've had bosses come up to me and say they were checking for cigarettes and try to put their hands on me or slap me on the butt. I've had to threaten them. I've had union guys walk out the [men's bathhouse] door naked, shaking their clothes. One constantly did it. He would walk out to shake his dusty work clothes with a shirt on and nothing on below. I'd be outside waiting to go underground. One day he said, "What do you think, Martha? Do you like what you see?" I looked at him and said, "With what you have to offer, you probably have to squat to pee like I do." The guys just cracked up laughing.

You do learn a lot about each other, especially if you're working with one or two people. Anything that was really personal I kept to myself. But if there were other problems, there were a few people that I trusted. Some of them talked to me about their wives, trying to understand them, or about the kids. They'll ask for your advice about what to do.

Some days, if you're working with one other guy hanging pipe in the returns, you'll talk and joke around with a bunch of them eating lunch. And you start listening to their problems at home. You just grow a relationship. Most people don't understand how you can have a friendship like that with men.

But when you get underground, you depend on people. Their eyes are your eyes also. You're watching out for their back, and they're watching out for yours. It's gradual, but pretty soon there's that bond. There's a few I wouldn't give the time of day to, but there's a handful that I would do anything in my power for if they needed my help.

I got along fairly well [with the men], but you always have that handful who want to do you wrong. I had a foreman who was pretty good about keeping me with people I could work with, who wouldn't try to hurt me.

There were men that we worked with who went out of their way to see if they could hurt you. If you had to lift a prop or a crib, they would drop their side, just to see what would happen. Or they would sling heavy stuff at you. You could tell which ones were out to hurt you, and you tried to stay away from them. But I had a pretty good foreman. He could see this, and he tried to keep all the women away from these men. But there were times when you had no choice but to work with them. I had a few bosses who were the same way.

Quite a few of the women ended up with back injuries from stuff like that, and most of them have quit. They said they just couldn't take it. The most we ever had was seven women, and they've just about all quit. Now there are three of us who are permanent and two that are temporary. I've always had the attitude, "I'm not going to do anything to make you happy." I'm going to stay around. I've said to some of the bosses, "I'll be here after you're gone." And I am still there, and most of them are gone. I'm doing what I have to do to live. That's my attitude.

I think you build an invisible force field around yourself. You never let anybody get that close. You watch what you say, how you say it. You have to be careful how you talk around the men. Not especially the cussing, but how you word things so they don't take them in a sexual way. Even joking. You learn who you can joke with. You have to think before you speak. They don't. They can say what's on their minds. But most of the guys I work with and get along with, they learn what you'll take and what you won't. If they cross that line, they know I'll come down on them.

It takes a while to learn. You're joking with them and then realize, I shouldn't have said that. And you think ahead, saying, no, I don't want to go down that road. You learn which ones you can joke with and how far you can go, and then you have to stop.

At first you really don't know what to do, and they get away with a lot more. Sometimes if you can't get it done through the men themselves, you can find other men who will take care of it, or you go to the company. But eventually you start handling it yourself. And if you can't, then you take the next step. Sometimes it takes a call to their wives. I've done that.

One man continually pestered me. Underground he was always manhandling me. Outside he made comments. One day I ran into his wife, and she said, "If he gives you any trouble, call me." And so I got fed up one day, and I called her. She said, "I think he's just playing with you, but there's no reason for that." After that I had no trouble with him.

I tell most of the women that go to work up there to get a nonlisted number right away. Certain wives whose husbands are out messing around seem to think you're the one to blame because you work with him. Not all the wives feel that way. You meet a lot through the union meetings, and the men will come up and introduce you. Wives get to know you, and they see that you're all right, that you're not out to get their husbands. One comment I make to the wives is that, "I work with him for eight hours. You don't see him as much as I do. If you did, you wouldn't want him around" [laughs]. But some men say, "I can talk to you now, but don't talk to me if you see me with my wife."

It was easier for me because some of the men's wives were my cousins. Some women at the mine had a reputation before they got up there, and these women still keep a distance from them, because they don't trust them. I tell these women [workers], "Well, you have to prove yourself to these women, and you'll be all right." But most of the women who are at our mine are not interested in getting involved with the union or anything else. Sometimes they're working six days a week, and who wants to give up their Sunday?

*She used her wages to raise her child and help her parents. The extended family took care of the elderly and the young.*

In my family, if you have money, it all goes in a pot and everybody lives off of it. You don't say, "This is mine." It's the way we were raised. I've always helped my folks, especially after my dad got hurt, because they lived on a fixed income. When my grandfather had black lung, we kept him until he died. And then my grandmother got Alzheimer's disease, and we kept her. You take care of them until their time is gone, and do the best you can. I look at my mom and my dad, and they're getting to that point now, and I think, I can't quit my job. I told my sister that she'd have to quit her job and take care of them, and I will pay because I make better money.

Because of the mining paycheck, I could support my child. I had money to help my parents. It gave me the freedom that I didn't have to depend on anybody but me. The money made me feel better about myself. If I want to give to charity, or go out and blow it, I can do it.

My folks helped with child care. They basically raised my son. I put a trailer in their backyard, which made it easier. I started off on graveyard shift, and then I moved to day shift. During all the years he was growing up I stayed on day shift. People would ask him what his mom did, and he would say, "She's a coal miner," and they would look at him dumb. One day he came home from school, and a girl was carrying his books. I said he was supposed to carry hers, and he said, "Why?" I think he's learned from my hard work.

One of my biggest problems has been my experience. The bosses they brought in knew I had fireboss papers. They thought they could just demand that I do their job and then threaten your job if you didn't do it. They threatened to pull my papers. The papers can only be pulled by the state that issued them. I had been hurting my ankle constantly firebossing for them. I thought, this is my body I'm destroying, and I'm not gonna do it anymore. They got a little bent out of shape about it.

A lot of guys have told me, "You know, Martha, I liked you a lot better when you first started." They call me a royal bitch now [laughs]. Before they could talk to me, and I was innocent, shy and sweet. That lasted five or six years. And I got walked over right and left. One day I just got fed up. Now I tell them exactly what I think, whether they like it or not. As women, we are now thinking for ourselves. We don't need them to do our thinking for us.

I told one boss about a place where there was a possible cave-in. And he said, "I can't believe you because you're a woman." And then it caved. Guys hated for me to know more than they did. I've got my mine foreman papers. That really irked them. They say, "Why did you get them?" And I say, "To prove that I knew what the heck I was doing."

When I first got there, I hid what I knew. As time went on, I thought, why shouldn't I use my brain? Our old CONSPEC machine was an old computer that used floppy disks. They were training one of the other women that was junior to me. I filed a grievance, and if it was up to the union, we would have lost it. But I fought it myself and won it. They gave me 13 hours to learn how to operate it. They gave the other girl two months, and the other guys a month and a half. I memorized it, and I got the highest grade out of all of them.

The first time I took the fireboss exam, I was rated assistant mine foreman. I took it a second time and was rated mine foreman. Everybody said, "You're just showing off." I said, "No, [I took it] just to prove that I knew something." On the book you use as a fireboss I would write in details. The guys said, "You don't need to write a book about it." Then we had a big blitz [inspection]. We got more than 400 citations, and the federal inspectors said, "This is what we want. Details." And I laughed.

I've had to file a lot of grievances. I've filed on bosses working, I've filed grievances on not being called out when I've been running machines. I've tried to file against my own union brothers for discrimination and harassment, and my local has always told me, "You can't file against a union brother." Then they'll turn around, and somebody will do something to me, and it's fine for them to go after me. I've had to threaten the company with grievances to get a bathhouse, to get portable potties—things that women need that they don't think about.

*She is worried about the increasing use of diesel-powered machines underground.*

We're having problems with diesel equipment. The company had wanted us to test equipment from a distance of four feet. The state backed us and required 18 inches for the tests, which makes a big difference on what you're breathing. We started out with diesel pickup trucks taking us underground. Our seam is six and a half to seven and a half feet high. Now we have these new little cars that can destroy your hearing, and they're experimenting with ram cars.

We're breathing more and more exhaust. There are times when you catch the company short-circuiting the air. They finally got to where they are going to put in drill holes for exhaust and one will be for a new fan.

Since we've used diesel, a lot of people have had trouble with allergies. More people seem to have a hard time breathing. Some women have developed breast cancer and had hysterectomies. The women were fine before they started working. They worked five to six years, and suddenly they all went in for hysterectomies. Is this what is going to happen to all of us? I went for a physical not long ago, and they told me, at the age of 41, I could be starting the change of life. The doctor seems to think it goes back to the mine. She says all the heavy work, all the stuff we're breathing, adds to stresses on the body.

I have 19 years in, with three and a half years on layoff I need to make up. Fifteen years is underground, and the rest is outside. I do know that bosses outside treat you better than the bosses underground. Outside you were treated like a person. Underground you're treated like a mule. Now I'm going back underground to work on the beltline. I chose swing shift, because they won't make me fireboss as much.

When I first started in the mine, I got hurt riding the belt. I've had the top come down and the ribs. I messed up my foot firebossing, because we had five and a half miles of belts and returns to make and you have to make it in three hours. You can trip and fall because you're having to move so fast. That was one reason I went outside. I thought, I'm not gonna fireboss any more, I won't have any legs left. The doctors told me that if I messed up my ankle one more time, they're going to have to do replacement surgery.

Underground is a whole different world. There's more respect for life and property. Outside they'll run those trucks and they don't care. Underground you're very careful with your cables. You're always watching out when you're on a machine. There's more of a sense of caring underground. Life is more valuable underground than it is outside. I always liked the underground. The work's harder, but the temperature is always the same. You get on equipment, and it doesn't go down the road sideways. Underground you make a lot closer friends. I worked in one section, and we were known as "the Undertakers" because every day we were burying some piece of equipment. You learn who you can trust, and you learn to depend on people. On the strip, you're pretty much on your own, and you just do whatever. You just float around.

*Martha Horner became active in the local union after receiving encouragement from the local president.*

At our mine we're mostly Hispanic workers from mining backgrounds. We've got about 350 people now, although we expect layoffs. P&M [Pittsburg & Midway] is a white company that is having to put up with Hispanic workers in northern New Mexico. Down around Gallup, P&M has to deal with the Indians.

All our local union officials were kids of miners, and we believe in working for the best of the local. There were a lot of times I didn't want to be active in the union. But [the local president] pulled me back in. I trusted him because I graduated with him, and I knew his brother and other people. Through them I started to believe in it again. The old-timers said, "Martha, come help us." I started to see a different viewpoint. If they went against me on things, I'd get mad, but I said, "I'm not gonna leave, I'm gonna be at every union meeting whether they want me there or not."

I dropped out for five or six years after they walked out on me. But for the last eight or nine years I've been pretty steady. [The president] talked me into going into the safety committee, the audit committee, and almost anything else he can drag me into. He talked me into coming to the women miners' conferences.

A few of the men will work with us because we do a lot for them, but the men still haven't put us on the same level. Most of them still think we're in there invading their jobs.

*Working long hours, she has little time for the community activities she once enjoyed. Being a miner has also made it more difficult to find an accepting man to share her life with.*

When you're working six days a week on swing shift, you don't have time for a social life. You go out late, or on a Sunday, or take time off. I used to go to the theater and get involved in different things. I belonged to the museum, but you can't get involved because you're working. And they need help, and I go down and give a little bit of time here and there. But sometimes you're so tired you don't have the energy to give.

I've become more confident, and I trust more in my own instincts about what I want to do. But I never got married because I never found a man I thought would treat my son right. I missed out on a lot of good men because I wouldn't give up my job, because men didn't like the idea of their future wife working underground.

You figure out what you want. You think, well, a man may be here now, but three or four years down the road he could be gone and I could be out of luck. And we live in a small town, and I'd be making more money than they would. Most of it was ego. They just couldn't handle the fact that I'd be bringing home most of the money.

Having a network of women miners has been a good thing. The Coal Employment Project has educated me and helped me with my own cases. The guys don't know much, and what they know they're not going to relay. I've found that with the safety committee. If I have to go on bimonthly inspections, there are some men who won't go with me. You have to fight to get involved in the major issues.

Without CEP, I don't know if we'd still even have these jobs. Because you need people that you can call and talk to and get some kind of support, or write a letter to. You don't have the answers. And you start feeling that there's nobody out here to help you. The pressure at times gets bad, and you want to say, "Forget it, I'm not doing this any more."

*Self-respect, she believes, is crucial for women's economic equality.*

I think we've made big strides, and if we continue, we're not going to be forgotten. But the men won't view it that way. They'll downplay our contribution. The companies may not want to hire any more young ones, but maybe we can force the government to start hiring and training women. I always ask at my company, and they look at me like, forget it. But eventually they're going to have to start training, because everybody is getting older.

I think we must teach women to have more respect for themselves, their bodies, and their minds. They need to take pride in themselves. To show girls growing up that you don't have to stay in a little job, you don't have to stay home and raise kids because a man says you're not good enough to go out and make a decent living. It may not be in the mining industry, it might be in another industry. But you have to say, "I'm as good as that man or better, and whatever it takes, I can do it."

# 11

# A WORKING LIFE
## Elizabeth Laird

*"Before the mines, I was bringing home $3.40 an hour at the spinning plant. And that place paid more than anywhere I'd ever worked before."*

*Affirmative action mandates opened doors for women who had already spent decades in the low-paid workforce. Elizabeth Laird worked for nearly 25 years in the textile industry in Alabama before going into the mines at an age when most workers are already contemplating early retirement. She had worked through World War II, raised a family, and survived personal tragedy when her son was paralyzed in a car accident. She and her husband were subsequently divorced. The experience challenged, and later strengthened, her natural optimism and spiritual faith.*

*She worked as an underground miner for 15 years, then retired after spending nearly 55 years in the nation's workforce. She remarried her husband and enjoys the retirement projects of restoring a farmhouse in southern Alabama and spending time with her grandchildren.*

*Our interviews took place in 1983 and 1994.*

I went in the mines when I was 54 years old. When I went to apply, the company people just looked at me. With all the younger women, I figured I had about one chance in a thousand. But it didn't take long for me to get on. I had a good work record, and friends. It was 1976, and the company was under pressure to hire women.

My work record helped me. It should have. I had been working for 40 years. I first went to work cleaning house when I was 15 years old, after my father died. He had dug coal with a pick and shovel down the road in Aldridge, Alabama. He worked from "can to can't," you know—from daylight until dark—and I was lucky to see him on weekends.

I was born in 1921 in Aldridge, a coal camp. My daddy's granddaddy was in the Civil War. My grandfather on my mother's side mined coal near

Elizabeth Laird

Birmingham. One of my grandmothers was part Indian, and my other grandmother worked in the cotton mill here in Cordova around 1901. The women were allowed to bring their babies to the mill and put them in a box when they were nursing. They set up a kindergarten next door for them when they were weaned.

My father was in his fifties and my mother was very young when they got married. He was making good money in the mines, but when the Depression hit, he only made around a dollar a day. He would work one day and then bring home the clacker, or scrip, that we would trade at the company store. Mother took in washing and ironing from the foremen and raised a big garden. We lived from day to day. Everybody did. Then my father's health failed, and he died at 77, when my youngest brother was four months old. So I left school and went to work to support the children.

By the time I was 20, I was working in the cotton mill in Cordova. I started in 1940 and worked almost 25 years as a weaver. During the war we worked seven days a week, making herringbone for fatigues and cloth for tent ducking. My pay started at $14 a week, eight hours a day. I started in 1941, and when it shut down after nearly 25 years, I was making about $89 a week.

The mill was straight production work. If you didn't make production, you didn't eat. You had to keep your looms running. Some of them were wide as a sheet and ran on electric power. I'd run the machines and tie knots and pull the threads through.

The mill had its danger, just like the mines. It was a strain on your eyes to keep your threads drawn in straight. The lint was bad. With the coal dust we have respirators, but with the lint we didn't. I lost two fingers in the mill. We were running 75 looms without gear guards, and I slipped on a wet floor and caught my hand in a gear. The company gave me $18 a week.

After that mill closed I went to work at a place in Jasper making golf bags for three and a half years. But they came out on strike, so I went to another cotton mill and worked there until it shut down after three years. So then I came back to the golf bag place, and it shut down. Finally I went to a spinning plant and worked there for ten months, and it about killed me.

Working as a spinner is the hardest thing I've ever done, because each spool had a brake on it off the floor about chair level and you had to brake it with your feet. So if you messed anything up, you had to hop around on one foot all night.

*Tragedy struck when her oldest son was paralyzed in a car wreck when he was 17 years old.*

My husband couldn't understand how it could have happened. I couldn't either, but I prayed a lot and made it through. Our son Ernie and this boy had been to Jasper to a church social. The other boy was driving, and they hit a culvert and were thrown about 70 feet in the air. The driver wasn't hurt, but Ernie was riding in the suicide seat. The doctors didn't promise us anything for the first two weeks. My husband couldn't absorb the shock. When the doctor told us Ernie's spine was severed, my husband didn't even hear it.

After the accident, I went back to work at the golf bag plant. I sewed zippers, I turned pockets, I trimmed, I beaded bags. It helped, because I didn't have time to sit still and worry.

The shock of the accident eventually brought on a divorce. After that I had to work extra, to get the kids raised. I was working two jobs because I had Ernie and another child in college and one at home. My husband paid child support on the youngest, but in Walker County that was nothing. What can you give your children on $20 a week? For five years I worked day shift in the mills and worked at a diner from five-thirty until two in the morning and then had to be back at the mill at seven.

The mines was the only place where I could work one job and have time with my youngest son. He was about 14 when I went underground. He was so proud! He was the first kid in school whose mother was a coal miner. The middle one didn't say much about it, but Ernie would smother whenever he thought about me going underground. He thought I might get in an accident.

Before the mines, I was bringing home $3.40 an hour at the spinning plant. And that place paid more than anywhere I'd ever worked before. In the mines I was making about $2,000 a month. Around here, there's nothing else that pays women like the mines. There's the Arrow shirt factory, and some little places that make things like car seats, but they pay minimum wage. There's a lot that's shut down.

*In the mine Elizabeth Laird worked for several years on ventilation in low coal, crawling to hang the heavy curtains used to direct the air flow. In a 1981 interview she reflected on her five years underground.*

The first day in the mines I was real nervous. But I enjoyed my work. Keeping the air up to the section was an important job, and I could work by myself. I was one of the first women hired at Mary Lee No. 1 mine, and after five years, I've been on all the equipment—the drill, the scoop, the miner, and the pin machine.

Some of the men thought I was crazy, because I worked so hard. What they didn't realize is that I had worked production for so long that I'd set my pace years before. I told them I was there for two things, quit time and payday. And quit time comes quicker when you keep busy.

My seniority isn't going to help me much to get one of the highest-paying jobs before I retire. I wish I'd gone into the mines earlier, but they weren't letting women in. It would have been a lot easier on me. I wouldn't have had to work two jobs and gone five years without a decent night's sleep.

On the section they call me "Mama Liz," and we get along fine. But it took a long time before they felt they could trust a woman. I'd told them before to watch a rock, and they'd tell me I didn't know what I was talking about. One time I told this boy to watch some top, and he didn't listen. A few days later two whole crosscuts fell in, and they still wouldn't listen. It took an accident to change that. One guy's shuttle car got juiced with electricity. He had taken the buggy partway through a ventilation curtain, and I heard him saying ugly words. That wasn't like him, so I went to him.

When I got there, he was getting voltage bad. It was throwing him around on the buggy. The wet curtain was touching the hot buggy, and the other buggy boy was being held down at the other end. I hollered to get the power off and for somebody to come help. And buddy, it wasn't two minutes before they were there. We got those boys on stretchers and got them out of there in good time. After that the men seemed to trust my judgment.

Safety is the most important thing when you're underground. I watch out for the men I work with. I make sure they get the air they need, and if the top is fixing to fall in, I signal them with my light.

I talk to all the men, and when I eat out a lot of them, married and single, will sit down with me. I know the men's wives. But I stay so busy, I don't really socialize with anybody.

Working all your life can hurt a person. I worked two stand-up jobs back to back for five years, and then I started scooting around on my knees in low coal. My knees wore out, and the doctor finally had to put new knee joints in. I had two operations. The only trouble it caused me was one Thanksgiving. I stood in a dress for two hours watching the floats. The wind was cold whipping around my legs, and my knees froze. I couldn't walk. So now I keep sheepskin bandages around my knees to keep them warm, even if it's 100 degrees. Six months after the operations I was back at work, but not in low coal, even though I liked it. We didn't have as many roof falls. It seemed safer.

Coal mining is the easiest job I've ever had. One thing I've been lucky about is my lungs. I've had no problems yet, but something may crop up. I'm sure I've got at least as much lint in my lungs as coal dust. I saw my X rays when I was hired into the mines. They were a little cloudy from the lint. The last time I saw them, they were black. But I was turned down for black lung benefits, and I don't know any women who have gotten them.

You never know what's going to happen. But whatever comes, I want to be prepared. We were brought up in the church, and I've stuck with it. I want to know I'm forgiven when I die.

Living right makes you feel good. So does being single. I had a happy marriage for 26 years, which is a lot longer than most people. It took me four years to get over the divorce, but now I'm perfectly happy. I like being my own boss, not having to report to somebody. I like to be independent, but I was never part of the women's movement. I never had time to think about it. I was too busy working.

*In 1990, at age 69, she remarried her husband. She retired from the mine the next year.*

I am very happy in my marriage, and with my children and grandchildren. I don't feel like I'm 73. For me, it's mind over matter. I always think about the future, because the past can drag you down. I have lived a long, long time just one day at a time.

When I first went in the mines, I only planned to work until I got Warren through school. But when he finished, it struck me that nobody needed me any more, and I started getting depressed. My doctor advised me to keep working, and I perked up and felt better. I retired a few months before my seventieth birthday, but I wish I was still working. I didn't have any responsibilities, but I worked every day. Before I got remarried I worked day shift and would go sit in cafés and talk to people in the evening. I closed the town down every night!

There have been a lot of hardships in my life, but I've come out all right. I don't regret anything. I live for tomorrow with faith in God and hope for the future. I have always told my children they can be or do anything they want to. You have to build your life up like a house, like something you want to live in for a long time to come.

# 12

# WALK IN BEAUTY
## Evelyn Luna (Evie Tsosie)

*"The community has changed toward us. Now I come across Navajo women who ask, "How did you ever get that job?" They say, "You must be really doing well." . . . In the beginning they would accuse us of messing around with their husbands, and now we get compliments."*

*The Navajo nation lies in a stark country of mesas and desert, beneath which lie thick coal seams. Peabody Coal Company opened a mammoth strip-mining operation on the sacred Black Mesa in the 1970s and employed hundreds of Navajo workers, including a handful of women. Evelyn Luna (who in 1996 changed her name to Evie Tsosie) works at the Kayenta mine, a few miles from the soaring spires and monoliths of Monument Valley, Arizona. The Kayenta and the neighboring Black Mesa mine supply coal to a huge generating station, which supplies electricity to several western states.*

*As a child, she learned the sacred traditions of the Dineh or Navajo people. In Navajo culture women occupy a position of power and respect, and family lines and property descend through the mother's clan. Her more recent legacy was mining. Luna's father toiled in a uranium mine, and she later married a copper miner. When a divorce left her with two young children to support, she went to work at a uranium mine. A few years later Peabody Coal Company offered her a job at its Kayenta mine. She took the job and then struggled with the conflict between her traditional beliefs and the need to earn a living on the reservation, which suffered from high unemployment.*

*In 1986 I drove through the long stretches of desert and sky on my first visit to the Kayenta mine. The car radio picked up the local station, which broadcast in Navajo. Circular hogans, used by traditional families, dotted the mesa with their single opening to the eastern sky. At the mine a towering coal stockpile dwarfed an abandoned hogan and nearby sheep pen. Driving through the sprawling mine with union officials, I saw an elderly Navajo woman in traditional dress herding sheep near the perimeter of the mine. Later Evelyn took me to her uncle's hogan for the first night of the squaw dance ceremony. As a bonfire shot*

Evelyn Luna (Evie Tsosie), with her daughter Ericka

*sparks into the starry night and cast a circle of light upon young and elderly faces, rich smells of fry bread filled the air, and dancers pounded the earth in a potent dance of healing.*

*In 1993 the Navajo women hosted the national women miners' conference in Gallup, New Mexico, and shared the history of their culture and the economic struggles faced by the reservation.*

I am a Navajo, and I was raised with traditional beliefs. One of those beliefs is that the woman should work inside the home. But I was left with two kids after a divorce, and I knew I had to get a job if I was going to make it. And I wanted to show my ex-husband I could do something more with my life than go on welfare.

The reservation is a hard place for a woman to get a job. A lot of Navajo women took training on heavy equipment, but there weren't many openings. Many were never hired. Those of us who were lucky enough to get hired are resented by the men. They think their own sisters should have our jobs. There are about 420 union people at the mine, and only nine are women.

Other than the mine, there aren't many jobs for women, and those pay minimum wage. So a lot of divorced women do end up on welfare.

Women on the reservation want jobs not to be equal with men but just to have enough money to live. The women's movement hasn't had much effect on us. We're isolated in many ways from what goes on in the outside world.

I was born at home, like most in our culture. We didn't have birth certificates. A witness—a midwife or relative—would report to the tribal census about when you were born. I had three older brothers and one older sister, and five younger sisters and one younger brother.

We follow a clan system. Your first clan is on your mother's side, and then you have your father's clan. In English, my mother's clan is the Towering House People, and my dad's is the Red House clan. Some family names reflect their clans, and some don't.

I was raised in the belief that my ancestors held all of nature to be sacred, and they lived their beliefs. I live in the tradition because it works for me, and the blessings are the core of it. Our medicine men are like your doctors. They are specialists. One will do one kind of blessing, and another will do something else.

Growing up and being taught by my parents, I felt that tradition was what we had to do. We had ceremonies done when we got sick or had bad dreams. We had prayers done for our stresses. Every four years or so you have to have certain ceremonies done, like the "good way" blessing and the womanhood blessing. We have the squaw dances. They last four nights, and they say you are affected by them. Like if you've seen a dead person that's not a Navajo, or touch the body, you have to have something done.

When I was young and in school we had a program where we spent a year away from the reservation. I went to Salt Lake City to live with a Mormon family, and they told me my Navajo traditions were bad and I shouldn't go to the ceremonies. But I couldn't give up what I had grown up believing. It made sense to me.

To the Navajo, everything in nature has a special meaning. You learn what Mother Earth exists for, and the sky, and there are prayers to give reverence to all parts of nature. The prayers begin, "Beauty before me, beauty behind me, beauty all around me."

*Copper and uranium mining were more familiar to her than coal mining, which expanded on the Black Mesa with the opening of Peabody Coal Company's surface mines.*

My father was a uranium miner, and my ex-husband worked in a copper mine at Morenci, Arizona. I'm sure that's part of the reason I became a miner.

First I went into the uranium mines and worked there for four years. That's probably why I didn't have any trouble getting hired by Peabody on the strip mine at Kayenta. The problem wasn't getting hired. The problems came later, with the men.

The men at the uranium mine were a lot easier to work with. There weren't as many young male chauvinists there. I don't know why. And the uranium mining company had hired a lot more women. Where I work now at Kayenta, there were three other women working on the main operation when I got there, and 400 men.

The uranium mine was underground, and I did laborers' work. I helped hoist materials like timbers and rock bolts and chain links and transported magazines of explosives. It was a big drift mine where you have a lot of room to walk around. They have cars on tracks to move the uranium out of the chute.

We got checked for radiation exposure by a device they put on our hardhats, like a temperature gauge that changed colors if you got more radiation. Once you got so much exposure, the company would pull you out of the underground and put you on the surface until your reading came back down.

When my dad worked they didn't do anything to check the radiation. That's why he's got so many problems now. He's been sick a long time with constant coughing, but he won't go to the doctor. It's a tradition among the Navajos. The men won't go, no matter how much they need it.

Sometimes when he's sleeping or he gets really bad, it's scary just to listen to him trying to breathe. He wheezes, like he's short on breath. And the doctor—the government doctor is who we have to go to—says there's nothing wrong. I think they don't want to pay all this money out to these sick people who worked in the uranium mines.

Nobody knows how bad uranium is on people. I was born at the time my dad was working at the uranium mine, and when I was young, in my twenties, I was tested for breast cancer, but luckily the lumps weren't cancerous.

I went to a government meeting on behalf of my dad. The government was promising to compensate uranium miners $100,000 apiece. But in order to get that compensation they have to pass so many tests that it seems like they have to die before they get it. That is very wrong. Why make these men suffer? They should never have promised something like this.

Some of the widows who got compensated told me they had to prove they were married to receive it. In our old time, they didn't have marriage licenses. They had traditional weddings, but they never thought of marriage licenses until recently, when the compensation came along and they had to be married. So now they have to go to a justice of the peace and have witnesses that they actually were married for so many years. Even then those widows don't get the whole benefit. They get 10 percent of it.

*When Luna began work as a miner in 1981, she grappled with conflicts rooted in Navajo tradition. She reflected on this problem in a 1986 interview.*

I liked the work at the uranium mine, but I wanted to be closer to my folks in Kayenta, and so when I took vacation in 1981, I put my application in, and they called me right away.

We mine coal on the Black Mesa, which is a sacred mountain in Navajo tradition. We call it the Mother Mountain. Digging out coal is seen as torturing Mother Earth, tearing out her inner organs. After the mine opened about 15 or 20 years ago, there was a lot of feeling against mining the Black Mesa. If something wasn't right on the reservation, if it hadn't rained, or if there were dust storms, many of the Indians felt it was due to the mining, how it had upset the balance of things.

By our tradition, it is worse for a woman to torture the earth than a man, because it's like torturing our own body, the body that gives life. I used to have bad feelings at work and felt like I shouldn't be there. Other Navajo people told me to have a blessing done by the medicine man, and I did, twice. It's necessary sometimes, because when you start feeling bad about violating a sacred mountain things start to happen. You start losing things, or you have credit problems. You need the blessing to straighten it out. The blessings work. That's what they're there for. They're not to play with.

At the Kayenta mine, it is common for Navajos to call a medicine man up there for a blessing. It has to be done in a private place on the mesa, and the medicine man performs the sacred ritual. You pay the medicine man individually for the blessing, which forgives the person for destroying the things of nature.

I don't have those bad feelings too much anymore. I have talked with other Navajo men about it, and they make me feel better by telling me that by driving a water truck, I'm not tearing up the mountain, I'm just wetting it down. If I was working on a dragline or on a dozer, it would be harder for me.

Coal mining has weakened the tradition of the Navajo, but people have to make a living, and the company has given us jobs. We've given up something, but we've gotten something in return.

Where I work now, the men don't like it that women are there. They complain you're not doing your job, even when you are. They say we get by with things because we flirt with the foremen.

One thing that makes it hard is, I am already tired by the time I get to work. I work afternoon shift, and I wash clothes, cook, clean house, wash dishes, and drive 65 miles before I even get to work. Then I work until midnight, get home after one in the morning, and can't get to sleep until about two. Lots of nights I don't get more than about four hours' sleep. It's not enough, but I get by.

There are problems on the reservation for any women who work at the mine. People say things to my kids. They say, "Your mom's rich. You think you're better than us because you can have anything you want." And neighbors try to make me look bad. We don't have cable TV on the reservation, so I bought a satellite dish, and people started throwing things at it and leaving their beer cans and wine bottles in my yard. They tore the fence down I'd just put up. I know it's jealousy, but it's hard to take.

Most of my free time, I'm with the kids. They are eight and ten now. I try to teach them the Navajo tradition, but a lot of the young ones don't have the beliefs. Some think Navajos spend too much money on traditional things. In my family one of my children speaks Navajo and the other doesn't. They have to stay alone at the house a lot now, because I can't find a babysitter. I try to make it up to them any way I can. There isn't much time for a personal life other than the kids. Even if I had the interest, you have to drive hundreds of miles if you want to have a good time and relax. By the time you get there, it's time to turn around and come back.

*Through her union she learned what her rights were on the job.*

I do try to go to all the union meetings. I'm the only woman who does. Even after 15 years the idea of a union isn't well accepted here. The men don't understand how much the union has done for us. I know if it wasn't for the United Mine Workers, I probably would have lost my job one way or the other. The company would have found a way to get rid of me.

It's important to get involved, to see what your rights are. I've tried to be active, but I've been pushed aside, even when I'm just trying to find out my rights. The other women say they don't have time for the union. But they know when the meeting is every month, and I feel like they could schedule their time so they could make it once in a while.

The first couple of years I worked at Kayenta, I didn't worry about my rights, or about saving money, or anything. I just goofed off and had a good time hitting the rodeos with my friends. But then I realized I had two children depending on me, and a job to protect, and I straightened up. It's important now for me to show people a woman can make it alone on the reservation.

I just hope they don't lay us all off. That's always a worry. Our mine was shut down once because a pipe busted somewhere in Nevada. When it comes down to it, you don't know what's going to happen. So you live day to day and hope for the best.

I've always been a withdrawn kind of person. I don't communicate a lot. I live in a commuted area, which is a part of the reservation where there are a lot of houses, and there is a lot of drinking, which gets frustrating.

When I was young I always stayed with my brothers when they were working on their trucks, just to learn what I could. They didn't want me there, but it didn't bother me. I picked a lot of things up. One time when I was in Salt Lake City with my girlfriend my truck broke down, and I ended up installing a new drive shaft and U-joints. I just listened to the sound, and it was real familiar.

On the job the older men don't give women such a hard time. They're better to the women. They've got daughters my age. The older men tell me to hold on and not listen to the young guys, because it's jealousy. Other people tell me to talk back, but that's not my way. I would blow up. That's why I never say anything, because I know my limits.

I don't know how long I'll last at Kayenta. So many days I get frustrated and think about quitting, but then I think about my house and my truck and all the other bills. And when I think about that, I get up and go back to work.

*By 1994 she had stayed on the job for 13 years and wanted to complete 20 years before retirement. She was 39 years old and a local union official. Her two older children were grown, and she was raising a two-year-old daughter.*

It seemed like the women were being ignored, so I got more involved with the union just to find out what was going on. I thought it was the only way to find out how the union functions and to know how to use the union to protect our jobs. I'm on my second term as a union officer, and there are lots of problems just keeping the local together. It's been a rough year.

I've found out a lot being a union official. Peabody is being investigated now because of their hiring practices on the reservation. People are complaining that the company isn't following the Navajo preference laws. Under those rules, you have to be a Navajo or be married to a Navajo before you can get a job on the reservation. If they can't find anyone qualified, then the companies can hire from the outside.

Before, I withdrew from everything and didn't care what happened as long as I had a job. But now I've opened up and begun to realize how the company is supposed to operate. The women are still having problems. We are still fighting to be heard, to be understood, and to be recognized by our union brothers. Sexual harassment is still an issue.

Through the union we repealed the right-to-work law on the reservation. We decided to repeal it because we had guys working at the mine who were scabbing on us. We wanted to make it a solidly union mine. We got delegates from different chapters and went to the tribal council, and there was a vote. With "right to work," we felt we were fighting for a contract for scabs to work. We got it repealed, but it hasn't really solved the problem because they're not enforcing it.

I'd like to get more people involved with the union and to get people to recognize that the union is there to protect them. If we didn't have a union, a lot of people would be out of a job. Our wages would be lower, and our benefits would be nowhere. We wouldn't last in our jobs. The union hasn't been supportive of the women, but if we didn't have a union at all, things would be a lot harder for us.

The Coal Employment Project has been important because we need somebody to go to if we can't go to the union. We feel like we have support there. I've learned a lot from the other women, and I've utilized it in how I've worked with the union. I'm sure that's why I've got more of a fight on my hands now, because I've opened up and spoken up on issues. Neither the company nor the union really likes the women to go to the women's conferences, because we learn too much.

*She noted that, over the years, the community has become more accepting of her lifestyle as a working single parent.*

My [older] children are 21 and 18, and I have a two-year-old. The older two are in school in Utah and Chicago, and I'm encouraging them to go on with their education. I want them to do better than me. I took my education for granted. I didn't go there to learn. And now I have a very hard time talking in public or explaining things so that other people understand.

Things have changed for the better at work. The men are more accepting of us being there. Our fights with them are different, although there still is a problem with harassment. And the community has changed toward us. Now I come across Navajo women who ask, "How did you ever get that job?" They say, "You must be really doing well." They feel that women who are working up there are well off. They think it's good that we're working. In the beginning they would accuse us of messing around with their husbands, and now we get compliments.

# 13

# WOMAN ENOUGH
## Janice Molineaux

*Money doesn't mean that much to me. Peace and happiness, that's what I want.*

*Mine safety has attracted the talents of women miners in many capacities: as members of local union safety committees, as company safety officials, as international union representatives, and as state and federal inspectors. Janice Molineaux now works as the only woman inspector in West Virginia's mine safety enforcement agency.*

*Molineaux's voice is rich in the melodies of her native Trinidad, although she has traveled a long way since her childhood as the daughter of a West Indian fisherman. An adventurous young woman, she struck out for Italy and the United States, where she has worked in nontraditional fields since 1977. After a layoff at her mine she enrolled in truck-driving school and went on to drive 18-wheelers across the country for three years.*

*As an inspector, Molineaux has her daily workplaces in the low-coal seams of McDowell County's small mines, through which she crawls to monitor dust, gas, and roof conditions, among other problem areas. Her job is protecting miners' health and safety.*

I knew nothing about coal mines before I came to West Virginia. I didn't even know what coal looked like, because in Trinidad we have no coal. We lived on the coast in a small fishing village called Carenage. My dad was a fisherman, and I was raised with one brother and one sister. Trinidad is an island of its own, but there are smaller islands along its coast where people had vacation homes. My dad also worked on one of the other islands. We would go down where he worked, and we would make a garden on the hillside.

We had to make a garden to eat. We ate a lot of roots, like cassava—the Mexicans call it yucca root. Down on the island there are a lot of fruits. So

Janice Molineaux.
Photo from Tom Johnson

while your mother's cooking and your father's out fishing, we would just swim or go fishing, too, or gather fruits—mangoes and papaw. We had sugar apples. Banana we used as a main dish. We had coconuts. We had cherry trees and orange trees and grapefruit trees. Plantains, cassava, and dasheen and sugarcane—we had food growing in our backyard to eat.

I had a great childhood: bathing in the rain, playing in the river, going fishing, going to the country. I didn't know anything about snow because our temperature averaged 85 degrees year-round. And with the trade winds blowing all the time, it never got too hot.

My father had a big net and three boats, and in the nighttime they put a bright light at the end of the jetty and the sardines came in. You have to entice the big fish to come in by throwing out the sardines. When the big fish come in, you throw out the seine, and that's how you catch the big fishes. That's how my daddy used to make his money.

The women stayed home. There was a song my dad used to sing to my mom:

Brown-skinned girl
Stay home and mind baby
Brown-skinned girl
Stay home and mind baby
I'm going away on a fishing boat
And if I don't come back
Stay home and mind baby.

That's how it was. I never knew much about my grandparents. I met two great-aunts and a great-uncle. They didn't speak much English. They spoke patois, a broken mix of French and English. My parents spoke the patois, too, but when we were growing up, we were under British rule then and had to learn the queen's English. They didn't teach us the patois. The elderly people would converse in patois, especially if they didn't want you to know what they were talking about.

Trinidad is about 30 miles from Venezuela. Culturally we always say we are the melting pot. We have a lot of Negroes, Indians, Chinese, and whites. It's a mixture of a lot of races, and the races intermarry. You could hardly find a pure Negro. They are a dark-skinned people, but there are a lot of light-complected black people. And some whites are so tanned they look bronze.

*On an island where class divisions were more obvious than differences in race, prejudice developed along economic lines.*

Over here you have the distinction between white and black. You have prejudices. In Trinidad you have prejudices against what we call the "bourgeoisie." They might be Negroes or black people, but they have money. It's not so much their race as who has money.

That's how the prejudices really go. I've seen it between black and black. One is real light-complected and has money, and the other is dark-skinned and don't have nothing. To me, the Indians are somewhat prejudiced against the black people. There are Indians that are darker than I am. They don't mix much. They work hard, and they farm. They have wealth.

As far as my own background, I think it was a mixture of Indian and French and Negro. My aunt who couldn't speak English was a Creole, French and Negro. She had long hair down her back that we call dugla hair. When you mix an Indian person with a black person, you mix straight hair with curly. The mixture between them is a wavy kind of hair, like my aunt's.

*The island of Trinidad was under British rule during her early childhood.*

I went to an all-girl school with the nuns. We were Catholic and British. We didn't get our independence until 1962. In school we had forms. After form five, you took your exams. If you passed, you could take a job with the government or go on to university. My schooling was strict. We had uniforms, and while in uniform we weren't allowed to speak to boys.

When I grew up most of the girls left to go overseas. If you wanted to better yourself, you went to America. At one time there was a rush of people leaving to do domestic work in the United States, where they could buy things you couldn't get in Trinidad. Now you can get what America has in Trinidad, but you pay for it. A pair of Michael Jordan sneakers costs about 150 U.S. dollars.

My mother told me that if I passed my exam they would send me anywhere I wanted to go. I was, as she said, hot-footed. I was adventurous. They had a program where you could go to Italy and help the mother take care of her kids. I worked for the government for a couple of months, and then I resigned, and I went to Italy, and then I got a visa and came to America.

I came to Boston and met my husband at a party. He was from Jenkinjones, West Virginia. He was going to a trade school to be an offset printer. After he graduated he wasn't making enough money, even with overtime, so he decided to go back home and work in the coal mines.

When you are in another country, you hear so much about America. You don't hear about the poor parts like West Virginia. I had never seen poor white people until I came here. In our country all the white people have money. You don't see them dirty. You don't see them going around in bare feet. To come here and see such poverty was a shock to me.

When you live overseas and you see country life in America, you see a big farmhouse with flat land spread out. They don't show you the back roads like McDowell County and Harlan, Kentucky.

We caught the bus from Boston and went in through Bluefield. I don't know really what I expected, but I didn't expect what I saw. You had to cross three mountains before you got to Jenkinjones. I got motion sickness. It seemed like we were going and going and never getting there. But when we finally got there I cried. I wanted to go home. I was expecting what I saw on television: nice homes, flat land. And what I found was a lot of old, run-down houses and coal heat, which I knew nothing about.

We lived with his grandmother. She raised him, and she didn't like me. The first thing she said was that I was using her son to stay in this country. That was the furthest thing from my mind. He'd had an old girlfriend who'd lived next door to him, and I was a stranger and a foreigner.

Another thing she didn't like was that when I heard music I would move my body and shake. In Trinidad we are used to dancing. She complained all the time, and one day I just couldn't take any more. I went back to Trinidad to get my green card, and when I got back my husband had a little house for us.

When I got back to Jenkinjones there wasn't much work available. All the men were coal miners. The women stayed home and raised children. I had no intention of being a housewife. I had to find a job. My mother always drilled into us, "Be independent." "You need a man," she said, "but you need a job so you can depend on yourself, too." I decided, if I'm going to work, I'm going to work where I can make some money. I'm not going to sell hamburgers for eight hours. So I decided to go into the mines. That was in 1973, before women had been hired.

The company wouldn't hire me because they said there weren't any women working in the mines. So I let it rest. I went to work at a Head Start center as a janitor, and I helped in the classroom. In 1977 I reapplied and got the job. I went to the mine, I spoke to the superintendent, and the next week I was hired.

At first I was kind of scared, because you don't really know what to expect. But after going underground the first time, it didn't bother me. The first day was moving a beltline. Being bent over for eight hours was pretty rough.

The coal seam was anywhere from 48 to 60 inches high. I went to work with the masons, as a helper. I stayed with them most of the time and ended up bidding on a mason job. Nobody wants to be a mason because it is one of the lowest-paying jobs and one of the hardest. We worked with 40-pound cinder block. I enjoyed it. I played softball then. I was building stoppings then, and I could hit home runs!

I enjoyed going to work. I never missed. The guys cracked a lot of jokes. The boss said, "You've got three strikes against you: you're a woman, you're black, and you're a wetback." I said, "I didn't come here illegal, so I'm not a wetback!" The guys used to crack dirty jokes, but I feel like I went into a man's world and have to accept certain things. But I told them to tone it down sometimes. If the jokes were funny, I would laugh. I would tell jokes, too. I used to get a lot of black jokes, and a lot of Polack jokes. If I didn't think it was funny, I would tell them I didn't like it.

You have to let them know how far they can go. If you don't, they'll go as far as you'll let them. I told them, "Don't touch me." But I worked with one buddy on mason work, and I didn't have much problem. Sometimes when I went on the section the guys would wrestle with me. One time they were unloading rock dust, and they held me down and filled my shirt full of warm rock dust. They didn't touch me, or look at me, or do anything out of the way. Another time, I was outside, they held me down and filled my mouth full of chewing tobacco. But the guys never showed me any harsh ways.

Some of the women flirted and let the men do their work for them. I held my end. I worked hard. You can put me up against any man. But the industry was going down, and the first layoff got most of the women around 1982. Before the layoff there were approximately 300 miners at the mine and close to 20 women. I got laid off, but I got called back in '85. I went to work in the warehouse for the company in '84. The union tried to kick me out of the union, but they couldn't, because the company paid me hourly wages. When they went back to work in '85 I went back underground. Then they shut it down a year later.

I went crazy being out of work. I went to Consolidation Coal Company to try to get a job. That company has a nonunion mine, and they don't hire many union people. They interviewed me, and I think it went well, but they never called me. They asked me what my goals were. I told them I wanted to be the superintendent one day [laughs]. If I can't improve myself, what is the sense of working?

After the layoff in 1986 I went to Trinidad for six months and came back and started college. My first husband and I had gotten a divorce. We were married for eight years, and it just died. The divorce didn't bother me. I just kept on working. And all the time I worked underground I was working on my mining engineering degree part-time. I worked midnight shift, went home and slept, and went to school in the evening, and then went back to work. Another time I took welding as a trade and became a certified welder.

I went back to college and studied construction engineering, because the coal industry was going down. I met an old boyfriend and fell madly in love. We got married and decided to drive a tractor-trailer rig after my graduation. I went to truck-driving school and drove a tractor-trailer for three years all over the country.

It was nice because you get to see so much of the country, but there is no way to have a family life. You're on the road on weekends and on holidays. It paid the bills, but I wanted to be home where I could make a garden, go to church, and be home in evening times. I applied for a state mine inspector job so I could be home. Then my husband left me, two days before I took the state exam. It was depressing, but I hung in there because I was determined. I wanted that job.

*In 1991 Janice Molineaux was hired as West Virginia's first woman mine inspector.*

To me, inspecting was another step up. Very few women take the exams, although they've been offering them for years. We had four women who took that exam out of nearly 100 people.

I inspect small mines. My job is safety. Frequently miners cannot get materials to make repairs. They will point out things to me that need fixing because that is the only way they can get it fixed. And the inspector, having the ability to write a violation or talk to the owners, can get it done more quickly and with more cooperation.

Some operators will work with inspectors. I've never had any harassment. I guess they're afraid to, because of my position. I have a lot of authority. I make it clear to the operators, "You can work with me; do as I ask." Being a woman, I ask them. But I am very clear: "I can ask you to take care of the problem, or I can make you do it. You have the choice." And I have good relationships with mine owners and bosses and the men who work in the mines.

In West Virginia some state laws are stricter than federal laws. But you can inspect a mine this week and go back next week and you find the same problems and more. There are some good ones. There is one mine that is almost perfect. Three brothers run it. You never see a piece of paper or a cigarette butt on their property. The belt is spotless. But going to other mines will just floor you. Most are low coal with mud knee-deep, and you crawl through mud to inspect the mine.

I'm crawling all the time! [laughs]. My mines are 30 to 40 inches high. We crawl the belts and the returns. Most of my mines are one-section, but they can go pretty deep. One day this week I crawled on my hands and knees for more than a mile. They had four beltlines.

My knees are all to pieces. I wear knee pads, but you have to crawl through mud and water. After your clothes get wet, it irritates your knees, so they get sore. You're carrying a five-pound rescuer, which after a mile feels like it weighs a ton. If you stay in low coal all the time, you tend to get used to the crawling. I don't go to the high mines hardly ever, so I'm used to the crawling. And you pace yourself. It's up to you to do your work in as much time as you choose.

In McDowell County there are about 100 small mines. The majority of them claim to be union mines. The men pay dues, but they have little union representation. Most of these small mines run rubber-track equipment in 36-inch coal. The conditions are muddy and wet. Some owners pay the guys a flat rate, and they don't have any insurance. I spoke with some of the guys. They know if they screw up there's always somebody else out there wanting that job. Most of these operations are nine to ten people. Some may have 30 people, but there are not many large mines left.

An inspector cannot baby-sit a mine seven days a week. While I'm there everything is fine, but when I leave it may go all to pieces. Some companies will neglect safety to run that coal. They're looking at that dollar sign. They're not looking at the consequences to a man's life of running an unsafe mine.

*Her enforcement power commands respect and protects her from gender-based harassment.*

In this position I get more respect from the men and from the bosses. In the mines the guys will say anything around you. As an inspector, [I find that] if they curse and I say something to them, it stops. They will apologize and make an earnest effort not to do it again. A lot of times all I have to do is ask them not to do something and they'll stop. Or if I'm having problems and I talk to the boss about it, he'll go right away and talk to the men, because they don't want me as an enemy.

The men may talk about you like a dog when you're not there, but while you're there you have respect. They come to you, they ask questions. I may not know it all, but I try to find out the answers. The mine operators ask questions. When you get there, they're attentive. I don't know if they're scared or what. Some of the guys on the section get very nervous when they see an inspector coming, because they know the power inspectors have.

A lot of the men in these small mines have never worked with women, because small operators don't hire women. The women coal miners were only in the big mines. Sometimes I might run into maybe a superintendent or a mine owner who gives problems, but the miners themselves tend to go along with me because they want the problems corrected.

I have an unlisted phone number, but I have to post it at all my mines. If the men have a problem, and they don't get cooperation from their employer, they can call me seven days a week, day or night. But it has to be strictly business. I have had calls. It's also there in case they have an accident after office hours.

*As an inspector, she can shut a mine down for safety violations.*

By the stroke of a pen I can shut a mine down, if I go in there and find imminent danger. If I go in and find smoking articles on the men, I can pull the crew out of the mine. Without a comprehensive program in place, they cannot work. I can shut them down maybe for a day or two. I've never had to do it, but you need to have that authority.

I am there to help, not to hurt. It's hard to make the owners understand that. Because from the time they say "inspector," you're a dreaded person. The owner thinks the inspector is there to put him out of business, to slow down production, to keep the men from doing what they have to do. But without an inspector, the coal mines would go all to hell. Even the miners will tell you that. If the inspector didn't come, the company wouldn't check the equipment, check the belts, they wouldn't take gas tests or keep up the curtain to keep the ventilation up.

*She was injured on the job and now fights chronic pain.*

As an inspector, I have to inspect coal trucks, too, and I had just finished inspecting a tractor-trailer and citing the driver for taking out his front brakes. I was standing between my car and another inspector's car, and the truck was behind the cars. His truck drifted down and hit my car, and I was caught between them. The tissue and the nerves of my left calf were damaged. It was bruised badly but nothing was broken, so the doctor sent me back to work. But it was swollen for at least four or five days afterwards, and I couldn't walk on it. The pain is terrible, even after a year, no matter if I'm crawling or just lying down. I'm trying therapy and shock treatments.

I am an implant to West Virginia, but now it's home. I'm used to the laissez-faire way of living. I feel comfortable there. You leave your door open. I buy my groceries and put them in the back of the truck and go off to another store. If anyone is desperate enough to steal my groceries, they can have them. Sometimes I get depressed—nowhere to go, no people to meet. But then, I can walk in my yard. I can make a garden. I can go walking in the mountains. I can go by the river and fish. I can have a dog. I'm thinking about getting a cougar. Can you see me walking through town with a cougar on a leash?

When you live by yourself, you can't depend on people. I cut my own grass, I tarred the roof. Either you do it, or you pay somebody to do it, if you don't have a man around. I change my tires, I change the oil in my truck. I think I am woman enough to do it all. You have to put your mind in it.

My first decision was coming to this country to better myself. Going underground, I asserted my independence and proved I could work side by side with a man. My failed relationships made me more determined, made me stronger. It made me want to achieve more. I've got to have something to focus my time and my energy on. I promised my boss I was going to go back to college to get my mining engineering degree.

A lot of men can't accept an independent woman. I don't see why they should be threatened. They should be glad if a woman is willing to work. A husband should be willing to share and be proud of what you're doing. Sometimes when I get down and depressed, I lose a little of my self-confidence. But you can't let yourself go down. I would tell myself, I am woman, I am strong, I can survive.

*She prefers inspecting to coal mining or truck driving because of the greater freedom on the job and thinks occasionally of returning to Trinidad.*

The pay was very good as a coal miner, compared to other jobs. It helped you to get what you wanted. You could buy a car, buy a home. You could live at a certain standard that other people can't. When you attain a certain standard, you don't want to go down. You're always striving to be up there or go better. It enabled me to take trips back to my country.

I liked the travel of the truck driving, and yet I want to be home. I like the freedom of being a mine inspector, working when I want, how I want, and be off when I need to. I think I prefer inspecting, because I have more freedom in that line of work, because I arrange my own schedule. If I want to work midnight shift, day shift, or evening shift, I can do it.

If the coal industry lasts, I would like to get with the federal government. I wouldn't mind going to a different state and see a different kind of mining. I would like to travel a little bit while working in the industry. If I can be an inspector and inspect different types of mines, I would do that, or work in an administrative position. In the federal agency there's more room for advancement than in a state agency.

If the coal mines went down, I would go back to driving a truck. The money is not bad, but money doesn't mean that much to me. Peace and happiness, that's what I want. I drove for some months by myself. You have to be careful and have a straight head on your shoulders. But there are a lot of women truckers now, and I like a challenge.

If I could get a job and make enough money, I would go back to Trinidad, because my family is there. I want to be around my dad and my sisters and my brother. I would go back, but it is depressed there, and sometimes there are shortages. There won't be any rice or any flour. Over here you can get anything. In Trinidad you can't afford to buy a T-bone steak. You can buy it, but how many families can afford $13 or $14 a pound for it? A head of lettuce is six dollars.

*In 1994 she attended her first national women miners' conference and left with a new understanding of the need for women to protect their interests.*

We need a strong organization. In McDowell County the big mines are closing down, and that's where the women were working. We have about 100 small mines, and there are no women in those mines. The only mines that have women at all are U.S. Steel, and in a couple of months they're going to have a big layoff, which will get more women. We need to organize. If we don't fight for what we need, we're not going to get it. They will take it away. They will say, women's place is in the home, and that the man needs the job. But the way family life is going now, women need the job, because they're the ones left to raise the children.

# A Different World: Struggle and Survival

Once across the portal, women miners faced other barriers: sexual harassment, isolation on the job, and the added workload of home and family. Although anecdotal evidence suggests that harassment by management was more prevalent, discrimination by coworkers was particularly difficult to address, particularly in unionized mines.

Women miners developed ways to cope with harassment. Patsy Fraley's union activism resulted in greater harassment. But it did not stop her from exposing corruption at a mine safety agency. Early in her 14-year mining career, Patricia Brown adopted a steely attitude at the mine mouth to protect her innate gentleness and compassion, but she later found ways to remove the disguise. Brenda Brock offers a close view of women's struggles in a non-UMWA mine in the early 1980s and tells of the effect of her pregnancy on her male coworkers. Sandra Barber didn't experience serious harassment problems until she rose into the higher ranks of the Mine Safety and Health Administration (MSHA). Rita Miller describes her efforts to combat discrimination on and off the job, and how she coped with repeated layoffs.

# 14

# A THOUSAND LIVES
## Patsy Fraley

*It was money and greed that cost me my children,*
*and that's what it took to get them back.*

*Patsy Fraley worked at one of the largest union mines in eastern Kentucky in the 1980s until she and her husband Tom lost their jobs when the mine was dismantled and sold to a contractor. Most miners in eastern Kentucky have two options: working in a "dog hole"—a small, often unsafe mine—or trying to piece together a living in some other line of work.*

*Patsy fought back against the job loss, just as she has challenged all the obstacles in her life. She has survived more than most people could endure: a troubled childhood, domestic abuse, the loss of her children in a custody battle, job harassment, unemployment, and battles with her employer and with her union. She won some major victories, such as exposing corrupt practices by mine safety officials and advancing to a position of leadership within her local union.*

*Our interviews took place over a period of two days in 1992. As he has done in every aspect of their shared coal-mining life, her husband Tom participated in the interview as a loving and supportive partner.*

After I got hired in 1977 I worked one shift, and then the men went on a three-month strike over the medical card. I was absolutely broke. I had gotten put out of my apartment with my baby, and the landlord said he didn't care where I went. Luckily, I borrowed some money and got me another place.

Right after I started at the mine, I was waiting at the elevator to go underground, and they stopped us and brought out a guy who had been electrocuted. They were doing mouth-to-mouth with him. He was already dead, but they kept working with him all the way up on the elevator. I couldn't believe it. He was 34 years old, he had three kids, and his wife was

Patsy Fraley.
Photo from Martha Duncan

expecting another one. The men were crying. They sent us home. I was so nervous I had to pull off the road and throw up. I cried and thought, "This is what happens to you." I prayed all the way to work the next day. The job was my only hope, because I had tried everything else to get my children back and to give them a better quality of life than I had.

When I was a little girl, my dad ran off and left us, and we about starved. My mother would go to the secondhand store and buy our clothes. She would starch them and stay up all night to iron them. But the other children would laugh at me.

In our town there were the very poor and the very rich. They were really middle-income, but to me they seemed rich. I had a lot of love, and my mother did her best. Still, I didn't want my children to have their feelings hurt like I did. If they fit in, I think they have a better chance.

My mom and dad belonged to the church, and then my dad went off with the lady preacher. It was the Church of God. I remember the big scandal. My dad had moved us from Ohio to West Virginia, and you had to live in the county for a year before you could get welfare. There was no such thing as food stamps or a medical card. My mother cleaned houses, took in ironing and washing, did baby-sitting, and bought furniture to move us to a nicer

Tom Fraley and his son Jeff.
Photo by Cosby Totten

place. I think she was starting to get cancer then and just never went to a doctor. She didn't know what was happening to her.

My mother died when I was 17, and before she died she asked me to raise my two brothers. I had gotten married a year earlier, and by the time I was 22 I had four kids to raise, plus my two brothers. My husband believed in keeping me barefoot and pregnant. He never allowed me to have any money. I wasn't allowed to work. I never will forget going to my high school graduation. I sat there with two children watching my class graduate and cried my eyes out.

But then I got on a program where they paid you so much to go to night school in typing and bookkeeping. I found a baby-sitter and was really feeling good about myself. But my husband resented it so much he ran the baby-sitter off after two weeks and busted up every dish in the house. The kids were screaming and crying, and he told me I could not go back. After that, the lady wouldn't baby-sit for me anymore.

I suffered a lot of physical damage from the abuse period. There weren't any abuse shelters when I was married. Because his family was well-known in the town, you could call the police on him, but it would be a joke. I was terrified to stay with him, and I was terrified to leave him.

I finally got my courage up and left with my kids. He stole them back at gunpoint. He swore on his mother's life that he would shoot them in front of me. I knew he was capable of it. He said, "You let me take them, or you watch them die." So he took them, and I stayed in Columbus so I would be there when the divorce came up. I hid out, and he found me and beat me and shot at me until finally I had to leave town.

I told my lawyer my address, and I called him every week when the divorce was coming up, because my husband had said he would take everything. We had a nice brick home and a swimming pool in the backyard. But I had no money at all. The checking account was in his name.

By the time I found out about the court date, it was too late. One Saturday I got a registered letter saying the divorce had come up the day before. Since I had not shown up, I lost my kids.

The next two years of my life I don't remember much about. My head was messed up. You can't afford a lawyer on a minimum-wage job. I had talked to a judge's wife, and she said all I needed was a stable home. But it takes money to have a stable home. I went to Florida and married my second husband and had my fifth child.

I was never satisfied without my other children, so I came back to West Virginia. For three years I worked at the VFW as a bartender. I hated that life because I was raised a Christian. I wanted to get out of it and find a job where I would earn as much money as my ex-husband did. On the radio I heard that a coal company had been forced to hire women. That was in January 1977. So a friend and I drove to mining school for a month and cashed in pop bottles for gas money.

I was determined that I was going to break this cycle of poverty. My mother struggled so hard. The women in her family had always had a hard time, working in a restaurant 12 to 14 hours a day for $15 a week. To my knowledge, only one of them ever finished high school. I was going to send my children to college. Everybody thought I was crazy, wanting to go to work in the mines, but it took money to get my children back.

Before this happened, I thought mining was the awfulest way in the world to make a living. I said, "None of my boys will ever go to work in the mines," because the only thing I could remember was dingy-looking clothes hanging on the line and all those dirty men. But desperation will change your mind, especially when it comes to your children.

When we finished mining school we went to every mine we knew to fill out applications. Half the companies wouldn't accept them. The other half laughed at us. Finally we called Island Creek, and they sent us on some bum runs, but then they called me to take a physical.

I was sent to Big Creek 2 in Sidney, Kentucky. It was a shaft mine, and at the time they didn't even have the elevator finished, so they dropped you down in a bucket that would hold six people. The men would get in that bucket, and they would horseplay, and that bucket would go sideways.

The first time I walked in the superintendent's office, he looked me up and down. He said, "You don't look like the women miners we're used to. Would you like to go out to dinner?" I said no, and he said, "Well, I'll tell you right now, we're not going to put up with no hanky-panky." I said, "If I was looking for a man, I would have stayed in the bar where I was at."

I started to work in July of 1977, and we were off on strike until September, and in December there was the contract strike. I was just about to get on my feet, and then we were out of work. I wanted to be active in the strike, but the men didn't want us on the picket line. We went up there, and they said, "Girls, go make us some sandwiches." I felt like, hey, I need to be here. But it was just me and another woman, and we didn't really push it.

It was a real struggle in the mines. It was physically demanding, but a lot of it was my fault. When I would shovel, I wouldn't have a dry thread left on me. For two months we loaded 400 bags of mortar mix at 90 pounds a bag. I lost 40 pounds. We had to handle 500 or 600 solid cinder blocks. The men laughed. They said they didn't do that much in a week. I thought if I worked real hard, I'd get ahead. I thought, if I bust rocks today, tomorrow I'll be on a piece of equipment. It didn't work that way, because they want to bust you, they want to run you off.

If you do 50 things better than a man, they'll say, "She tries." If you come up with a better way to do things, you can't just say, "This is the way it ought to be done," because they'd say, "Who gave you bossing papers?" So instead I would say, "Well, I don't *know*, but I think *maybe* if we try it this way, it *might* work." Then the boss would say, "Well, it *could*." And then about half an hour later he would say, "I think we ought to do it this way," like he came up with the idea. And then I'd praise him for what a good idea it was.

When none of them could date me, a bunch of them drove down on Saturday night and watched my house. They started the rumor I was gay. You can't win. I tried to make them think I was a boy. I had long hair, and I put it up under my cap. When I first got hired, I had frosted hair, and I tried to downplay that. I'd heard about rapes. They laughed in the dinner hole at a story about two men raping a black girl underground. Then they'd laugh about women they'd been out with. I said, "I'll never be the talk of the dinner hole."

But you had to worry about what people in the community thought about you. They talked about me when I worked at the VFW, and when I went in the mines it was worse. I met a couple of wives who were supportive, but the rest of them thought we were after their husbands. I told one, "I stand on my feet to make my living. I don't lay on my back." That's come a long way. More wives accept women miners now.

I've found that most older men respect you more than the younger ones. The older men have seen hard times, and they saw we were there trying to feed our kids. But the younger men came in and reaped all the benefits that the older men had struggled for in the union. They came out of high school, got a job through their daddy, and spent their money partying. And they had no respect for women.

The men resented the fact that women were making that kind of money. They think you should stay at home, take your beatings, and not try to better yourself. All of a sudden everyone resents you because you're trying to be somebody.

When I first got the job in the mines, naturally I joined the union. But I thought, I don't need a union, because I'm going to do my job. I wasn't against it. I was green about it. My grandfather was in the union, but I wasn't close to him. My dad was in the union, but he left us. I wasn't raised union, but later I fought for it and believed in it so strongly, it was like I'd been union all my life.

At Big Creek 2 we had guys who would strike over the least little thing. That hurt the union. If one guy was wronged, even if they didn't like the guy, they stuck together. They filed grievances. Everyone went to the union meetings. One time another woman and I were shoveling the tailpiece, and the boss cussed us out because we weren't shoveling fast enough for him. The men heard him and shut the place down.

After I'd been working three months this boss told me to go up and hang curtain in this place. I started up through there, and the pin machine was backing out. They said, "The boss pulled us out. It's getting ready to fall." I went back to him and said, "Why are you sending me up there?" He said, "That pin machine costs thousands of dollars, and it's hard to replace." I said, "You know, I've got five kids that think I can't be replaced." He said, "Charlie, you've got to take care of the equipment." The more they tried to run me off, the more determined I was to stay. I got the respect I wanted and finally got a buggy [shuttlecar] job. My buggy was like a Cadillac. The electrician fixed my seat up close, and all the lights worked.

I've worked fast foods and on an assembly line in a plastic factory. If you're standing on an assembly line or flipping burgers, what are you learning? Nothing. But in the mines it's a different world. It is fascinating to see little leaves in there that's been compressed, and where it sparkles, where there's been logs and animals in there. Mining was exciting. You stayed alert. I believe you could work in a coal mine for 60 years and still learn something new. It was dangerous, it was bad on your health, but I *loved* the actual work.

I used to say, when the men got on that elevator and went halfway down, they turned from human beings into animals. The men change. Your personality changes. I think it's where the men are scared, and they can't act like they're scared, so they put on a macho front. They get *wild*. Dirty talk and jokes. I liked the jokes. It made the time go faster, and it put the danger to the back of your mind.

I loved it. But I didn't like the stress, and having other women look down on you. Someone told me one time—she was on welfare, and she had four children, and she had her nails polished, with nice clothes—and she said she wouldn't damage her body by working in there.

I wanted better. I was tired of poverty. I wanted to break the cycle of my family. I felt like the women in our family were put in this category, and they never could rise above it. They were slaves to men, some were abused, and most didn't have an education. I thought, if I can bring us up one level and my girls and sons can go up another level, we'll pull ourselves out of this.

By 1978 I had saved up enough money to get a lawyer, and I called up my husband and told him I was coming to get my children. They were in Bellevue, Ohio. He told me that he'd blow my brains out if I came. I told him to load his gun because I made just as much money as he did. I went and got a motel and picked my children up. Once he found out I was making as much money as he was, he didn't fight me because he didn't want to spend the money. But it took me seven years to get my children back.

They shut Big Creek 2 at the end of 1979 because it was run as an experiment by the government. They transferred me to Big Creek 1 at Turkey Creek. When I first went back it was hard, because I had joined the church and I was used to cussing underground. Everybody tried to joke with me, and I'd go around the rib and laugh. If they got real dirty, I'd just walk off. You have to accept them for what they are. You need to stand up for yourself, but you need to know when and how to do it.

They shut Big Creek 1 down, too, and I was out of work for almost a year. I had paneled for [had union transfer rights to] the Pevler complex in Martin County, and they called me as a buggy operator.

When I was saved and joined the church, my faith was real strong. But when I went to work at the Pevler mine they put me on second shift, and I had to move to Kentucky. We were working six days a week, and I couldn't make it to church.

At Pevler I don't believe they knew they had a union. We would never eat with a boss at Big Creek, but at Pevler they'd sit there with a boss and talk to him just like he was a family member. And then we'd get up from lunch, and he'd put it to you.

I felt like I'd walked into another world as far as safety. There was no such thing as a safety meeting. Half the buggies didn't have brakes or lights. At Big Creek 2, if the fan was off, we came out. But at Pevler, if there wasn't any air in there, you'd go ahead and work. We always knew when the federal inspectors were coming because everything would be fixed up. If we did have a good inspector, the company got rid of him. We were living in a twilight zone. I don't think there had been a grievance filed there for two years before I came.

I told them they needed to unionize that place. I started filing grievances. I was hired as a buggy operator, but they wouldn't let me on a buggy. The first job they gave me was retrieving rails. These rails were sopped up in mud, and here I am trying to hand-lift them up on the flatcar. I was having a lot of female trouble. I was bleeding all the time. At the first mine I worked at the boss had understood and made sure I wasn't bruting it. At Pevler it was the opposite extreme.

At Pevler the struggle to prove myself started all over again. One man told me, "You ought to find you a good man and get out of the mines." I said, "A so-called 'good man' is what put me here." One man told me I couldn't work there and be a Christian. I was shocked. I said, "If I quit my job and raise my kids on welfare, I can be a Christian, but if I work in the mines, I can't?" He told me I couldn't go to church and wear pants. But I didn't listen.

I was raised in the church but had strayed away from it because of the abusive situation and my mother dying. She suffered so much and died so young, and I resented God for that. It took me a long time to understand you need faith. I figure I've had enough hell on this earth. I sure don't want to go to hell when I die.

After two years in the mines I was physically wore out. I would come in from work hungry, but I'd be too tired to eat. I would lie on the floor. I'd stop and pick up my son, and my back hurt me so bad that he would walk on my back. I'd take a shower, and by that time I'd be ready to go to bed, but he wasn't. He wanted to eat and go places. I tried to be involved, but it was hard because I was so tired.

It was hard to be active in the schools, and it was impossible to keep my house the way I wanted it. Everybody said, "What do you do with all that money?" If you're paying for five children in school out of one payday, you're not going to have much left over. I've never had a new car. At one time I had two in college and three in school. That takes money. Of course, I splurged some, because I'd never had money like that. I wanted to buy a home. But even though the man at the bank told me I made more than he did, even though I had excellent credit, he would not give me a loan.

When I first met [my husband] Thomas, I was kind of scared of him. He had this beard, but he had kind eyes. I was dating this other guy at the mine, and then we broke up and I dreaded going back. The rest of the men flirted and carried on, but Thomas had respect. He was one of the hardest workers there, and they were going to brute me so they put him with me.

Then my car tore up. You had to work the first month without pay, and I didn't have money to make my lunch. I called the bank for a $500 loan to get me to the first paycheck. I had paid off two loans at this bank, but they would not loan me the money.

So I left my little boy at the baby-sitter's all week, because I didn't have any food. I was putting on my coveralls in the bathhouse, and I just started crying, and Thomas came by. I said, "I have a job, but I can't get to it." He loaned me money, and he said he'd take my car to his brother-in-law's to get it fixed, and I could drive his car home. He got my car fixed, and I paid him back at my first payday. From July until February, we were friends. I told him I wouldn't date him as long as he drank, and he quit.

Then I got real sick and had a hysterectomy. It was in December of '81, and the surgery was in January. Right after that I had bleeding ulcers and lost three and a half pints of blood. The stress from the job—my boss was so cruel he wouldn't let us go to the bathroom—plus my physical condition was too much. Through Thomas, the men took up a collection. Thomas kept coming to see me in the hospital, and we would talk.

I thought it over, and we started dating and got married in May. I said to him, "Thomas, my children are my whole life, because I lost them. I don't think it would be fair for me to marry anybody." My daughter said, "Anybody who would marry someone with five kids must be crazy." It's been a good marriage. It's been rough at times. The bosses would say to the men, "Do you know how much money they're making together?" There were brothers working on the same shift. I never begrudged nobody what they made if they worked for it. But we had my three children at home, and his son came, and we needed money.

The men would make smart remarks to Thomas. They'd say, "Why don't you make her quit?" I didn't want to quit work, because I'm very independent. I didn't think it was fair that Thomas had to pay all the bills.

At work I filed a grievance because the company was hiring younger men and putting them on day shift when I was begging to go. We weren't even allowed to see our time cards. Another woman and I found out that the men were getting overtime and almost everybody was getting top rates except us.

We got together with the girls down at the tipple. One of them said she'd been sexually harassed. Somebody knew about the Coal Employment Project. There was eight of us to start out with in the early eighties. Our women were scattered out, and each one was going through her own hell. We started calling them and saying, "Hey, are you having any problems?" Our support group started when we met with Betty Jean [Hall, a lawyer and the CEP director] to see if we had a discrimination case. We started it as a gripe session. Betty Jean told us we didn't just need to tell war stories, we needed to do something to try to *change* the way women were being discriminated against. We kept notes, and we got proof of discrimination and kickbacks.

We had us a good case going, but we couldn't find a lawyer in Kentucky, and then two girls in the office who were helping us got scared. They were feeling pressure from the big bosses. We were getting pressure, too. The men refused to work with us because they thought we were keeping notes on them. They resented Thomas for sticking by us women. They told him he had to squat to pee and all that kind of thing. The boss was putting him on bad jobs, and I felt guilty.

The rumor went around that me and Thomas were trying to run the mine. The men hated Thomas. They called him pussy-whipped and a wimp, but we were just trying to go by the contract. Thomas came out one day to take a truck off a steep hill. Somebody had messed with the brakes, and he ran off the road and almost wrecked.

After the case started, the word was out that this boss was going to get us hurt, since he couldn't fire us. The bosses told the men we were ratting on them, and the men said if we got buried up, they couldn't help it. When we heard this, we went to the federal Mine Health and Safety Administration, because we had proof of what was going on.

Joyce and I took a sick day and went to MSHA and tried to explain the problem. We felt like our lives were in danger. The company had put her in a place that was falling in on day shift, and me in the same place on second shift. We couldn't even holler, because there was no phone. The union wouldn't help us. If a man complained, they'd take care of it. With the women it was, "Bless their little stupid hearts. Wonder what they're quarreling about now," even if it was a contract issue.

We walked into MSHA, and they acted like they were scared to death. They said we needed a lawyer. We said it was a safety issue. They said they'd have to get hold of the district head of MSHA.

We were scared to go to work. That night I got a call from this lady who said she worked at MSHA. She wouldn't give her name. She said, "[The company] is one of their biggest payoffs."

The district man called us. The agency had written up our complaint completely differently than we had told them. He recommended we get a lawyer through the union. Then I got real scared. I heard about a federal judge who had started an investigation of payoffs and safety problems. We hung tough, but it was wearing us down. The judge wanted us to come to Washington, so we went. The lady who called said she was working with the judge. Later somebody broke into her house and stole papers, because she had access to a lot of information. She said she had heard bosses tell MSHA inspectors on the phone, "There will be tires waiting on you, or two fifths of liquor." So not only did I have to worry about the company and the men, but I had MSHA to deal with.

Two other women miners and I were on CNN [Cable News Network]. The reporter came and did an exposé on company payoffs to MSHA. There was an MSHA employee who told about how he had witnessed payoffs and then left his job for fear of his life. But nothing else happened. The judge retired, and it just went away.

In the discrimination case, we couldn't find a lawyer in the state. By then we were all laid off and had no money. To go to federal court would cost $16,000. We ended up settling out of court for $3,200 apiece. But we wouldn't sign a gag order. My phone bill was almost that much. I did feel it was a victory not to have a gag order.

Through the case we got started trying to help other women. We just had our little support group, and then we had the [national women miners'] conference here in 1986. We built support in the community so they'd have a better outlook and support for women miners. I think the conference and the parental leave campaign got us more respect. We wanted to show them we weren't out to steal their husbands and that we cared about the community. We started the fight for parental leave—now it's "family leave"—way back in 1979. We wrote our congressmen and Representative Pat Schroeder [D-Colo.], and then she decided to jump on it.

As I got involved with women's organizations, I had some problems with dealing with their lawyers. When we talked about mining, we used a language they had no idea about. And they would come up with words I'd never heard of.

I felt so inferior when they would ask me questions I didn't understand. I would say yes, too ashamed to tell them I didn't know what they were talking about. They kept pushing me into meetings with lawyers, and I'd come away depressed. They were trying to build me up, but they didn't. I'd like to have the education of a lawyer so I could interact with different people. I limited myself years ago. Some of it was my fault, some of it wasn't.

Then CEP called and wanted me to go to Canada. I thought, they're trying to make a silk purse out of a sow's ear. They were pushing me to be something I didn't want to be. But because they'd done so much for me, I felt like I owed them.

I resent some women's organizations, because they get grants that are supposed to help women and they don't. I'm for equal pay. That may mean I'm a feminist. I don't know. I'm for survival. I don't consider myself an activist, but I believe in justice and doing what needs to be done.

*Patsy Fraley suffered several back and neck injuries working underground.*

I've hurt my back three times in the mines. The first time I sprained my back, and I was off six weeks. I went back, and then they shut the mines down. The second time I was setting timbers, and my back messed up again. The third and final time my buggy was sticking in high gear. There were no brakes and no lights. The inspector wouldn't help me out. I drove the buggy through a big water hole in high gear, and I hit the canopy at the back of my neck. I ruptured two disks. What made me mad was that the other buggy driver would lie up on the power box. The next day he told Thomas, "Ha! I knew I'd bust her!"

That's why I was so determined to go back and show them. I was off two years and went through physical therapy. I went back in '86 and worked one day. They didn't want me back. They put me on the beltline, carrying rock dust and throwing it across the beltline. By the end of the day I could barely walk. I had no circulation in my legs. I almost paralyzed myself.

I went through a deep depression after that, because I had to eat my words. If they'd let me run my buggy, I could have done it. They knew my back couldn't handle all that rock dust. If a man in the clique had come back after an injury, they'd have started him out slow. When I went back and couldn't make it, it felt like such a defeat.

Both Thomas and I were always fighting the company. We didn't have what you'd call a normal marriage. We were on different shifts. I'd come up at night and tell him what they did. We were tore up all the time, and we didn't have quality time with the kids.

Our time was limited anyway, but we were always in a battle to keep a job to where we could raise them. We were always in a battle with the company or the men or the union. We were juiced up fighting over this, or down and depressed over that. We shared the job, we shared the problems, then he got hurt and I got hurt, and we fought with this company and then with the union.

A marriage is adjusting to different stages. We had to adjust to the part where we worked at the same place but different shifts. Tom had to adjust to suddenly having three kids. We had to adjust to working on the same shift. We had to adjust to the part where I got hurt and stayed home and he went to work.

I resented the homemaker's life. It is a thankless job. I've been a coal miner, and being a homemaker was harder, because you're taken for granted. I used to say I would hire somebody to pack my lunch because I got so tired of packing my own. I love to cook, and I like making things Tom likes, because I know what it is to work and come home. I used to tell the men at work I wanted a wife to have my supper fixed when I came home. When a woman comes home she goes to work again. When men come home it's a different story.

When I was at home, mentally I was with [Tom] every day at work. I'd ask, "Are you setting your jacks?" I worried the whole time. In a way I wished I was like those other women and didn't know what went on. Other women, I know, are scared, but I *know* the dangers. I know the hassles. And he's had jobs in these little nonunion mines where they worked him 16 hours a day, and he'd come home, take a shower, sleep two hours, and go right back again.

It's been such a struggle. I don't know if we'll fall apart when everything goes smoother. I don't know if we'd get bored. I would like to have a little peace. Mentally, I'm drained. To talk about it I have the fight, but when you leave here I'll collapse.

*Thomas Fraley listened closely. Born in 1949, he is a slightly built man with a dark beard and gentle eyes. He told his story in slowly considered words.*

The first I ever heard about women in the mines was in 1974, when I first went in the mines at a small coal company. There was a woman who had applied for a job before me. When I filled out my application, the superintendent said, "Backdate that, because there's a woman that wants to get hired." They didn't hire her. But she got a job at another mine and bossed for Island Creek for a long time.

My dad worked in the mines at Inland Steel for a long time before he got hurt. My grandpa on my mother's side delivered house coal on a sled made out of two-by-four posts. Back in the thirties they heated with coal and they cooked with coal.

In about 1949 my dad got mashed up in the mines. My uncle got mashed up, too, and went to Michigan. After my dad became disabled, he sold beans and corn to the grocery for our school clothes and shoes. That's the way my dad worked for the last 12 years of his life. I really admired my dad.

After he died in '61, we went to Michigan, and I stayed four years in Detroit. My oldest sister had moved up there to get a job. I think I was 13 or 14, and I got me a paper route for the *Detroit Free Press*, and my mother worked at a White Tower restaurant. It paid very little, but my mother was a strong woman. She'd have to take a bus all the way to Dearborn just to get to work.

The army got me in February of 1968. I was in the eleventh grade, but I was failing. So the army drafted me, and I pulled my two years, went to Vietnam, caught hepatitis, and was sent back to the United States.

After Vietnam I went to New York and got a job for a few months, and went to Arizona, hoping for better money. That didn't happen, so I started working on the railroad in Baltimore. A group of us would drive back and forth every weekend to Kentucky. Whenever I got laid off I came back home. They called us back to Michigan, and we built railroad lines. As long as we had a car to drive we was all right.

In 1974 I got a job through a friend at a nonunion mine. It was a killer job. They started me out at five dollars an hour. It was a family job. Still, it put food on the table. I worked there for three years, and finally I got up to seven dollars an hour.

That company treated their men like slaves. They had me there eight hours a day continuously shoveling on my knees. You'd have to lay down to take a doggone drink of water. Some places the coal got down to 28 inches high. Even crawling, you couldn't hardly move.

I got on at Island Creek in '77 and worked until '82. After the Vietnam era, coal really started moving. We had a good time out of the coal, all the way up to 1980. It was a steady paycheck.

When I started there they had a woman, but she quit. And then Joyce Goble worked there, and Patsy came in '81. The men that I worked with were what you'd call male chauvinist pigs. I've heard men talk: "I can outrun her." They'd be unsafe a-doing it. There's men that wants to put pressure on a woman in the mines. I just can't understand it. I've always considered the women buddies as long as they do their job, and I never did see one who didn't, because they have to prove themselves more than a man.

In 1982 me and Patsy got married. I'd known her almost a whole year. Whenever we were dating I would take a drink of whiskey and do other stuff. If I hadn't married that woman, I would probably have been six foot under. She changed my life.

When we got married there was remarks at the mine about how much money we made. The men said, "That's not right." And I said, "Boys, doggone it, she earns her money just like you do."

We'd only see each other at shift change, for maybe five or ten minutes. I'd get me a little peck of sugar when I was going off the shift. I had a CB at that time in the car, and I'd talk to her on her way home.

I'd be at home in the evening, and the kids would be driving me crazy. Patsy would get off work at 11 and get home about 12, and she'd take a shower and finally get in bed about one o'clock. At 10 or 11 o'clock she'd get up and fix something to eat for me and the kids. She always did have our supper fixed before she went off to work. She was amazing.

I'd get up about five-thirty and would get myself off to work. I never expected Patsy to get up and fix my bucket. I've heard a lot of men say, "I've got to have a breakfast before I go to work." If the woman is working, why put that much more pressure on her? There's a lot of men don't even fix coffee or make themselves a sandwich.

I helped with cleaning the house, but I'll be the first to admit, I ain't no cook. But Patsy makes one mean pone of cornbread. Me and Patsy got married in '82, and I really love it. Working in the mines together, there's a whole lot more to share. Like I've always told her, "I hope you live to be a hundred and four, and I live to be a hundred." That's the way I feel about Patsy. I don't think there's anybody in this doggone country that could love a woman as much as I love her.

I think it's an equal partnership between me and Patsy. There's a lot of men working in the mines, and their wife might be a secretary at a law firm making $4.75 an hour. I've heard men say, "If something happens, I'll take the house because I made the money to pay for it." Like her money wasn't a supporting wage. If we was to get divorced, I think it should be split fifty-fifty. I'd say that she's contributed more to this place than I have.

At the mine I stood up for the women's rights. I just thought it was wrong the way they done the women. If they filed a grievance, I would be a witness for them. When Patsy and Martha and Joyce was in on this discrimination case, I almost had a wreck because somebody took a wrench, messed with the brake lines. And the bosses put me on bad jobs. Safety has gone downhill since 1980. The companies want more production. If they asked me to do something unsafe, I'd tell them, "Go suck an egg."

*Thomas was active with Patsy in the campaign for a family leave law initiated by women miners through the Coal Employment Project.*

If a woman wants to work, she should work. But you want somebody taking care of the kids who will be good to them. That's why family leave is so important. It's something everybody needs, just like health insurance. I've got a cousin over in Germany. His wife had a baby, and the government paid him, plus she gets off work for about three years to be with the baby. They've got family leave, and it's a whole lot smaller country. He said he'd never come back to the United States.

I've not worked any since October of 1990. I had been working in Knott County in a little nonunion mine. One Saturday I was the only roof bolter there, and I was putting up four-foot rosin pins by myself. I hurt my back about five o'clock and continued to work until nine-thirty that night. I had to crawl around just to finish the shift. I came home and spent five days in the hospital. My disks are messed up. I've been trying to get compensation. They awarded me 55 percent, but I've still not got the first dime. The justice system sucks, if you ask me.

In 1984 Patsy got hurt in the mines, and she draws two little checks a month, each of them $132.60 or something like that. That's all the money we've got, and our land payment is $229 a month. But Patsy can take a dollar and make two of them. She knows what we can afford and what we can't.

We haven't had health insurance since 1988. I think we need national health insurance like what Canada has got, or a little better. Right here in Johnson County, 40 to 50 percent don't have no health insurance. Patsy's had a lot of problems with her stomach. And whenever you try to get insurance, it's a "preexisting condition," and they do not cover it. There's something wrong with the system here.

In '84 they were shutting down mines left and right because the company was cutting costs. Island Creek was one of the best companies I ever worked for until it made a deal with the Italian government and the company went after the union.

At the time we were about the only union mine outside of Floyd County. Without a union, a company will work a person to near retirement and lay them off. That's why I believe in the union so strong.

A lot of people say, "I ain't going to work in nothing unless it's got windows." But I loved mining. I'd go right back if I could. It's a challenge. I think everybody should be able to go in and visit.

I don't see much of a future for coal miners around here. Why can't the government take all those tax dollars and put scrubbers on the power plants instead of sending up men in space? I don't ever intend to be up there in space. I've probably been spaced out a couple of times, but it only cost a couple of dollars [laughs].

*Patsy comes back into the room, and Tom leaves for the grocery store. She picks up the conversation.*

I was depressed after I'd stayed home, but then I started getting real involved with the union again, and that got me out of it because I got mad. In 1985 the company was bringing in nonunion truck drivers while we were working just two and three days a week. That happened when Island Creek sold out to another company.

Our mine was a big complex, and with 1,100 members in the local. You couldn't keep up with what was going on, and pretty soon the local started cutting their own deals. I'd try to file a grievance, and the union would say I couldn't, for no good reason. The company started shutting down union mines and punched a hole around the side and ran it nonunion.

I tried to get everybody to understand. I said, "Boys, don't let them do this. If you do, you'll be going out of the holler, as the old saying goes, a-talking to yourself about what went wrong." And finally only 20 people were left working union up there, out of more than 1,100 active members.

I always carried my contract book with me. I would file a grievance, and half the time the boss would scare the men so bad that the men who asked me to file it wouldn't even show up as witnesses. But I still won grievances. They can write a contract and it looks fair, but when they start having arbitration on it, one word can throw the whole thing out.

In 1987 Thomas got laid off and the mines shut down. I was already off hurt. But we knew this other contracted-out mine was working. The arbitrator ruled the contractor should put us back to work. By contract he couldn't put in a nonunion mine on union coal lands. But the company started fighting it.

The national contract expired in December of 1987, but people kept working without a contract. The company brought in guards and sandbags, and we weren't even on strike. Our company had pulled out of national bargaining, so we weren't covered anyway.

Then the international [union] forgot about us. They let nonunion people work our jobs. And we couldn't file a grievance because our contract had expired.

*The only chance laid-off workers had to get their jobs back was the arbitration case filed while the contract was still in effect.*

A magistrate ruled against us on the case, but we won a circuit court appeal. The company took it to the Supreme Court, and they refused to hear it. Then it went back in front of a federal judge.

By 1990 we'd been out for three years and still hadn't seen a new contract. Some people were still working without a contract, and we were pushing on this case to get our jobs back. There were families breaking up and people losing their homes and cars. We had no contract and no jobs. And every day we seen these people going up and taking our work. We thought about striking, but the international wouldn't recognize the strike. They told us the only people who could draw strike pay were the 20 active employees. They said we didn't have any community support. So we had a rally to show them they were wrong. After the rally the international sent down a contract proposal with nothing in it for the laid-off workers. They were pitting the working employees against the laid-off workers.

I persuaded the laid-off workers to vote down the first contract. The international union representative met with the laid-off separately from the working people. I went over to the meeting and stayed all day. One of the provisions of the proposal was that we would give up the case, which was our only chance of getting back to work.

I walked through the meeting room to go to the bathroom. The union representative said, "Boys, if it wasn't for that little sister right there, you all wouldn't have a job." I said, "If you vote for this, you won't have a job for long." The working men did vote for it, but I got busy and got hold of all the laid-off men, strung out all over the country, and we got together and voted it down.

The international sent us another contract, and it still had no provision for laid-off miners. They had language saying the company "intended" to hire a few more people, but nowhere in that contract did it say it was a union mine. It had no pay rates. People were desperate. Thomas worked this little nonunion job where they hired people a month before their first payday and then they wouldn't pay. One mine owed him $3,000 in back wages.

The international union wanted to force a vote real fast, and everybody was getting excited. I said, "Wait a minute, boys." I was on the voting committee. Nobody else would work on it, just me and Martha and Thomas. The UMWA vice president came down and presented the contract. He says, "This is the best we can do." I said, "We heard that about the last one." We voted it down by a very slim margin. I even swayed some of the working miners to vote against it.

Before Christmas the international vice president sent for me and some of the local union officers to come to Lexington. When we got there, he said he got us a contract, but we had to drop our arbitration case. He said they were going to give the laid-off $2,500 apiece if we agreed to drop the case. I've had lawyers tell me the case was worth $15 million. If we had won, the company would have to pay 60-some people back pay. From July 1987, when the arbitrator ruled in our favor, we were owed back pay.

It was a few days before Christmas. They wanted to have the vote right then and said they'd send our checks overnight to get there on Saturday. They said, "Patsy, you and Thomas will get $5,000 right before Christmas. Wouldn't that make a wonderful Christmas present?" I asked the vice president why he sent for me. He said, "Patsy, everybody listens to you. You carry a lot of weight. All I'm asking is for you to consider it real good before you present it." But it wasn't my job to present it. It was the district's job, with an international representative. I wasn't even a local officer. They were asking me to do that because I had fought it.

I patiently listened. I started asking questions about the contract language, and he admitted the mine the company wanted to open was not even going to be union. I said, "There is no way that I'm going to stand and tell these people to vote to go to work in a nonunion mine."

But a lot of people were swayed by the money before Christmas. Everybody said, "Patsy, how are you going to vote?" I said, "I'm going to vote no." The union representative asked me how it was going. I said I thought it would be a tie or close. Sure enough, it was a tie, 29 to 29, and they had to have another election in January. After everybody settled down I told them, "Boys, that $2,500 is an insult. You could make that in two weeks if those people up there weren't doing your jobs." So there was a lot that swayed back and voted it down.

Then some of the working miners got mad because they didn't get offered the $2,500. And some of them voted it down. They [the international union] thought they were pitting the working miners against the laid-off, and it backfired. So we voted it down for the fourth time.

In February they implemented a contract without a vote, which goes against our union constitution. We didn't find out until April. They claimed it was the same as the 1988 agreement, but there was a little side letter that let them keep the contractors and hire new people. I was fighting mad. The union representative said, "Patsy, this is the '88 agreement. It's already been voted on." I said, "It is *not* the '88 agreement. This provision is not in the '88 agreement. You tell the international officers that we *demand* the right to vote on this contract." Three weeks later the boss of the complex introduces himself to the local president and says we have a new contract. We filed charges with the [National] Labor [Relations] Board [NLRB] against the union, because we have the right to vote on our contract. The union put the screws to us, and we had no say-so. To me, that's dictatorship.

It was a hard fight, because around here you've got either union lawyers or company lawyers. No lawyer wants to fight both of them. We needed someone to fight the company because they did this to us, and the union for letting them do it.

*In 1992 the international union settled the arbitration case. The settlement amounted to a few thousand dollars apiece before taxes. As part of the settlement, the NLRB charges were dropped.*

I don't think anybody has the right to give away people's jobs when they're paying a union to represent them and they believed in it and fought for it. Some of the men want to drag on your skirt tails and let you do the fighting for them. But you can't just give up. If you don't fight back, it's your fault. But if you can, always get a lick or two in. But sometimes I wish I didn't have that philosophy, because it's awful hard.

I always thought the union would stick by us, and it didn't. When I needed help the most, they let me down. They're just not doing what I paid them my union dues to do. I supported them by filing grievances over safety issues and getting people's jobs back when they was laid off. I did my part, but they didn't do their part.

I have a lot of regrets because I put everything in the job, and the union. The worst thing a woman can do is to put her self-worth in a job. Nobody knew how much that mining job meant to me. I thought, I'll show them that I am worth something. The paycheck was the answer to my problems. It was money and greed that cost me my children, and that's what it took to get them back. I was breaking the cycle, I was bringing myself up, my kids were going to college. And it was a hard job, a scary job. I had to fight the men and the company.

I was doing it for the children, but in some ways I wasn't there when they needed me because I was too busy fighting for a job and working and getting involved in the union. I shouldn't have let that happen, because I don't have a job now, and we don't have much of a union no more.

People who know me now can't believe I ever lived in an abusive situation. My stomach trouble now came from that, because I was never allowed to state my opinion. I was not a human being. I didn't know who I was. All I knew was that I was a mother, and an abused person, and I wasn't fit for anything. But when I got that job, it was everything. That was my answer. But it was nothing but a hassle.

I should not have let the union and my job interfere with my faith and the worship of the Lord. This is where that old union blood comes in. I found a church I thought I could stick with. We'd had an informational picket line up to stop the nonunions, and one of the men who went to the church crossed the picket line. I was so upset with this man that I quit going to church. We went around and shook hands with everybody, and I couldn't shake his hand. I haven't been back since. But the way the world is now, everybody's against everybody. If I picked a church where there wasn't a scab going to it, I wouldn't have a church to go to.

As far as women miners, most of them around here are laid off. Women miners came up through CEP—I have to give CEP a lot of credit, not the union—because they hung tough with us. But a lot of these women I talk to will never go back, because it was such a hassle. I'm too old to get another mining job. And some women say, "I could work minimum wage and see more peace."

Right now I don't know of any mines that are hiring women. It's sad. And most of the women who worked in the seventies are my age. Now I don't think I could pass the physical. It's time for the next generation. It makes me resentful in a way, because I loved mining work. You may come out crippled, or with black lung—if you're lucky and they don't carry you out—but in the mines you have the best benefits there are, and you have a retirement plan. We women who've made the sacrifice, what are we going to end up with? For the time that I worked, I'll never reap the benefits.

What it comes down to is self-preservation. I gave so much to causes and my children and my husband, I didn't take care of myself. I've got emphysema and black lung. After all we've been through, we can't afford to take care of ourselves. We don't have the money for eyeglasses. We are getting farsighted and can't hardly see to read. I went to Wal-Mart, and the glasses were $16, and we didn't have it. So then I went to Big Lots and found these little plastic hot-pink glasses for two dollars, so I got those. And his doctor gives Thomas medicine samples for his back. The compensation was supposed to pay it, but they don't.

My life is about over with, but before I leave this world and travel to the next one, I'd like to know my kids are established and they've got a home. And what am I going to leave them? We've worked for what we've got, but it's not much. Of course, nobody ever left me anything.

I'm a fighter. But when you're financially embarrassed [laughs], you can't concentrate on doing good work, or fighting for this guy down the road, because you're too busy trying to survive. Thomas and I used to be considered middle-class. We're way below that now. Ten years ago we made $40,000. We lived on $7,000 last year.

I've always had to struggle. For every step I've taken forwards, I've taken 30 back. One sickness could wipe up out. And it's not just us. It's everybody. But through all this, I have grown. Before, I would never have stood up for women's rights. After the brainwashing I got for ten years from my first husband, I didn't know where to put my frustrations out. I drank and I partied and I hated the whole world. And then I thought, he's still destroying you.

With the mines, I thought I was just getting a job, with good money. I didn't know it was going to take my soul and my life. It aged me. My body's busted. Still, sometimes I can outwork my girls. And before, I didn't own my own home. Now we own this place, and we're paying on land next door. I'm disappointed, because I always believed in the American dream. I thought, if you work hard and don't waste your money and don't drink and don't run around and are good to your children, you'll have a nice home and have medical coverage and have your retirement. But that's not the way it works.

Looking back, I feel like I've lived a thousand lives. I've gone through almost every kind of situation that a person could be in and experienced things a person shouldn't have to go through. But it's made me the person I am today. It made me realize you can't fight the world, and you can't go around being bitter.

Part of me likes the person I've become. I still have some bitterness, but I like the fact that I have grown, that I have learned to stand up for myself instead of being submissive. I like that fight in me, but I don't like what it did to me physically and mentally. Women can do a lot more than they think they can. You can reach down in there, honey, and get it where you don't think there's any more left.

*Her "thousand lives" presented a series of obstacles that challenged her faith and threatened her happiness. But she was not defeated.*

I was raised in poverty, and that was one life. My dad left us, and that was another life. Then I got pregnant and had to get married at 16 years old. Then my mother died when she was 40, from cancer. Living with an abusive husband was another life. Being a mother, that was another life. Then I kind of lost it when I lost my kids. So I started drinking and partying, which I'm not proud of, but that's part of my process. I worked in bars. Then I married this little rock star, and that was a different life. He was a good little feller, but he was in Nam, and he had that post-stress syndrome, and he would flip out.

I went to Florida and stayed four years, but there was always something tugging me back to the hills. Then I was Patsy again, but I was no good because I gave my kids up. I came back broke, with 11 cents and a baby and no mother to come home to. I had to work in the bar to make it. Going to work in the mines was another life, and learning to stand up for myself. Then I got saved, and that was another life, and being married to Thomas is another one.

Working in the mines will make you strong, but it also humbles you down to realize that any breath could be your last. Especially when you've got a little bit of money, you still need to have faith and have God in your life. When you're young, you think you'll live forever. Going down in the mines opened my mind, but it took a toll. I don't know if the good outweighs the bad. I really don't.

# 15

# MY OTHER LIFE
## Patricia Brown

*"Women are not taken seriously by the unions. That needs to be changed."*

*At the time of our first interview in 1983, Patricia Brown was facing the problem of sexual harassment as one of two black women working in a large underground mine in Alabama. In her Bessemer home she had the quiet self-assurance of a doctor, a teacher. She was careful with words. At the mine driftmouth her manner changed. She appeared larger, more powerful, defiant of any sign of the harassment that had marred her daily life. She acknowledged the behavior shift by saying simply, "Coal mining is my other life."*

*Ten years later she had reconciled her two lives. She had earned the acceptance of her coal crew and dedicated her life to the church. She was laid off shortly before our second interview in 1994; besides the loss of her job, she was also dealing with her children's departure from home. She was maintaining her strength and compassion and said she felt "blessed." Her desire now is to find work in child care.*

I was not prepared for the coal mines. My brother-in-law tried to warn me. He said, "Pat, you don't know it now, but the men are going to make passes, they're going to talk about you." He said I was going to have to harden myself.

I am a widow. My husband got killed in a car wreck. I try to block it from my mind. It was terrifying. I was almost 18. We had two kids, and I was pregnant with my third. It was rough, but it may have helped me prepare for what I've gone through at the mines. I didn't have anybody to lean on but myself.

For nine years I had worked as a preschool instructor, around very young children. When I went in the mine I had to make a complete turnaround and become a whole new person. With kids, you have to be gentle. My life changed completely. Before, I was making $7,000 a year. Now I make

Patricia Brown

$19,000, but the stresses working with kids are nothing compared to what I have now.

I went underground in 1980 because I wanted a better life for my children. I wanted them to go to college. I was in my early thirties. My family didn't want me to do it. They had heard about the danger and the harassment, but I make my own decisions. I studied coal mining at Walker Tech, and when I went in, I was aware of what was happening. They tried to scare us, but because of mining school, I was more aware.

In 1980 coal companies were just getting into the women's equality thing in Alabama, and we happened to be fortunate. There was one group of women hired before us, and that gave us a cushion, because the men had gotten a little bit used to women miners. They gave the first women a hard time.

But really, they did us the same way. We had men curse us out as soon as we got into the mine. Some would call me "black bitch" and other things. They would talk about a woman's period and make you feel as low as a dog.

For instance, this black guy—and I've had white ones do it, too—said women had no business in the mines. He said, "If I ever get to work with you, I'm gonna work your ass off, and it won't be any use for you to go crying to the boss." They'd make threats. But I hardened myself. I made the transition. It didn't take me long to learn their ways.

For instance, I have a big bust. I never thought about it until I got down there. They made me feel ashamed. They would tease and comment and write things on the walls. They drew pictures. Some guy on the shift before me would go down and draw pictures of my body, and somebody else would come and tell me, so everybody would get a big laugh. Mostly they did it to see how I would react to it. At one point I thought about getting an operation. But then I decided to ignore them. If they find something that upsets you, everybody in the mine will do it to you until you crack.

Of course, it did bother me. The first time I saw it, I went home and cried. There have been many tears shed over things they've done and said down in that coal mine. I did have this one black guy who was very supportive. If he went down before me, he would spray-paint the pictures off so I couldn't see them.

The worst times I had were on the track, and not on a working section. That was when they drew the pictures, and the men decided they wanted to get rid of me. It came from jealousy mostly. They didn't like a woman coming in taking a job a man could have. And they could not accept a black woman taking that job.

I just worked harder. I didn't want conflict. But working harder made them mad. They thought I was trying to show them up. That wasn't the case, of course. I was just determined that they were not going to break me.

*She received threats and felt she had no effective remedies, and no one in whom she could confide.*

They would take my bucket, or my jacket, and they would put them in a hangman's noose, to give me a message. My life was threatened. One man tried to run me down with a track car more than once. It was frightening. I was crossing the track with a heavy piece of water pipe, and this guy ran the track car right at me. I barely made it to the other side. I found out later he was planning to run me down and make it look like an accident.

Most of my coworkers were white, and I didn't feel comfortable talking to them. We knew where we stood as black people in the union, and you really didn't feel like you had a voice. Bosses, especially, we would never tell, because you get branded if you go to the company. One of the things you didn't do was go to the company. I didn't want to be branded on top of being a woman and a black.

I couldn't talk to my children because it was so frightening for them just to have their mom in the mines. They heard about the danger that was down there, and I was all they had. To tell them that someone was trying to hurt me would have been devastating to them. So I sheltered them from a lot of things. Some things I talked to my daughter about, but I didn't share the details. I tried to prepare them in case something happened to me.

It was a very stressful time. When I would get upset it would come out in such a way that I was hostile at home. It put a great strain on my children. I would go home at night and wake them up and make them do things that were unnecessary, because I was angry about something that happened at work.

In the end, though, I gained strength from my family. That's the only way I survived in the mines. Later I also gained strength from my church. I had to gain inner strength because I didn't have much outside strength to draw on. But thank God, I finally learned to cope with those guys. I let them know I'm stronger than they are, or more intelligent. I've got the reputation now that I can get them off my back.

One thing about coal mining for women, you can't go in there and just be yourself. You have to always put on an act. For me, it's a tough act. When I hit that bottom and get on the portabus, if somebody punches me, I punch back. You have to stay on the defensive all the time. The men will go for weeks doing things until they find something that gets to you. They're like little kids. They'll do it until their attention span turns to something else, but you have to constantly be on your guard.

And they try to feel on you. I've had some walk up and pinch me on the butt or try to grab my breasts, but I stopped that. I told them that if I ever caught one of them putting their hands on me, it would be over for them, and I meant it.

Whatever kind of behavior you start with them, you're going to have to continue. If you start off letting them do things to you, you're going to have to deal with it not only on your section but all over the mine. Word gets around.

I never went to the union about it, because then the company brands you a troublemaker. If I had gone to them with a discrimination grievance, it would have meant more discrimination.

I do feel more secure having the union. I don't think I would like nonunion work. Unless things changed, I would stick with the union. We have our gripes with them. They don't seem to care like they used to. The UMWA always was a caring union, but that seems to be changing.

I'm not the type to harp on racial stuff, but in certain instances I have been treated differently because I was black. Certain things I was given to do, the white women weren't. There were two white women hired when I was, and they would put them together shoveling and put me by myself to do the same amount of work. They would get a pallet of rock dust to unload together, and I would have to unload one myself. I would have to haul concrete blocks, whereas they would be shoveling a little around the feeder. There was a difference.

There were some [white men] who wouldn't talk to you if you were black. There was plenty of them who wouldn't speak to me at all. There were two of us black women down there. We felt like sometimes we got a raw deal, but we didn't think there was anything we can do about it.

Another thing we had to contend with was the Ku Klux Klan. We knew who some of them were. Some boasted about it, but it doesn't have the power it used to. We say, "To hell with the Ku Klux." This one Ku Klux on our section is friendly toward me. We get along fine on the section. He don't seem to show any prejudice.

I drive alone to work. I have to. The only other people who work down there from this area are white men. There are no women from here working, and I never tried to ride with the white men. Their wives wouldn't have approved of us riding with their husbands. We've heard the way they talk about us. Women coal miners have a bad reputation outside. The wives resent the fact we spend more time with their husbands than they do, and that we can sit down and have a better conversation with them than their own wives can.

I started out riding with this one guy, but I was afraid of the gossip. I was working hoot owl shift. Maybe I was jumping the gun, but I told him we shouldn't ride together, because people would talk.

Overall things have gotten better for blacks around here. I don't know whether civil rights did it, or whether it was the passage of time. Some whites won't ever accept blacks. At the mine I have to keep my cool. They think if you raise racial issues, you're a troublemaker. You always have to hold it in. Me and the white guys kid sometimes. One calls me "Tar Baby" down there, and I call him "Redneck Honky."

In some cases I get treated better by the white males than by the blacks. The black men get resentful if I talk to white guys. They think I'm not supposed to. But I have to work with white guys every day. Why should I not talk to them? The black guys say ugly things to me, that I'm messing around with the white guys. The same thing happens to the white women. The white guys give them a hard time if they talk to the blacks. This one guy in the Ku Klux got very upset with one of the white women.

No matter if you're black or white, if you're a woman, you're the center of attention. The men watch you. They gossip about your personal life. There are a lot of people doing things that might not please other people, but they don't get punished for it. But the men watch the women like a hawk, to see if a woman is friendly to certain men more than once.

I wouldn't go out with any of the guys at the mine. Right now I don't have time for a social life. I go to church. But I don't want a coal miner boyfriend, unless he worked at another mine.

Another kick in the butt to me is what people outside say about women miners. Any guy wants his girlfriend to be the best in the world, but they hear stories about women going down there and stripping, and all they want you for is one thing. I don't tell anybody that I work in the mines.

I do tell the ones I want to know better, but I still don't tell everything that goes on in the mines. The guy I date would think I contributed to the harassment. One male friend told me not to wear T-shirts with any writing on them, because that drew attention. What he didn't understand is they're going to harass a woman no matter what she does.

Before going underground, the men that I came into contact with were nothing but gentlemen. I trusted them. But working with these guys gave me a whole new outlook. I tried not to, but I couldn't help but look at men on the outside the way I look at them underground, and I see them as putting women down. I have to look at men twice now before I can accept anything positive about them.

Right after I first got hired, I was placed on a curtain job, on a section with a rough reputation. The guy before me had been disqualified by the boss after one week. A lot of guys in the mine resented that I had been placed on a section while the two white women were shoveling mud. The white men got hostile. But the company went by seniority, and that's why I got placed on that job.

The boss had disqualified the other guy for pure meanness, and he had gotten away with it. I had a rough time with that boss. He made passes. He wanted to feel on me. He tried to kiss me. I told him to keep his hands off, and he tried to make it hard for me.

Bosses have their ways. He would send me out into the old works to get supplies. The boss tried to disqualify me after two weeks, but it didn't stick. Some other bosses came down there to observe me, all unknown to me, and they decided that I should be given the 120 days allowed in the contract to learn the job.

We've had at least five bosses, maybe more. On our section bosses don't stay long. We are a strong-willed group. We won't be pushed around, and we stick together. Some say we're the worst section in the mine. Some say we're the best.

I think we're the strongest. I imagine it was the hardest to get accepted into. There are two black guys on the section, but I am the only woman. It took four or five months before they took me into the crew. They saw I was there to do my job.

I really like the guys. We all fit in. On the section we talk about our families and our personal lives, but not about sex. I feel proud because some of them come to talk to me about family problems when they wouldn't go to anybody else in the mine. They know I don't gossip.

But still, even though we get along, we have fights and curse at each other. We stop speaking for a while, then we go back to speaking. Me and this one guy stopped speaking for about three months. We had a big fight. He called me a "black bitch," and I overheard him. I threatened him.

Another thing is that I have never eaten with the crew. I couldn't stand to eat with them, they are so vulgar. You'd be sitting there eating, and they'll bring up something foul. I used to have the weakest stomach in the world, and when that happened, I would go off so they wouldn't see me change color.

That's why I love ventilation work. I am by myself a lot and have some quiet time. I can be off all night and do my work and sometimes not see them the whole night. I prefer it that way. I've always been the quiet type. I was the oldest one in the family, and I had to do everything, and that annoyed me and put me in a shell. There were six children at home at that time, and I got more and more withdrawn, until I just about left the human race completely. That was in my teens. Even now I don't visit people. I don't really have any friends. I just can't get close to many people.

I don't think seriously about marriage, because I have two daughters and I had a stepfather who made some passes at me. I put a stop to that. I threatened him. It is bad to brand all stepfathers as evil, but it doesn't hurt to be careful. I would never make my daughters go through that.

Maybe I'll get married when my kids are grown. I always planned to marry at 55 so I could be retired and travel and live it up. Even if I don't get married, I want to do the things I never had a chance to do. But as it is now, I plan to stay in the mines for 20 years.

I worked second shift most of the time. I also worked on owl shift, which I loved because it gave me time with the kids. But I was rotated off and put on permanent evening shift. It was the least desirable shift for me as a single parent. All my children were school-age. I had them in child care, and I had a person who would come to the house until my oldest daughter got old enough to take care of the family. I also lived close to my sisters and brothers, which helped. In a mine you can't easily communicate with your children. But the paycheck gave them a better life. I bought a house, and they could go to college. It gave them a good foundation, but they lost something, too. We lost a lot of time together.

People around here have a bad attitude about coal mining. I've had people tell me I wasn't supposed to wear pants. And the church thinks that if you've got a good job, you're supposed to pour out the money. I do what I can, but I have a family to provide for. If I didn't have a family, I wouldn't think about going into the mines in the first place.

I want a better life for my kids. I wouldn't want them in the coal mines, especially my daughters. I feel that my boys could take care of themselves, but the girls . . . of course, they'd have a good role model!

I am so glad women are learning to be independent. The more independent the better. The women's movement has helped a great deal. If I had to get out there and march to get a job for another woman, I would do it. I'm more of a women's libber than I was. I strongly believe in giving women a chance. We deserve it. We've been trampled on all these years.

After all I've been through, I think I can do anything I set my mind to do. I have gained confidence. I've learned a lot about men—and women. The coal mines have taught me to be tough and hard.

But I've lost a lot healthwise. I feel the congestion in my lungs. I did quit smoking about six months ago. I also have a back problem. And I have scars where rock has fallen out of the roof and hit me.

I worry about getting killed down there, but I try to push it from my mind, because it would drive me crazy working alone so much. I have had some close calls. One time I had just stepped two feet back and a rock as big as half this room fell in. I didn't see it coming. Then another time the whole place we just moved out of fell in. It was very close.

I just want to live to see my children grown. I think about that a lot when I'm down there working. I wonder if fate has it that I'll get killed down there, and I worry about leaving the kids. I feel guilty sometimes that I've denied them so much of my time, but even if I did get killed in the mine, I think my kids would have been better off for me going on.

We talk about it. I let them know it's dangerous and tell them I could go in tonight and they might never see me again. I tell them to learn to stand on their own two feet. I try to teach them values. And I appreciate each day I go in there and come out alive.

*In 1994 I saw her again in her Bessemer home. She had recently been laid off after working underground for 14 years. I asked if things had improved with her coworkers as the years passed.*

Things are worse in one way. The company can do more what they please, because there are fewer of us. They don't have to answer to anybody concerning minorities and women. But I have grown, and I know there are other outlets if you have a problem. As black people, we still don't have a channel through the union. If issues come up, we don't bother to carry them to the union.

The only encounter I had with the local union taught me where they stood.

I had bid on a coal drill job. I was the only person at the mine who bid on it, and the company did not want to give it to me. I filed a grievance, but when the time came for the meeting with management, I had no representation. My elected representatives who were supposed to present my case didn't say a word. Another union person came and spoke up.

The company fought me tooth and nail on that job, and nobody even wanted it! Some higher-ups in the company made them give it to me, but nobody even told me when the job was awarded to me. It was months later when I was finally told.

The company discriminated on job bidding, and the union wouldn't fight it. Most of the time, when blacks bid on a job, the company would say the job was under investigation and wouldn't give it. Some of them were awarded the jobs, but a lot weren't.

As unions stand, there seems to be so little hope. But if they could start all over again and restructure themselves and commit themselves to representing everybody, it would work. I used to go to union meetings quite often, but then when I got more involved in my church, I had to choose between the union and my church, and I chose my church activities.

I think women have really been shortchanged by the unions. If you carry a problem to them, they act like you don't know what you're talking about. Women are not taken seriously by the unions. That needs to be changed.

I've been hanging curtain almost since I've been in the mines, but I did not fail to learn other jobs. At one time I was a coal drill operator. Then they eliminated the coal drill. So I bid back to curtain on a different section. I've been on a section since then. I have operated the roof bolter and the miner. I can operate just about everything down there.

The last section I was in was 32 inches high. I have worked in 28-inch coal. It's terrible. You crawl to do everything. But I want to move faster, so I would always duck-walk, or bend. You have to be in pretty good shape and keep your weight down. It's doubly hard if you're in wet, muddy places. It's a constant struggle.

Coal dust was a problem because I was allergic to dust. I learned to love a respirator, but in low coal you're sweating a lot, and a respirator will rub your face raw. It's hard to keep them on.

I was only on the track a few months, thank God. Then I was placed on a section with a radical group of men. The crew was made up of the boss and ten men. They were mostly white. They weren't so hard on people who came, but they were a mean section that was hard on bosses. I think we went through ten bosses or more during the time I was there.

They weeded out bosses and people they didn't like. Finally it came down to the core group, and these people finally accepted me. You had to really analyze them and learn how to fit into their minds. I'm sorry to say it, but I adapted to the language and gave it back to them if they told me to go to hell. That really wasn't a part of me. I was going to church, but I adapted in order to survive.

You had to let them know first of all that you were there to do a job. I let them know we could tease, we could talk, but they could not touch me. I threatened them. They laid in wait for me, and they would tear up things. But after I really made it clear I would not take this from bosses or anybody else, they finally accepted it. It was a hard fight all the way, to show them I would not be broken down.

I became a part of that group. I learned to fit in. And finally I was accepted as myself, not by doing things they did or thinking the way they thought. I eventually earned their respect. It took at least three years. I determined that they were not going to break me. So they would have to accept me, because I was not going to change to their status quo, because I was going to be the person I was. I was there to do a job, and I did my job, and I wasn't going to change my personality or beliefs or anything to fit in the group. I decided to do that after looking over things for so many years in the mines. I decided I was not going to be changed.

We worked in places where the roof was tremendously bad, and we had rockfalls every night. It was stressful, and you became close and looked out for one another. I have warned the guys about the top and saved lives. Normally they wouldn't listen to a woman. I have told them an area was dangerous and several crosscuts fell in. After that they started to listen.

A curtain person is usually about the last person working in a place. I could hear the movement in the roof. I could see it chipping and falling around me. So I had a chance to warn them. We've had people hit by falling rock several times, but nothing serious. One man had his finger cut off. He almost went into shock. They sewed it back on, but he never got the feeling back. I've been lucky. I've never been seriously hurt. But I had two back injuries that have brought on arthritis in my back, which is very painful.

I worked on that section nine years. We were close-knit. They got to the point where they would defend me from anybody. They were my protectors. I learned to love them, and they learned to love me. If I would seem displeased about anything, they would come to my rescue.

I'd worked with white women before, but I'd never worked with white men. The family guys, especially, who had wives and kids, I could identify with. We would talk about our families. They would confide in me. I think it was a new experience for them. They really didn't know how to accept a black woman, because they'd never worked with one before. And I've met some of the wives. The guys with whom I've gotten to be close have introduced their wives, and I became close to some of them.

*Brown's layoff coincided with her last child's leaving home.*

Losing my job and my children at the same time was hard. With layoffs, I believe in seniority, but something needs to be changed to protect women and minorities. It's a no-win situation. You don't have any job security; it's a "last hired, first fired" system. You feel more vulnerable. I've never been without the children in the house. We shared so much. Right now I'm having to go through this alone, except for talking to them long-distance, when you can't really talk. I wanted them to be independent, of course. I wanted to give them a good foundation, and with this job I gave them a good education. Mainly I wanted them to be more stable in life than I was. To have stability in the household enabled them to go on with their lives and have their own families.

My oldest daughter graduated from college and is a computer operator. My second daughter went to college and is working as a medical clerk. My next-oldest son works for UPS. My youngest son is in the navy and has three children.

With the health problems and the humiliation and all the things I had to deal with, I wouldn't want to go through it again. I would like to work again in day care, where you can be loving and giving. In the mines you have to be hard and tough just to survive. I did learn to love my coworkers on my crew, and they loved me, but it took years. Some people never get to that point.

I've grown as a person from the struggles I've had in the mines. I've grown through my spiritual life. Both have made me less timid and fearful. I think women miners have earned respect in the industry. We have contributed in a positive way to the industry. The coal mines are closing down, but if there is any chance for younger women to follow us in the future, I would like to help them. But the coal industry and the union need to be more sensitive to the needs of women. We should be given a chance and not be humiliated and degraded for stepping forward to take a job. As women, we deserve a better chance than we've been given.

# 16

# AFTER THE STORM
## Brenda Brock

*"When women went underground the men's hero identity was kind of destroyed. . . . People thought if women could go underground, mining couldn't be as hard and as ugly as they thought. Instead of saying, "This is great, these women are heroes, too," we were minimized, our jobs were trivialized. We weren't looked up to. We were looked down on as sluts, as pigs. Why couldn't we just be good women who worked hard?"*

*Brenda Brock crouched to lift a 40-pound sack of limestone rock dust to her shoulder, then carried it to the loaded mine conveyor belt, which rumbled by in the darkness. Seven months pregnant, she was small-boned and strong. Her belly protruded in a neat paunch over her leather mining belt, fitted with a self-rescuer and a square battery pack with a cord hooked to the light on her hardhat like a black umbilical cord.*

*Brenda's pregnancy had been a contentious issue at the workplace. As a single woman at a large non-UMWA mine in Harlan County, Kentucky, she had fended off harassment from coworkers when the news first leaked out, but the men's behavior changed as her pregnancy progressed. Some of her coworkers grew openly protective as the delivery date neared.*

*Working alone, she shoveled spilled coal onto the moving conveyor belt while I took photographs, the camera's light jabbing the darkness. I commented on the irony of being pregnant inside a mountain whose portal led to the daylight world—a womb within a womb. She laughingly agreed. As I rode out later down the main entry, the portal was a pinprick of light that grew larger until we were thrust into blinding sunlight.*

*Later that night one of Brenda's female coworkers joined us at Brenda's frame house, and we turned on the tape recorder. A torrent of stories swirled, rushed, and eddied while we drank countless cups of coffee and cigarette smoke curled over our heads. Five hours later we ran out of cassette tapes and blinked, as if*

Brenda Brock, seven months pregnant, Eastover Mine,
Harlan County, Kentucky

*coming up for air. The transcripts of the first interviews reflect that intensity and the staccato rhythms as Brenda and her friend interrupted each other.*

*That was in May 1980. In August Brenda invited me to join her family to await the birth of her first child. In the languid August days she rocked on her porch and took cool baths. She named her daughter Israel. When Brenda returned to work, she began packing another piece of equipment to take to her solitary outpost underground—a breast pump.*

*We kept in touch over the years as she moved first to West Virginia and then to Arizona. She continued in nontraditional work until the early 1990s, when she left her aircraft refueler's job to study child development. After graduating with a bachelor's degree in child development and women's studies, Brenda started a construction business with her sisters and other women friends in Tempe, Arizona.*

It takes a certain breed of women to go underground. I think we're women who aren't satisfied with the role we've been put in all our lives. We're bored with it. When women first went underground, we were a big threat to the men because we were messing up their playhouse. They knew the system. They controlled it. They were afraid we'd go in there and tear their playhouse down. And we did, sometimes without ever meaning to.

When you're underground, you're underground. When you're on the surface, you're on the surface. They are two completely different worlds. What happens underground, you don't take outside. We didn't know that going in, or how hard it was to keep them separate.

It's hard being a woman miner around the other miners' wives. The wives have never had the threat of a woman working with their husbands. It makes them feel inferior, especially when you're talking about work. When the wife is sitting there, she's totally excluded. I don't like being in that situation, because some of the wives get jealous. They think about how dark it is in that mine when you're in there with their husbands.

One time I went to this place where there was a live band, and this woman was giving me the evil eye. I worked with her husband every day. She was well known for going out on him; she even had a steady boyfriend. So she asked me in front of her husband and everybody if I'd ever gone out with him. I said no. And I said, "I'm sick of people never giving women miners the benefit of the doubt."

So then she asks me, "Has he ever asked you out?" And him standing right there! I said, "Oh yes. He's asked me out plenty of times." His eyes got big as 50-cent pieces, and his mouth dropped open. It was obvious what he'd been telling her. After that she hated me twice as much. Maybe it was because I told her that there was nothing about her husband I wanted, or would go through the problems of taking him away from her to get.

*In obtaining a mining job, Brock renewed a family tradition.*

My dad's father worked all his life as a miner in eastern Kentucky. Dad worked in the mines as a young man and then moved to northern California to work as a logger. Later he moved us to Michigan, where he worked for General Motors at its truck and bus plant until he retired.

At the time he and Mom married, being a coal miner wasn't the kind of job anybody wanted. People were ashamed to say their dad was a coal miner. They saw their fathers and brothers dying in the mines, and they wanted to get as far away as they could. And in the fifties and sixties, nobody was working. Oil and gas took the coal market, and everybody left for the big cities in Michigan and Ohio.

I grew up baby-sitting in Michigan for 50 cents an hour, and then I worked in a restaurant for a dollar an hour plus tips. When they wouldn't let me take mechanics in high school, I quit and got married. My dad and mom were divorced, and life at home was hard. When I was 19 I realized I didn't want to be married and that it was unfair to me and him both. So I stuck my thumb out and hitchhiked all around the United States. I kept on moving and ended up in eastern Kentucky.

I came down here to tell my grandparents good-bye, because I knew I'd never see them again. I went to Leslie County and met an old boyfriend and fell madly in love and stayed that way for about four years. But I knew it wasn't going to work and that I needed a job and some money if I wanted to pick up and leave. Then the thought came to me: Go to the strip jobs. They don't have any women. Everybody laughed at me, but I had a feeling. Then my cousin asked me, "Are you serious?" And I said, "Yeah. I do value life, and I'm starving, and I've borrowed money to live on." So he says, "Eastover's hiring 15 women."

My sister and I went up there, and we hid down in the floorboards of the car so we wouldn't be turned back at the gate. About a month after the interview, they called my little sister and hired her at one mine. She worked for six months. The men beat her with capboards [wooden wedges]. She'd come home bruised up. They treated each other that way, and she was no exception. One day some of them—including the boss on the section—knocked her down and plastered her with hickies. She came in from work that day a wreck. She said, "Look what the men done to me."

She went to the woman who was head of personnel, and the woman acted like it was her fault. My sister blew up at her, and the woman said, "You will never work for this company again or for anybody else around here either." They blackballed her, and she couldn't buy a job around here after that. She didn't have much of a union, just a little company union, that wouldn't take up for her. So she gave up. She quit. But if it hadn't been for Tammy, I wouldn't know what I know today about my rights. She went through so much, and she was only 18. She was the first woman at that mine.

When I got hired I went in to look at the equipment. I was so small that nothing fit me. I couldn't work that day. I had to go out and buy a pair of little-boy boots and put some steel toes on them. They were a boy's size three and a half. I walked into the mine office, and everybody just stood there and stared at me. Nobody said a word. Finally the company guy told me to go with a man who had just gotten out of prison for killing his wife. But I was in such a daze that he didn't scare me.

This boss took me up one return entry and down another all night. We were pulling a roof-bolting machine. They dragged me everywhere and nearly walked me smack dab to death. But I stuck right at his heels. He must have speeded up his pace because I've never seen him work that hard since!

I was pretty well educated about men, I thought, until I went underground. Then I realized I didn't know as much as I thought. I did fine with them in my own life, or in public. But working with them underground was a whole different show. One foreman always let me know he'd like to take me out, but I told him I'd never go out with him because he was a boss. After that we got along all right.

But there was another foreman who made me stay right on his heels all the time. He wouldn't let me talk to anybody on the section and would put me working away from the crew. He never came out and asked me out. He just let me know he was available, that he was divorced. Then one night he came in, and I thought, he's acting just like a man who's had a bad fight with his wife!

Sometimes there would be ten men sitting around discussing something I would know something about, and they would totally overlook me. Like they were thinking, why is she here? Why is she trying to talk to us?

*As she gained experience, she became more outspoken on the job.*

I'm not company, and everybody knows it. But for a year and a half I didn't speak up for nothing. Then one morning I got out of bed and thought, they've got me brainwashed. So I decided, I'm going to have a voice in this. I'm going to speak up.

The men told me I was just making it hard on myself. But it couldn't have gotten no harder. When you've worked a year and a half in the old sections by yourself, it mentally gets to you. Then I started hearing that the company was out to get me. The men said, "You'd better watch that guy there, he's known for hurting people." That worries you, because some of them will try to hurt you intentionally if the company wants to get rid of you.

The rumor got out one day that another woman and I were going to be put on a track crew with a guy known to hurt people, and that he was going to get one of us hurt. We stayed up all night talking about it. It's easy to get hurt lifting those rails. And this guy was known for being fast, doing it so you could never prove it. Well, it didn't happen, maybe because we stuck together, but we were scared to death. Then the company tried to split us up, get us mad at each other, which wasn't hard. They put her on first shift before me, although I'd been there longer.

On third shift I worked in water up under my armpits, and there was nothing you could do but try to pump it out. I stayed wet for six months and danced on hot cables. I thought all the sections were like that. This guy came up and said, "I'd bet you'd die if you saw a dry section." And I said, "You mean there are dry sections?" And this was after ten months underground.

After that I begged them to let me go on repair work, and finally they put me with this repairman who told the company he couldn't work with a hard-on. So I was put with a sorry, sloppy repairman, where I couldn't learn anything.

Since we had a small company union, the men didn't want us women to get involved. They wanted to get rid of me and Pat because we were radicals to the point that we would speak up. One man said to me, "I don't want a damn woman in there, knowing everything I know." And then a company man said, "If you do get a woman in there, get one who can be controlled."

They call you radical because they want to brainwash you to not be democratic in a company like the one I worked for. I really believe that one reason the company didn't want to hire women is that when the big strike happened at their other mine in Harlan County in the early 1970s [to organize the mine under the United Mine Workers of America], they saw what the women got done.

When the company interviewed me they asked me what I would have done if I was on the other side of that picket line full of women. Those women weren't afraid of nothing. They got up and said, "Our husbands can't do it all, because they've got injunctions against them." But those women didn't work for that company. They weren't controlled by anybody.

I've thought a lot about a strike. Our little union never struck. Even when the company cut back on medical benefits, the union just sat back and took it. There's four of us women who have been attending union meetings. At one meeting I made a motion that, due to the special problems of women working underground, we should have a woman representative. And the men told me to hush. I said, "Don't you think we need one?" And they said no. We didn't even vote on it. They just slid over it to something else, and we couldn't do much with just the four of us there.

The company then started a get-together for women who worked at the mine, and after a couple of meetings I realized that two of us were the only two underground women there. Everybody else was company. I saw we didn't have any business being there. They just wanted to know everything that was going on underground.

At the last union meeting there was the company president, the superintendent, and the general foreman present. That shows you just how company-owned the union is. The president of the local couldn't answer any of our questions. My friend jumped up and said, "If you can't do the job, why don't we get a real union?" And everybody there cheered her on. The men said, "Yeah, let's drop them!" But the minute she said that, a union representative came up and changed the subject and broke the meeting up. So we lost the thread of our discussion.

*When Brenda became pregnant in 1980 she wanted to keep working underground as long as possible.*

When I got pregnant I decided to use the pregnancy as a way to get the women together. I wanted the women to know that the men had changed union policy about donations for sickness. When a person is off five weeks they get an automatic $1,100 donation from the union. But the men told me, "We're not giving no five dollars for a baby we didn't have the pleasure to make."

When I told my boss I was pregnant, he jumped up and said, "You can't work! You'll be up here with morning sickness!" And I said, "I've puked for two months and you haven't even noticed." At that time I had just been put on a belthead that would spill 15 to 20 tons of coal a day, and I was cleaning it up with a shovel. Before I knew I was pregnant, I'd feel my side hurt whenever I lifted that shovel over my head. And every time I'd sit down I'd fall asleep.

So I went in about a week later and told everybody. The men were shocked. They said I should quit. And they talked bad about me. So I said, "If any one of you wants to talk about me, fine, but I can make your life perfectly miserable because I'll go straight to your wife and tell her it's yours. And for nine months you'll be miserable because they won't be able to see the kid, and for the next two years your wife will be staring at it, then for the rest of their lives they'll be wondering." When I told the men that, it shut them up.

One of the men who had been trying to get me to go out with him kept giving me a hard time about my pregnancy. He said, "Brenda, if you'd stay out of the backseat of people's cars, you wouldn't be like this." Well, he had given me a watch for Christmas. So I told him the next time I saw him and his wife at the steak house, I'd bring out that big gold watch and say, "Honey, you'd better take this back because it's not working." And then tell him how sorry I am to hear about his divorce, and ask, "Is this your girlfriend?" Well, he got wind of it and apologized to me five times in front of all the men on the section. I began to realize that you can use their wives to protect yourself from harassment.

My pregnancy has protected me all the way around. The men don't verbally harass me like they used to, because they've all got wives who have had children. They know the kind of moods you can get into. But sometimes I get worried about the dust and the effect it's having on the baby. I just started wearing a respirator, and when I looked at how black the filter was at the end of the shift, I was shocked. I could feel the difference. But I worry about the dust I breathed before I wore it. And then I smoke, and that worries me, so I smoke more. I do feel handicapped down there being pregnant. I can't do anything like I used to. Used to be I could shovel for eight straight hours.

Sometimes I don't know whether people think I'm gay or what. They can't understand why I don't have a man. They're always saying, "Brenda, why don't you just get you a man and settle down?" Like you go out shopping for them.

When you're young you're just looking for someone to be in love with. After I realized I didn't need a man for money and realized what I did need one for, I was lost. Partly for sex, partly for companionship. And I started to see what it meant to be alone. I mean, I had everything I'd ever wanted. It had taken me 25 years. And I was sitting there in tears because I was alone.

At first all these characters knocked on my door. They all knew I lived alone. The men at work said, "What in the world do you do about your sex drive?" But what I miss are the little things. I'd give anything in the world to get up to a breakfast just once in a while, or have my lunch packed for me just once. It doesn't sound like much, but it is.

I'd say some of the wives would like to have this job themselves. Most of the men have got their wives brainwashed against us, to think we're whores and home-breakers, and that the underground is no place for a woman. They want to keep their wives at home. We're a threat to the men because we're evidence to their wives that they could, if they ever got the notion, get a job and leave them behind. But I do think that every wife should go underground just to see what it is like, with the thought that her husband is doing it every day. She would definitely learn how to treat him.

At first coal mining was just a job. But now I wouldn't ever want to do nothing else. It started out as money, but as time went on I began to realize just what I did have. I've never felt this confident, this independent.

When I first went in I asked a blue million questions. And I begged to get on equipment. And then this man said to me, "Brenda, you don't know what it's like. I've been on this bolt machine for five years. You've never had this feeling of having to do this one thing eight hours a day for the rest of your life." And he was right. Now I'm starting to get that feeling.

*After a layoff in 1984, she lived in West Virginia and Arizona. Eight years later she returned to eastern Kentucky to visit friends and relatives in eastern Kentucky. In a motel room in Harlan she pored over the transcript from a decade earlier, and we turned on the tape recorder. I asked her how she felt after reading her words from a decade earlier.*

Angry. Very angry. I feel a river of emotions. It's tough to go back and think about it, because I didn't know what to do then about harassing situations, and I still don't know what to do. I'd like to believe that we've come far enough that companies want to deal with this so it's fair and just to everyone. But it's not. It's just like it was when we went underground, except it's not just there, like I used to think it was. It's on the best jobs, with the best management, and it happens to some of the best women. I think it happens to men, too, and they're even more scared, because they feel more threatened about their jobs.

Almost all the jobs I've had have been "nontraditional." I've experienced harassment on every job I've been on, except for one right after the mines when I was a construction crew leader in Michigan. We built a 101-unit senior citizens' apartment complex. They hired me as a general laborer and a few weeks later put me on a forklift. The boss realized that I could figure out how to run almost any piece of equipment. That was the only job I ever had in my life where the boss treated me with respect, gave me tons of responsibilities, bragged on me, and all the men around me commented on what a good worker I was.

After a while he came to me and asked if I knew some people I could hire as general laborers. I would be a crew leader, and I would hire and fire them. I hired all women. I went into Flint, and the economy was really bad. All the women I knew were looking for jobs. So I hired them, and we got rave reviews for being the best workers and the best crew there. We did all the loading and unloading and carried drywall and helped the mudders. We could do anything, and I was really proud of that crew, because we busted butt. It was the only job I've ever been on where the bosses gave credit where credit was due. It was wonderful. It only lasted a year.

Then I moved to West Virginia and lived in a log cabin, and then moved on to Arizona. I worked in a K-Mart until the weather cooled off and then went out and worked as an oiler on a backhoe. I was the operator's helper. I was the one in the ditch with the shovel and the compactor. Eventually I went back to school and then got a job with the city of Phoenix as an aircraft refueler. There were only two other women on that job, and it was a huge job site. I was the first woman fueler they had hired. We also did maintenance. If a beacon light went out, we climbed the beacon and put the light back in. I used the mowers and backhoe for dirt work. We waxed floors and did all the cleaning.

On that job I was physically assaulted by a high-level manager. He walked up behind me, put one arm around the waist, and grabbed me with the other hand between the legs. They gave him two weeks off. Later I found out there were three other charges against him on file.

When he did that, I told him I had worked with coal miners and nothing like this had ever happened. But after a few days I remembered that when I first went underground, it did happen on midnight shift. Somebody grabbed me in the same way, and I had completely blocked that from my memory until the other incident happened.

Then I got even angrier, because it brought back so much of the pain and the things the guys did to us. Every time at the mine that something happened and we would complain, the men would accuse us of wanting to file suit and not wanting to work. Of course, if anything happened to them, they'd run for a lawyer. It was okay for them, but it was never okay for us to be mad or hurt, because it wasn't okay that we were even there. I don't think that's changed in Harlan County. If it had, there would be women still working there.

We were afraid to fight. We were afraid to file grievances. When they accused us of wanting to file suit, we'd say we didn't want a lawsuit, we don't want anything that we should have had. The question was always, what are you doing here? That shouldn't have even been a question. We were there for the same reasons the men were. We wanted the money and the benefits.

We were working hard underground to prove that, yes, we could do the job and that, yes, we were worthy of the money we were paid. We were so into trying to make the men see the good side of us that we became like the men in order to relate. We had to become like them because they weren't going to relate to us in any other way.

One day I was talking to my mom on the phone, and I said, "Yeah, us men at work . . . ," and it really hit me. I was upset about that. It scared me. I realized I had lost my identity, and I had to work consciously to get it back. That happened gradually, too. One man said, "Brenda, you're just like one of the guys." I went home and thought, no, I'm not going to let this happen again. So I sat the men down and said, "No, I'm not one of the guys. I don't want to be introduced as one of the guys."

I think that a lot of the men's resentment was because we weren't frightened. We were happy. We thought, this is great. The benefits, the security, the money. We thought we could work and live and be somebody. We were happy about it. I don't know about the other women. But the men looked very unhappy. I'm sure they were threatened, and that they didn't feel the same way about their jobs as we did. I saw them driving new trucks and living in decent homes. I saw them with their boats and material things. But most of them seemed depressed. They didn't seem happy with themselves. And I wondered how much of that had to do with having to work underground.

For years the wives and girlfriends did not understand what it was like underground, and they treated their men like legends. The women didn't know that some of the men mostly sat around all day. If their husband was a foreman, he probably didn't do anything except gas the headings. They looked at the men like heroes.

When women went underground the men's hero identity was kind of destroyed. They looked at us and said, "You're not a hero, women can't be heroes." They convinced the women of that. People thought if women could go underground, mining couldn't be as hard and as ugly as they thought. Instead of saying, "This is great, these women are heroes, too," we were minimized, our jobs were trivialized. We weren't looked up to. We were looked down on as sluts, as pigs. We were called lesbians. Why couldn't we just be good women who worked hard? That's how the men were viewed. Why couldn't people have felt that way about us?

In the end even we began to feel that way about ourselves, once we realized everyone hated us because we were underground. A lot of us lived hard. We would go out in public, and people would treat us like dirt. Now, years later, some people in Harlan County say, "Oh, you were one of those women. Well, there's no more women underground." And they laugh.

What they mean by that response is that we didn't mean anything when we were hired, and we don't mean anything now. They are saying hiring women miners was just a phase, and it only happened because of some fluke in civil rights when the companies had to hire women. People have said to me, "The Bible says women should stay in the home." They have distorted the Bible to keep women down.

There is a desperate attempt to put women back where we were. Some people feel they have to force women back because we're losing our family unit. I don't see it that way. I see the family unit changing. It has to change, but men have a lot of fear about it. They are frightened that we're losing the cohesiveness of what America is supposed to stand for.

We have to grow. When I grew up, *Leave It to Beaver* was showing June Cleaver as the perfect, at-home wife. In fact, she was an actress working full-time making that show. She wasn't at home feeding her babies. And yet we sat there and totally forgot that she was a workingwoman.

We don't have strong role models for blended families. But the positive potential now is that we can teach our girls and boys that we can work and live in harmony and be happy. But the distribution of power has to change. Always before, the power was with the man. He decided how much mama got and how much baby got.

But now women are realizing that they're born with a certain power and that we have given it away too easily instead of keeping it and using it. When we give our power away, we really mess up the harmony.

*By the time of our final interview in 1992 she had just left the last of the nontraditional jobs she held for 15 years and returned to school to study child development and women's studies.*

Nontraditional work has given me a lot of confidence. I was always a novelty on every job site I went on. I used to think, I'm only here because they had to hire a woman. Then I thought about it, and said, I'm doing the same thing the men are by saying I'm not here because I'm good or because of my experience.

When I went underground, I felt that way because it was a fact. Women were there only because the companies had to hire us. But what I learned in six and a half years of operating heavy equipment, and all that we learned underground, has carried me all these years. I'm very thankful for it. I'm glad I'm not still in the mines. I'm glad I was delivered out of there by a bad economy. If the slump hadn't hit, I'd probably still be there, or be hurt, or disabled, like a lot of the men are I used to work with. They are crippled and wait for a settlement.

I didn't want to continue doing nontraditional work, staying in that rut, because that's how I would die. I had been in nontraditional work for 15 years, since I was 22, and I found myself beating my head against the same old walls. I thought it was time to leave and move on with my life.

When I went to the airport in Phoenix, I knew I was hired for my experience. And my attitude changed. One day I woke up and realized I was worthy, that I was skilled. I'm not worried now about getting a job, because I've got some marketable skills. But I want to move to a different level. I want to work with nontraditional work, but I don't want to be within it. I want to work for change, for women and children. I've paid my dues in the men's world of manual labor. It's time to leave.

My part in this process is to get educated enough to help change the world. I'm not a feminist in wanting to get rid of the male race. I want it to equal out. I want white men to understand that they're not the only ones who are important. We all have an equal part in this system. And if we're not allowed to contribute, there will be no harmony. In our lives I believe we are supposed to put out energy into the system, and until we're free to do that, nothing will work out right.

I remember thinking I would never leave these mountains [in eastern Kentucky] because I felt a spiritual connection here. The energy can be very positive or very, very negative. I think there is a positive and negative side to everything, and your mental state determines what you hook into.

I believe we all have things to accomplish in our lives. Whether we do our work depends on whether our self-esteem is in place to do it. I don't think I'm special. But I'm willing to go through the storms and the trials and tribulations. It is scary, but I know I'm going to make it and the storms will pass and I will learn a lot from them. Everything clears when the storms are over.

It took me a long time to realize that we make choices that shape everything that happens to us, even the negative things. Nothing is anybody else's fault. We're here to learn what our purpose is, and I feel like I'm on my way. In part I can thank coal mining for that.

# 17

# THE FRUIT OF OUR LABOR
## Sandra Bailey Barber

*I always say, coal mining didn't cause my divorce—it made it possible.*

*Sandra Bailey Barber entered a large Kentucky mine in 1975, not long after mine portals were forced open by affirmative action mandates. In 1980 she became the first woman coal miner hired as a federal mine safety inspector by the Mine Safety and Health Administration (MSHA).*

*Our first interview took place in 1981 at her home near Mayking, Kentucky. We stretched out to warm ourselves in front of the Franklin stove and turned on the tape recorder. As the years passed, Sandy and I corresponded and had long talks at women miners' conferences. I admired her compassion, intelligence, and determination.*

*In the late 1980s she encountered a severe problem with sexual harassment from a MSHA supervisor. She was finally forced to file suit and subsequently endured a draining legal battle. During those years she struggled in an increasingly bitter and publicized battle that focused attention on sexual harassment. Eventually the agency discharged the official and her tort action was upheld.*

*Our second interview took place in early 1994, when she reflected on her earlier experiences and the effect of the legal struggle. Her children were grown and gone, she had remarried, and she had won her lawsuit, a victory that came at a heavy cost.*

My parents were blue-collar workers, and we lived in Cincinnati, Ohio. My mother came from England as an immigrant to this country with two young children. Her father was a union plasterer, a master craftsman. Her mother was very traditional. My grandparents on my father's side were blue-collar workers. My father, strangely enough, was in law enforcement, which I find myself in now. He was a guard on death row at the Ohio state pen when he died.

There was social activism and unionism in my grandfather on my mother's

Sandra Bailey Barber.
Photo by Bob Underwood, MSHA, courtesy of Sandra Barber

side. It seemed to skip a generation, and skipped most of the people in my family, and it came out in me. Before I knew about these things, I just thought I was different from the rest of the family.

The first things I can remember as a real young person were class distinctions. We were inner-city poor, but we lived in a neighborhood just a little bit better than where we belonged. There were activities like "Zoo Day" and "Symphony Day" that all the kids had to go to, but I had to sit in the office as if I were being punished because I didn't have the two-dollar fee to go. I had to miss school once because I didn't have any shoes to wear.

When I was about 12 there was a black fellow in our class that people kept calling "nigger," and there was an Appalachian young man, too, who was made fun of because of his accent. I was a city girl then and never knew I would end up living in the mountains. But I remember I thought that was cruel. I think things started happening inside me about justice and injustice.

I never saw the connection between unions and fighting injustice, but I've since come to believe that's what unions are about, justice for all people. I have been a waitress, school bus driver, and I've worked in a shoe factory and in a school lunchroom. The first time I got interested in unions was when I worked in a shoe factory. Grievances were a joke. Working conditions were a joke. Having a fan would be an unreasonable thing to ask, and there were no optional Saturdays. I worked there when I got married, and I asked my foreman if I could have Friday off to get married. He said no. You knew if you ever spoke up they wouldn't hesitate to replace you. I was in a union, but that union was no help.

I first came to the mountains to see my mother. She had married someone who had a home place in the mountains, and it was being sold for back taxes on the courthouse steps. After I was married and having problems, I came to the mountains for six months to think things over. I stayed, and my husband eventually came here, too. We stayed together in a very stormy marriage and then divorced. He died of a heart attack at 31. My four children were all less than ten.

When I first moved to the mountains, I'd see men in the stores that I thought were black people at first. Then I realized they were coal miners and there really were very few black people here at all. I first heard about women getting into mining because of some litigation that was going on. I'd read about it in the paper, and I'd heard people talk in the grocery store lines.

I believe the reason I was hired was that I kept applying and because I hadn't sued them. They knew they were going to have to hire somebody, and rather than risk the life of one of the local belles, they hired me. I also had some work history in management in industry, and I kept coming back and begging for a job. I was 28 years old.

It was a large company with about five mines. They hired two women before me. They'd hire one or two at the different mines, and we were kept very isolated from each other. I knew there were a few others out there, but at first it didn't occur to me that I might want to know them.

The women they did hire from time to time usually didn't stay. Sometimes I think companies hired women they believed would fail. They either couldn't do the job physically or they really were just there to look for a husband. I don't think it helped me that of the two women before me, one took a token management job within a year and the other married her foreman. Everyone thought I was there for some ulterior motive.

I needed money to support the family. That was my motivation. I had no designs on being the first, or famous. I just wanted to earn a living, and there was nothing else in the mountains that paid a living wage. There wasn't then, almost 20 years ago, and there isn't now.

*As one of the first women hired in her area, and as an outsider, she faced strong community resistance to her occupational choice.*

Not being from here, I didn't realize the social stigmas attached to women miners. I didn't know what a disgrace it was going to be. My kids didn't realize it either. They learned afterwards when people made fun of them at school. I'd hear comments from them about people saying I was just there to get a man. People didn't talk too much to me, but my mother told me that people had told her she should get her daughter out of the mines, and she should basically be ashamed. My perception was that we were thought of as tramps.

Then there was the sermon preached against me in the church. They didn't use my name, but I was the only woman coal miner in the community. People came up and told me that the preacher said it was sinful, women stepping out of their place, wearing pants, and trying to take men's jobs. I was the issue, and they pointed at me to tell these other women not to go astray. The fact that my husband and I did finally divorce probably just cemented the idea in their heads. But I always say, coal mining didn't cause my divorce—it made it possible.

At first I was afraid of the grueling work, afraid I couldn't do it, although I'd always felt that I was emotionally and physically strong. I was the third woman to work at our mine, but in the state I'm one of the only ones who stayed. At first nobody thought we would fit in with the job. The first women really worked hard to prove they could do it. Not just do it but excel. I killed myself proving it.

Where I worked at, the management let me learn any equipment I wanted to, but I had to use my lunchtime to do it. They didn't stop me. They even encouraged me, but I understand some women weren't that lucky.

I lived so far from work that it took well over an hour to get there. Out of desperation, I bought a van and had seats built in it and built up a car pool of people. I rode to work every night with 11 fellows, and I think those wives were really outstanding women. I wonder now if one of the reasons they allowed it is that they thought, well, she's not going to be drinking or doing anything with all of them. They knew I worked regular, and that they'd go to work all the time. Our bathhouse was three miles away from the portal and the men's was 50 feet. Sometimes I would be dressed for work and not even know the men had called a strike and gone home. Being female, you weren't privy to bathhouse discussions.

At the time I wasn't thinking about women's issues, but I did talk to the men sometimes. They told me many times they'd strike to get a [women's] bathhouse. But I never let that happen.

I'd worked about a year and a half when a women's first aid team got started by women at the company's mines. We competed with other teams from the area. The year before some of the wives made their husbands quit when women came to the practices. They decided that mixed groups wouldn't work because we had to touch each other at arterial pressure points.

The company used us to an extent with the first aid team, saying, "Look what we're doing, we're recognizing women." As it turned out, it was great for us, too. Getting to know the other women was wonderful. We had been isolated from each other at different mines within a general 50-mile radius. We really worked hard. There was such team spirit.

We should have won first place. As it was, we won second. One of the judges told me we deserved first but they were afraid to give it to a women's team the first year we competed. I would have stayed with it if it hadn't been so demanding on my family time. I had no time for myself. All the time I had was for my family, the laundry, the evening meal and homework and all those domestic things.

I stayed on hoot owl shift nearly the whole time. I was lucky because I adjusted so well. It's still the shift I would prefer to work, because it gives you time with your family. At that time it was important to me to have my children's meal on the table when they got home from school. I felt it was something a good mother should do. The kids were young then, and they'd go to bed at eight-thirty or nine. I'd leave for work at ten. I'd get home about a quarter of ten in the morning, take something out of the freezer for supper, go to bed, and get up at three and start supper and have the evening with the kids.

The income from mining was wonderful for my children. We could take three-day driving vacations. They never had to sit in the office when the rest of the school went on an outing. They could think about college. We actually got passports and went to Europe one year on miners' vacation. I'm a good saver and a good manager. Once a month we could go shopping and buy clothes that I didn't have as a child. We had a reasonably late-model car that could be depended on to start. The income made remarkable changes in our lives. But some of the kids were embarrassed over what I did for a living. Once I asked one of my daughters to go get some wood and coal for the fire, and she made the comment, "I'm not a coal miner."

I figure I've worked with hundreds of different men in the mines, and of that number only two have told me to my face I shouldn't be there. Some would make comments like, "You could make more on your back." My answer to that was, "Any time you'll give me a regular paycheck and supply my hospitalization and benefits, we'll talk." I didn't mean that, but I wanted to show them you take home more than a paycheck. If a man pushed hard, my favorite thing to say was, "I want you desperately. I've been in love with you for years. But I want more than an affair. I want marriage." But honestly, being without a mate most of the time I was working, I was surprised that nobody got any better-looking to me. They were still my brothers in the union, and my friends.

After spending some time in this area, I think the wives realize they and their husbands are limited socially. The wives may be afraid their husbands will go crazy with female contact. Or maybe they are threatened because they have no control over another woman's behavior. And the wives may have no idea how her husband will react, because he's never been in a position of dealing with women on the job. I think some of the wives' feelings are well founded. A lot of it does depend on what the woman miner is going to do.

Before, in divorces, I always identified with the woman. But now I've seen how men suffer over being separated from their children, and how difficult it is for them to readjust, whether they were at fault or not.

I asked one man I worked with if he wouldn't rather a woman stay with him because she wanted to, not because she needed to for the bread and butter. He'd been married several times and had said he wouldn't want a woman coal miner because they made so much they wouldn't need you and wouldn't do what you said.

On the other hand, why should a man, just because he is married and has children, have to work every day of his life to support them? Why isn't it equally the woman's responsibility? Maybe it's because I know how heavy that burden can be, of having human beings totally dependent on you for every bite of food they eat, every shoe they put on their foot. I think it's too heavy a burden for any man or woman alone.

*She developed an interest in mine safety and participated in the 1977–78 national contract strike.*

At first I was so ignorant about the mines that I didn't know the danger. It took maybe a month, and then one night I realized there were a thousand ways to get killed in a coal mine. Just by touching the trolley wire. Just by putting your hand in a belt roller. I started studying up at home on pamphlets the federal inspector would leave around.

I'd see people run through unbolted places sometimes for shortcuts. I'd say, "We could have been off work for your funeral tomorrow." But I used to take chances, too. One night I was running the scoop through an unbolted place, and the mine roof fell in just after I went through it. That made an impression.

I became more safety-conscious and was appointed to the union safety committee. Being on it was really frustrating, because you have all kinds of responsibility and no authority to enforce the law. Sometimes I felt like I was in the middle with everybody pulling in different directions, but it was a good education.

During the 1977–78 strike I was never a radical, although whenever someone is active, some people will consider them radical. The uncertainty of those times, of rarely getting in a full work week, was hard on a budget. But working three or even two days a week was a lot more money than I'd ever earned. I think the pay was $48 a day. It seemed like a fortune to me. We lived very well on it.

[During the strike] I traveled with picketers and fixed sandwiches when they stopped on the road. I would know where they were going to be at certain times, and took food to different locations. I guess I probably was doing things that auxiliary groups do now, but nobody was doing it then. I never did anything illegal. I was just involved. Some of the men thought they might end up in jail just for picketing and didn't want women exposed to that.

I was laid off for a time in 1977. That was a hard time financially. When we were about to go stir-crazy, we loaded up our kids and went off for a weekend away. We'd been involved in lobbying for some legislation, and we went to Frankfort in buses. We always took our children with us. It was a social outlet, and we felt like we were doing something useful.

Going into the mines changed me in a way that has a word: organization. I don't mean just union organizing. Seeing that people together can accomplish things was something I really didn't understand before.

When I first heard about the Coal Employment Project about 17 years ago, I thought, I don't want to be part of any group. I didn't get involved until several years later. I think it was dynamic and wonderful and the best thing that ever happened to women in the industry. We learned from each other, not just about the industry but everything about us.

*In 1980 Barber was hired as a federal mine inspector with the Mine Safety and Health Administration. She was the first woman miner hired in that position.*

During the years I was at the mine, inspectors would come around looking for women applicants. I took them seriously, and by that time I was involved in safety issues at the mine. I started applying at MSHA, but not until I got laid off did I really pursue it. I used to call them up once a month. I had it on my calendar. In my naive, rough miner way, I would say, "Aren't you ready to hire a woman yet?" I meant that to break the ice, but the MSHA district manager might have been getting sensitive that I might file an EEO complaint.

When the government hired me, most of the women resented it because I was hired off the street making significantly more than them. Of course, they were mostly clerical people. The men had concerns about traveling with a woman. A few people refused to travel with me because of their wives' objections.

At first I couldn't settle into working with the government. I had achieved this pie in the sky I had been working for, and suddenly I was without goals. Sometimes I had to fill in for the receptionist, and I got real edgy and felt confined.

It's hard for people to understand that being underground does not feel confined at all. You have so much independence as a miner, to do the job, not necessarily at your own pace, but to choose your priorities as to what to do first, with very little supervision. You can see the fruit of your labor. In an office you can't see anything visible you've done. I miss watching coal production. It is beautiful to see coal being cut.

I don't think anything is quite like working underground. You're just like family, especially if you're with the same crew for a long time, like I was. The old-boy system is alive and well everywhere, of course. But I don't think anything can compare to any group of people who work very closely together in extreme circumstances, be it mining or fighting fires or police work or anything that's very intense.

*As an inspector, she joined the bureaucracy of a federal agency to enforce safety laws in both small and large mines in eastern Kentucky.*

Within MSHA there are clerical support people, lab support people, technical support people, the inspector—the backbone of the agency upon which everybody else is dependent—the inspector supervisors, the specialist supervisors, the subdistrict managers (who are sort of like a vice president), then the district managers. Above them it eventually leads up to the U.S. Secretary of Labor.

Once [operators] realized that I knew a bit of the business and what I was talking about, it was to my advantage. They overestimated my knowledge in the beginning, not only about mining but about the broad picture of the business. I had a lot of qualifications not easily come by—mine foreman certification, electrical certification, EMT, a blaster's certificate.

I think that with a lot of people who achieve, it's their lack of self-confidence that drives them on. That was true with me. I'm more confident now in myself than I've ever been, but it was my lack of confidence that drove me so hard and made me portray such a strong image.

I was a very firm enforcer. The district manager told me one time I was the second heaviest issuer [of citations] in the district. Surprisingly, I didn't have operators threatening to beat me up or run me off the property. I think it was because when I took actions I was very careful to explain why. I never stepped out of the Jeep beating my chest and acting like it was my coal mine. I believe in treating people with respect.

Inspectors have awesome powers, but I used them carefully. And I was determined that I was going to succeed, because nobody thought I would. I think they thought I would quit or be shuffled into a specialist position. Inspecting is the most physically demanding job in the agency. At that time I was in real good shape physically, and I was energetic. I inspected for around eight or nine years. There was a layoff, and I was hired back in a different position, as an education and training specialist. I didn't enjoy that job. It was one of those that they thought I'd fit into so well. But as soon as an inspector position opened up, I took it.

I prefer inspection. I like to make a working person's life better. I did take some special assignments. I was a special investigator for a while and really enjoyed that. Women are natural-born sleuths. For a year I was on a pilot program to collect civil penalties issued against mine operators for safety violations. I went back to inspection work, and then they posted 30 positions to train people to be supervisors. The trainees would be placed throughout the nation as people retired and as needed. There were 27 chosen, and I understand there were over 300 applicants. I completed that one-year program with six months' classroom training and six months' on-the-job training. At the end of that year many of those people were placed in supervisory positions. Those of us who were not placed were extended for another year, and we acted as temporary supervisors in different places in the nation.

*Around that time she married a childhood sweetheart.*

I had known Leon since we were about 14. But when my parents divorced, I moved to Columbus to live with my father, and Leon lived in Cincinnati. One weekend he came up to get me with some friends, young men and women. We had a romance. And 25 or so years later, we got reacquainted and we got married, and it's been wonderful.

*In the late 1980s she began having trouble with a supervisor.*

In the training that the government gave, they emphasized truthfulness and honesty. I believed everything they taught me. When I happened upon this person who tried to make my life unlivable and tried to tie my advancement to going along with his deviant behavior, I told the truth, and the nightmare began.

He was an official who was all but worshiped in the upper tiers of the agency. He was well respected in the community. And he was also a jerk. Each person in the program had a mentor whom they were to emulate and follow in the ways of how to be a good leader and manager. And he was mine.

He began by touching my hand as we talked at the desk, and standing too close, and draping his arm over my shoulder with dangling fingers. I did everything I could to avoid him and asked for assignments away from him so I wouldn't have to be in close contact with him. It seems like the only time I got those assignments was when he was punishing me for not being receptive to his advances. One day, on three days' notice, he sent me to Illinois. And I think it was because I had rebuffed him.

His behavior got progressively more serious over the period of a year, until finally he exposed himself in front of me in a vehicle and grabbed my breast and thigh. And all the time he was talking about that God gave him these hormones and it was all right. I truly believe he would have hurt someone physically in time, because I was in fear on that day that he would hurt me.

Among the many things that have amazed me is how many people within MSHA knew about his deviant behavior and did nothing. It amazes me how many women didn't tell. Only six women testified at the hearing on his dismissal. But I think there were many more women affected during his 18 years with the government. From what I understand now, his behavior was escalating and becoming more overt as the years went by. I think he'd still be working today if he'd only said he was sorry. But he attacked my credibility and my children, with threats and harassment. It was a nightmare.

Then he hired a private detective agency, and detectives started cruising by the house, going to neighbors, inquiring about one of my daughters. Our neighbors were called for information about where exactly she lived. She became very fearful. She suddenly started getting telephone complaints at her work, and she'd never had any problem before. She'd been there more than five years.

It was very stressful on my marriage. We separated for a few days. It was very difficult. Leon knew the details of the ways this man had touched me, but he was steadfast that my telling was the right thing to do.

If there were one person in government that I would never have chosen to tangle with politically over this issue, it would have been this man. He was strong on the national level of the agency. He was probably the best public speaker I've ever seen. But he was sick. All he ever talked about was religion, even as he was doing and saying some of the sexual things he did. He always let you know how strong he was within the agency.

On the day of the worst incident, I went back to the office and confided in someone who was in the same tier of the agency as I was. I asked him not to tell. Believe me, he didn't want to tell anyone, because this was a secret nobody wanted to know. Then almost immediately I was sent to Illinois.

When the incident first happened, I pretended to be sick and hid in the bedroom. And then a death happened in the family, so my weeping and crying was attributed to the death. I called in to the office after the death, and my harasser mentioned to me a job vacancy away from him. I knew that he meant that I hadn't cooperated, so I'd better get the hell away. I spent the night thinking I was going to be forced to bid on a job away from my home. And I thought, what will the next hurdle be? For two or three days I was a wreck. I was angry at what he thought he could do to me and get away with.

*Her crisis coincided with the televised hearings on the Supreme Court nomination of Clarence Thomas in 1991, when the issue of sexual harassment gained national media attention.*

My husband and I were watching the [Clarence Thomas confirmation] hearings on TV, and I remember thinking, if this woman thinks she was sexually harassed, she ought to spend a few minutes in my shoes. I wondered why [Anita Hill] hadn't come forward against Clarence Thomas earlier. I said to myself, I haven't told for almost a month. I'm never going to tell. I didn't want the personal attack that I saw happening there, and I knew that to tackle someone like this was a foolhardy move. I didn't think I was being brave in telling. But I thought, if I didn't tell, he would eventually hurt someone. I couldn't stand that on my conscience. I was afraid that one day it would happen to another woman.

My husband and I were talking one evening, and we both said we were sick to death of watching the Thomas hearings, that this happened ten years ago and why is she bringing it up now. I turned the TV off and said, "What if this had happened to her a month ago? Would you believe it then?" He said, "Yeah, I'd believe it then." And then I said, "A lot worse than that happened to me a month ago." That's when I told him.

It had been 57 days since the incident. I didn't know how or who or what avenue I would use, but I knew I was going to tell. I went to look for the district manager, but he wasn't in. I talked to the person acting in his place. I broke down and told him quite a bit. We talked for a couple of hours. He said, "The district manager will be here in an hour. Will you tell him?" I said, "Yeah, the die is cast now." As soon as I started spilling it, I felt relief. Anyway, I asked him to give the district manager a little background, because I hardly knew him and it was so embarrassing.

An hour or so later I was called to the district manager's office, thinking that he'd been given a little information. He hadn't been, so I had to start from scratch. MSHA immediately started an investigation of my charges. Pretty soon I was transferred away. Ultimately I ended up back as an inspector with a huge cut in pay. I was bumped back down to buck private and spent 11 months there. Of the supervisor trainees, I was the only one that was moved back to being an underground, crawl-around inspector.

I didn't complain. By then I realized what I was tangling with, and that I probably would be lucky to be employed before this was over. I suspected there were other women who had been harassed by the supervisor. I knew of one or two. I thought they would come forward. One did, a black woman whose incident had been almost ten years before. It did give credibility to the fact that he'd done this before. But it took quite a long time for the identities of the other women to creep out. They were afraid of him and worried about the humiliation and the stigma.

After that the agency sent me back to the office I'd come from. They could have shipped me out west. But my harasser continued in his position for several months. Finally he left, supposedly on sick leave. It's hard to keep a secret in an agency, and it was coming out that there were other women who had been harassed. Not many people talked to me at the office, because I had toppled their idol and they turned on me.

My husband asked for a meeting with the agency, because they wouldn't tell me how the investigation was proceeding. The district manager did meet with us and said that we needed legal advice. I was completely startled. I didn't want to sue my employer. I thought, why should I need a lawyer when their man was in the wrong? We did call an attorney friend, who agreed to take the case.

After that I decided that everything was the lawyer's problem. I quit shriveling up. I filed EEO complaints. I filed reprisal EEO complaints because of retaliation from what I call his henchmen. Another management person tried to get illegal access to my personnel files so that they might know everything in the world about me. I also filed a tort claim against the harasser personally. I filed suit against the private detective agency.

I still had no problem enforcing, but there was one manager of a big company who laughed to my face about this. He'd seen it in the paper. I said, "Well, since you brought it up, what is the perception out there?" He said, "You're going to be out of work, and he'll be upheld." It was difficult to have to hear that. But I just kept doing my business and not discussing it.

*The agency eventually discharged the offender. He appealed the discharge and lost.*

It took more than a year, and his discharge was upheld. I was certainly not the only reason. There were other women who testified against him. There were his threats of retaliation. He contacted people to relay legal threats, saying he was going to sue me for $10 million. An administrator chose to take the daring step of actually discharging him for this behavior. Throughout the whole thing, I did have fears that MSHA would try to sweep it all under the rug. At the same time I think people realized, if that happened, they might have seen me on the Oprah show, which would have been embarrassing for the agency.

Once I came forward, I absolutely could not turn back. If MSHA had not pursued it, I would have found ways to help them want to pursue it. Because once I had put everything on the line, I couldn't just back off. I had a lot of feelings of shame, thinking, why couldn't I stop him before it got this bad? I realize now that when somebody is sick, you can't make them understand.

You can't stop them. I really believe he was drunk with power. He'd only been in that position for a year and a half. You're not a special person just because you have a powerful position. My mother went to a court hearing and said, "After what I've seen, I realize now nobody could have stopped him."

My lawyers were very competent. We ended up with a large monetary judgment against him as the result of a tort claim, but I've not yet been able to collect it. I don't know how we would have made it without legal counsel.

*The legal battle came at a high emotional cost.*

I'm badly burned. I don't know if I would ever advise another woman to tell. She should be prepared to go work elsewhere if she does, unless she's really into torture [laughs]. I wouldn't step back now, but it's been very difficult.

Nobody wants to deal with this kind of problem. It takes time away from the job. It took time away from enforcing safety laws. The man was discharged, and I was bumped back to inspector for almost a year. Then a vacancy developed in the office I worked in, and I bid on a supervisor's job and got it.

My case has had some ripple effects. On the positive side, it's brought publicity to the problem of sexual harassment. On the negative side, the agency might think it's dangerous to let women in from the industry and go back to hiring women who have been sort of molded coming up through the ranks of the government. I think women from the industry are good for the agency. You need a healthy mix.

*On a routine assignment she revisited a mine office where she had applied for a job 20 years before.*

I had forgotten the incident, and it came back to me. I put in an application, and they laughed me out of the place. It was weird. This is 20 years later. Now I laugh to think of myself walking into a coal company and asking for a job, as if I might get it.

I think this generation's children will always remember some of the struggles. But there's nothing like experience. It will take a taste of having doors slammed in your face to rile you up. The struggle will get easier each time. You might take one step forward and two steps back for a while, like the civil rights movement. The progress can be very slow.

I believe now that I can do anything, if I want it bad enough. If I want to drive a BMW like the guy who harassed me, I could buy one if that was all I wanted to do. If you absolutely zero in on what you want, you can do it. On the downside, I've learned some things I wish I hadn't about how people can disappoint you.

One thing I've learned from working with government is that having a life plan makes a difference, even if your plan fails. Then you have plan B. I've tried to teach that to my children. Nothing will happen if you don't help make it happen. If you don't like the way things are and you're not willing to work to change them, then shut up. Either get involved or be quiet. I think it's much better to get involved.

Before the harassment happened, I had hoped to become a subdistrict manager for MSHA. I was going to classes in the evenings and on Saturdays to further my education. But for a while I gave up everything but pursuing the case. And now I wonder if the agency will ever be receptive to me. They did what was right, they did what they had to do. I think they were aware that I would not let it be forgotten. But I don't think I'll ever escape the stigma. And this episode robbed me of time with my husband to pursue things together.

Things are going really well in the office I supervise. Some of the men have acted surprised at some of the initiatives I've started, like asking them for their input at staff meetings. I really believe in the team concept. I could apply for a promotion when vacancies develop. I know I could do it, but I'm not that interested in putting the energy into what it would take to make other people believe it. I know I'd be an asset, and that's enough right now.

You really have to be better than good, being a female in a male industry. To be considered the same, you have to be better. And I'm getting tired, I'm getting older, I've got over 18 years in the coal industry. For now, I want to repay my husband for all he's given me through this rather than chasing down a position.

Women have had an impact, and not just on the mining industry. There are more women attorneys, more women in medicine. But mining is a good example, because if a woman can be a coal miner, surely she can be an attorney. I don't know what is going to happen in the future, as we reach retirement age. But mining has given my children the opportunity to go to college. It's broadened their horizons, to be able to travel. We have had an impact. Who knows, maybe my great-great-grandchildren will say, "She was a coal miner."

# 18

# A LONG HARD ROAD

## Rita Miller

*"It's not an easy life, but nobody broke my arm and put me in it."*

*After graduating from high school in the 1970s, Rita Miller pursued her father's suggestion that she find a high-wage coal-mining job. As a new employee, she worked at a nonunion mine in western Kentucky during the turbulent national coal strike in 1977–78. Since then she has worked at two other mines, always pushing for jobs on production runs at the coal face. She has endured several layoffs and describes the evolution of her attitude toward periods of unemployment. She has also done nontraditional work as a volunteer firefighter.*

My mother's daddy lived in Clay, Kentucky. He was a coal miner, and all of her brothers were coal miners. There was seven of them. One got killed in the mines, and one is disabled now. My daddy was a coal miner. There was six of us, and I was next to oldest. I have an older brother, a younger brother, and three sisters.

My mother and daddy were divorced in 1974, and when the divorce came, me and my older brother took responsibility for helping Mom out. I was a junior in high school. For as long as I can remember, I baby-sat for five or ten dollars a weekend, and I did boys' work around the house. I made my own way and paid my own bills. I've always been self-sufficient. I never had a role model. I've never had nobody.

I enrolled at Madisonville Community College to be a phys-ed major. Then Daddy talked to me about a co-op program, where you went to school for six months and worked at a mine for three months. My daddy said, "The coal mines are where the money's at, Rita."

But when co-op time came, it was a bad deal. There were 12 guys and four or five girls, and only two [girls] got to co-op. Three of us seen right

Rita Miller

then that this old world is cruel, and if you ain't gonna play the game, it's dog eat dog. And the ones who played the game got the co-op jobs. So instead of finishing out our mining degree, the rest of us went straight to work.

In 1977 there had been some lawsuits around Kentucky. In a month's span 12 women put in applications at one nonunion company. The woman I was working for was related to the superintendent. I went to apply for the job at eight o'clock that morning, and by that afternoon I had the job.

There was nine women working there. I was kept outside in the safety department, but I wanted to be underground. I wanted to get dirty, I wanted to be down there where the manual labor was. I wanted to be a coal miner.

After about two months they told me I could work picking gob up off the belt, where coal was coming out of the mines. It was just breaking dark one night when the supply-house man came and told me I had a phone call from my sister. I knew better, because we didn't have a phone. But I also knew there was a party going on. [A mine official] got on the phone, and he invited me over. I went, dirty and black.

The conversation came around to, what do you think about our little game? I said, "Well, it's y'all's game." He said, "You can be a part of this game." And I said, "No, I don't want to be a part of it." He said, "Just hear me out." He said, "You don't have to get dirty, you can wear the finest clothes, you can have anything you want. All you have to do is be where I need you to be when I call you."

I said, "No, I wasn't brought up that way." He said, "I have no problem with that. As far as I'm concerned, everything stands as is." I said, "I appreciate you taking me into consideration" [laughs]. And that was the last I heard from him.

Shortly after that I wanted to buy my first new car, and I needed a thousand dollars. This man was known to loan employees money and deduct it from their checks. I was scared because of the episode, but two days later he wrote me a thousand-dollar check, and I bought my first new car.

As far as physical harassment, I never had that. Mental I did have. I put up with it for about two months, and like a lot of people, I thought it would go away.

Finally I said to one man, "The next time you try to harass me like this, I'm gonna own that big fine Lincoln your wife's driving, I'm gonna own that brand-new fifty-thousand-dollar house you've just built, and I'm gonna have every dime you ever earned in my back pocket." I didn't have any problem after that [laughs].

That year I had about seven different bosses and six or seven jobs. The women who didn't want to go underground quit. Some of the others shoveled belts or sat at headers. As far as being at the face of coal, at that time there was not a woman that mined coal at the mine.

When I first went in, the miners' wives thought women miners were a threat to their marriage. I have no doubt when the men got home their wives would want to know if they'd worked with any women, and what they looked like, and what went on.

I was never introduced to their wives on the street until the last three or four years. The men would dodge you. Some would introduce you at mine picnics, but the majority wouldn't. I've confronted quite a few of them and said, "If you can't speak to me when I'm out on the street, don't talk to me down here." If the wives had been introduced to the women instead of hearing them talked about in the home in certain ways, they wouldn't have felt so badly toward us. But I put myself in the wife's place. If I had been married to a coal miner and heard that three or four women were on that same section, it would leave your mind to wondering what they could get by with.

Quite a few men I worked with said they'd never let their wives work in the coal mines. And I don't believe it was because of the danger or what was going on, but the idea that she can get the same paycheck.

Before women went in the coal mines, the men did some pretty dirty things to each other. They would go to the extreme of completely stripping each other down and wearing their asses completely out, or greasing them top to bottom, or hanging them up on crossovers, putting grease on private parts, nasty things that men would do to men, grabbing each other and carrying on like a bunch of two-year-old kids.

*In 1977 the United Mine Workers of America called a national strike that lasted 111 days. The nonunion mine where she worked was targeted by roving union pickets.*

When the '77–'78 strike first came about, the company flew a helicopter to a grocery store in Providence to load up with food to bring back to the mine. It was the first dealing I'd ever had with a strike, and I was scared. I was ignorant of what was going on. I didn't know the difference between union and nonunion.

One Saturday I was working and noticed a lot of men were moving around outside. A boss came in and told us to go home. A company helicopter had spotted a caravan of roving pickets from Harlan County coming down the Western Kentucky Parkway.

For three days no one entered or left the mine because of the Harlan County pickets. The superintendent told everybody, "I don't want you risking your life. I don't want you to go to no extremes. If you're uncomfortable with it, you've got a job." Five or six guys told him they wouldn't be back until it was over. It was a new experience to the women, and we didn't want to miss work. We were sent to pick up tenpenny nails scattered all over at the mine entrance.

Five or six of us would meet at a turning point and drive by to see if there was two [pickets] or 50. If there were a few, we'd go through. If there were a lot of people there, we'd turn around and go back home. One morning there was cars stretching for three miles on both sides of the mine entrance. As far as workers going through the picket lines, I think that was tried once and the man's van was turned over.

It wasn't unusual to walk in the mine office and see men sleeping everywhere. They stayed at the mine 24 hours a day. I've never seen so many guns in all my life. I had one at home, but I didn't keep it with me. I didn't go to those extremes. One guy was killed. He was supposedly a security guard. The way the papers read, him and his buddy was riding together and the gun fell off the dashboard and shot him.

During the strike I learned a lot about the union and understood what it stood for. Afterwards I felt strong about it. People came to my house, and I signed papers that I would vote union. But I was ignorant about job bids, the pay scale, job qualifications, and shift differentials. As far as learning about the coal mines, when I joined the union is when I learned what coal mining was.

After the strike the company changed superintendents. The man who took over was a male chauvinist, and he did away with the majority of the women who worked outside. He took me out of the bathhouse and put me underground on third shift, rock dusting. I worked straight midnight shift for a year and a half.

On third shift you didn't have many bosses looking over your shoulder. The men I worked with treated me as an equal. So I had the opportunity to learn how to drive a shuttle car, pin, run a cutter, and drill. I've always been an observer, and I paid attention. I got underground, but at the company at that time, if you were a woman, you didn't mine coal. Then a black man left third shift and was put on a coal run, shooting. I felt like they were discriminating against me, and that's when I set my sights on other things.

After two and a half years I left to work at a union mine. The men were as curious about nonunion workers as we were about union workers. I got six months in before I got laid off again. It wasn't the union's fault, but I'd worked two and a half years and didn't worry about my job in a nonunion mine, and I quit and went to a union mine and six months later I had no job. It left a bad taste in my mouth.

That layoff took a bad toll on me. I was around 20 years old, I'd worked as long as I could remember, and there I was with no job. I was young, and I didn't really understand the world. I'd had a taste of the big money, and that's what I wanted. I didn't have very many people that I could talk to, to find out where the mining jobs were.

I sold vacuum cleaners, just whatever I could to make a dollar. I worked for a woman who had rental property, doing floors, plumbing, electricity, remodeling, carpeting, paneling. She had a trailer that was unoccupied and needed work, so I fixed it up and moved in.

After I was laid off from the nonunion mine, I got curious that the company started hiring inexperienced 18- and 19-year-old boys. In one year's time, three of them got killed. After that happened I thought, I could have been helping production for the company, and there might be one of those boys still alive.

I decided to update my application. I had performed my job. They had no reason why they shouldn't hire me back. They hired me in because they had to meet a quota on female employment, but they had no reason to discriminate against me.

*Miller ultimately joined a class-action sex discrimination case against the mining company.*

One woman went to the mine with a friend to update their applications. The superintendent told her that they weren't hiring. They walked out of the office, and there was a young man in the office, and he told them he was going in for an interview. So they waited, and when he came out he had a job. I felt they humiliated me, they discriminated against me, they caused me mental anguish and stress, as far as being unemployed and knowing good and well that I could be out there making that money. All six of us signed papers agreeing that if this was settled, we would never reapply at the mining company. But the lawsuit didn't go through because of lack of evidence.

After a year-and-a-half layoff, I went to work at another union mine. I had to stay out [away from] the face, timbering and bratticing. I've never liked dead work. I want to be at the face. I got a year in, and then I got laid off again. The second layoff lasted from about '82 to '84.

When unemployment ran out, I worked at McDonald's. That lasted about three days. Before I went to work in the coal mines, I'd been a waitress for a long time. So I thought, fast food, I can do that. But it was too fast. They had me back there on the grill flipping hamburgers, and everybody was hollering and screaming. In two weeks' time I put in 47 applications. All the union mines had employees laid off, and there weren't many small, nonunion mines open.

I got a job pumping gas, but I'd always heard that a bought lesson's better than a taught lesson. When I was young it didn't take me long to figure out that I could jack up a car and change that tire cheaper than calling a wrecker to do it. At the gas station I learned quite a bit. My boss trusted me.

When Green River Coal opened up, they serviced their vehicles there. The daughter of the company president came to get gas, and I asked her when they were handing out applications. At the time I was making about $4.75 an hour at the gas station, and the wages at the mine was around $12 or $13 an hour. I went to the mine and sat in my Volkswagen all night long and like to froze to death. But I got my application.

I started at Green River Coal Company in 1984. I was hired in as a general laborer on dead work. But one of the bosses knew I had experience at the face, and he put me going to the empty units to get them ready for production. My main function was roof bolting. I enjoyed that job. But walking in behind a unit, to set it up for another unit, is like cleaning a two-year-old's bedroom. It's a disaster. Then I got on a timbering job, and I was working with a black man, and we worked well together. We could set 125 timbers during a shift. Then I went on second shift, and the mine foreman sent me as a roof bolter on production runs. Then I bid on a shuttlecar job and got it.

I've seen union mines where bosses would not have a woman working for them. All the women would be sitting at the headers or shoveling belt. One woman asked me to teach her to drive the shuttlecar. The boss got mad. He told her under no circumstances was she to get on any piece of equipment on his coal run. But I pinned and drove a shuttlecar and drilled and shot.

Then I decided I wanted to run the loader. One of the best men I've ever worked with taught me. He showed me all the precautions and set out to make a loader-runner out of me. But when a loader job came up, there was confusion as to which boss I'd be working for. There was one boss who was a male chauvinist. One of the women was having a problem with him. In any union mine no single boss can fire you, which is one good thing about a union mine. I think that's what frustrated him with her. He did the same thing with some of the men. There's quite a few that didn't like him.

Just because we were women we didn't stick together in all situations. We had our own opinions. Out of the six of us that had been there a long time, two of us was outspoken. If we had something on our mind, we said it. We've had some heavy discussions in the bathhouse.

*In 1989 ten miners died in an explosion at a mine owned by her former nonunion employer in western Kentucky.*

When the mine had that explosion, my safety wits came back to me. I found out right fast that men were less caring about safety than women were. They took care of themselves. I went out of my way to make sure the curtains were hung, and I had a few go-rounds with men tearing the curtains down and not putting them back up.

When I worked there, if it hadn't been for one woman on the section, one guy probably would have bled to death. He was running a shuttlecar and came around a rib [tunnel wall] with his right leg or right arm hanging out, and he crushed his arm and his leg both. The guy had this habit of always hollering and carrying on. The woman who was on the unit was shoveling. They heard the car hit the rib and heard him squall out. She looked at the loader operator and said, "You reckon he's hurt?" He said, "Why, no. He's just up there carrying on."

When the car didn't crank back up, she said, "I believe we better go check on him." She could hear him moaning. There was her and the loader operator and maybe one of the other car drivers. They just went blank. She reached down and grabbed a pressure point under his arm. They got him on the stretcher, and when she let go of the pressure points he like to passed out, so she went to the hospital with him. He was lying in her lap, and she had his life in her hands. She hadn't been underground three months. I don't know if the men could have handled that situation, but she was the first to think.

At Green River we've had maybe three or four fatal accidents. One guy got crushed on a loader, another guy got his leg cut off in a jeep, another guy got his leg cut off by a scoop. I've never been on the scene. I've been scared before, as far as thinking somebody got stuck underneath a fall, or some rock had fallen.

I fear a rockfall. I remember one time I was eating dinner and it fell where I had just left from. There've been a lot of close calls.

Coal mining's rough. It's different. You've got your good days, and you've got your bad days. One time I remember getting ready to go underground and I had an eerie, eerie feeling. Nothing happened, but that's the only time I remember just having a bad feeling. I went in because I told myself, if I don't, I may not come back. It's like driving a car and running into somebody and killing them. Are you going to get back in that car and drive it? I thought, could I go back in this hole if something happened?

After ten years in the mines there's not many jobs I haven't done. I've run the loader. It's got a large cable on it, and it's strenuous. I thought the shuttlecar was the easiest job. I didn't like working on a belt by myself. I've run every piece of equipment there is in a coal mine. There's no job I couldn't do because I was a woman. Maybe a man could pack a bag of rock dust weighing 75 pounds for five crosscuts where I couldn't pack it but three. But as far as the equipment I've run, I've been complimented by quite a few bosses on the way that I ran it. In the coal-mining world, it don't take a lot of book sense. It takes common knowledge about what you should or shouldn't do.

*At this time, Rita shared her home with a woman who had been her partner for 14 years.*

Whether you're coal mining or whatever, being gay is a very soft issue. One fellow at Pyro harassed me as a way of trying to find out, because he'd never seen me with any men and he thought, if I can get with her, it will ease my mind about whether she is or not.

Two of the girls I went to school with worked at the mine I first was hired at. One of them knew for sure, because we were roommates when we went to college. We were just friends. But the superintendent at the mine didn't like me because I was living with his secretary.

At the other mine one black guy tried what I call the old-fashioned approach, confronting me with names, like "queer" and "fag," to see how I would react. A few girls at that mine tried to petition me out of the bathhouse. That didn't work. But you have to understand that it is an uncomfortable situation for me as well as for them. I could tell, whenever we'd get into the bathhouse, that they would wait until I left, or if they did take a shower, they made sure I didn't see them. Which was neither here nor there for me.

*Harassment on the job was based on sexual orientation more than gender.*

Being gay and working was more of a harassment worry to me than sexual harassment. If I was in a situation where somebody made me mad, I would have to override my feelings because I had this fear of them confronting me with something nasty about my lifestyle. And that did flip me off, because the way I see it is that you don't pay my bills, you don't buy my groceries, and you don't worry about how I live. I didn't have much sexual harassment. I don't know if it was my looks or whether they knew I wouldn't take it.

Some women I could talk to about it. One time I came out of the mine and found a three-page letter from one of the women who was gay and didn't know very many people who were. At one mine, I was the only woman who worked underground, so I didn't have any showering problems. When I went to a mine with more women, I didn't shower because I was uncomfortable. I just changed at the house. But I went to work at another mine 60 miles from home, and I rode with three or four others. There was no way I could come home dirty. There were three women at that mine who knew I was gay. It's never been a threat to me except at first. I thought, after I told the people I really cared about [and they] adjusted to it, why couldn't anybody else?

It was easier for me if someone did know, because I didn't have to watch how I talked and how I was. The last year at Green River I carried on a conversation about my family just like anyone else. If they asked who S—— was, I'd say she's my girlfriend. I'd gotten very comfortable in a seven-and-a-half-year period of time. I had guys that I could talk to, and one woman has become a real close friend. As far as harassment goes, working at Green River was about the easiest place I ever worked.

I don't think that being gay is due to my raising. I was a tomboy when I grew up, I did boys' work around the house. I was even married at one time. Not for long—six months. I was around 17. He was eight years older. I got a divorce right after high school, but it wasn't because I was gay, it was because he had two kids and he didn't do right by them. It didn't take me long to decide that wasn't what I wanted and to get out of it.

One time we were working second shift, and there was me and two other girls and two guys. We were sitting out in the parking lot drinking, and this one guy was getting out of hand. I rode to work with the other guy, and he knew about me, but he wasn't going to go around mouthing about it. But I finally told the other guy, "Look, I've got to go. I've got a woman waiting for me at home." The next day I thought, oh Lord, but we came back to work and everything was fine.

It's more of a hardship on people that I'm friends with, because they get asked about me. My true friends have always said, "If you want to know about Rita, ask her." One girl got a lot of flack from the guys because we were going to the women's conference together. And she told me about this one man who was causing her problems. When we got back I confronted him, and he turned and ran.

And then there's the men who live around me. It amazes them to see me working on our place. I put the foundation on this house. Anything that goes wrong—the cars, the yard work, in the house—I do it. I've always been one to admit if I can't do something, but I've always been one to try it and see.

S—— and I have been together for 14 years. Shortly after we got together, we had some problems with our neighbors in the trailer park. I had to come home one night because they were acting so childish. They were sitting out and throwing rocks at our trailer. There was a seven- and nine-year-old in our house, listening to all this. I went to the door, and I had a gun in one hand and my mining belt in the other, and I invited them over.

I finally got the police out, after nearly two hours. They came and knocked on my door and told me to stay in my house. I hadn't set foot out of my house, but they were allowed to sit there, totally intoxicated. It was a big issue because it was my hometown. S—— has always been open. And she made it quite clear that it wasn't no problem, and she wasn't gonna make it a problem. We'd come from her mom's house, and a man pulled up, drunk, shooting off all this nasty stuff. It didn't take him long to realize he'd better keep his mouth shut.

It's true I get looks that if I dressed different, or walked different, I might not get. But that's always been me. It's not an easy life, but nobody broke my arm and put me in it. I put myself in it.

A few of my nieces and nephews have had a hard time. One time my niece was waiting on me. And her mom said, "Well, maybe Aunt Rita's gone to see her boyfriend." She said—and her five years old—"Aunt Rita ain't got no boyfriend. She's got Aunt S——." Some of them have a hard time believing I'm a she and not a he. They'll say, "Aunt Rita ain't no girl. Aunt Rita's a boy." But that's just stuff you have to deal with.

*In November 1991 the Green River mine shut down and laid off the majority of the union workforce.*

When the rumors first started, the company said there was too much sulfur in the coal. Then they said there was too much chlorine. But I'm still optimistic. I think it's economic, and the economy can change.

I don't think this particular mine will reopen, because it's union. It wouldn't surprise me if, somewhere in that vicinity, they didn't go in there and build another shaft and open a nonunion mine.

I knew it was coming, but I didn't want to believe it. And it's not just coal mines. It's everybody. This layoff is not as depressing as the last one. Maybe it's due to the fact that I'm not the only one and it's happening all over the world.

When I was a kid I never was one to watch the news. I'd come home, and Mama and Daddy would turn on the news, and I'd go outside. But in the last two or three years I've paid more attention. It's my generation. It's me now. I've got to know what's going on in the world. Ten years ago it didn't dawn on me that one of these days I would care. But now everybody is in a hole.

I never paid attention to politics at home, but I always knew I was a Democrat. When I started voting, Mama always told me names. Just Democrats. Just like if you put a set of rich people on one side of the table and a set of poor people on one side of the table, and they lay their money out and decide what they're going to eat for dinner, who's going to get the best?

I've heard my daddy and mama both talk about the Depression that they were in. I've got some of the ration stamps that he was given back in those days. I never thought much about it. Before, I could draw a paycheck and make it on about a hundred dollars every two weeks, whereas now, if I was to get out every day to look for a job, it would take a hundred dollars to pay for the gas that it would cost me to look for a job, plus necessities at home like milk and bread. The cost of living has gotten so outrageous that a person can't make it.

This layoff feels different. I've got just about what I've always wanted. I've got some money saved, unemployment pays better, and I haven't let myself go overboard financially. I'll be able to make it. Before, I couldn't make it on unemployment. I had to find a job. I've still got bills, but I'm more prepared as far as savings.

In layoffs before, if you didn't make a payment, they were ready to come after you. But my credit is outstanding. If it gets to the point where I can't make a house payment, or I can't make a car payment, I'm gonna call them and say, "Look, either you help me to figure out a way and work with me, or you can come get it."

Twenty years ago you worked and what did it get you? There was nothing to show for it. Well, now there's nothing that I can think of that I need or want. If I lose it, I can get it again. I've had it once, I can get it again. As long as the good Lord's willing, I'll have it. During my first layoff I about ended up on the crazy ward. I turned into an alcoholic. I didn't care about nobody or nothing. But I ain't gonna do it this time.

One thing that does bother me is my pension. I lack two years getting my ten years in at a union mine. But I told S——, with her two girls and her mother and my mother, I said, between the four or five of us, if worse comes to worse, the house isn't big, but with what everybody's got, we could survive. We may have to start over, but I'm just hell-bent and bound that it's not gonna get the best of me.

I'm fortunate that I don't have a big family I have to support. My mother does depend on me. I've always been a giving person to my sisters, my brothers, and quite a few friends. I've never had to ask anybody for anything, and I won't. If it comes down to losing my house, I won't put the burden on one of them to try to make my house payment. If I lose it, I lose it. I told S—— what scared me the worst was the kids. She's got two girls, and there are their babies' diapers. It's no great expense, five or ten dollars for this or that, but it mounts up.

I still haven't given up on coal mining. I enjoy it, but I also like working on cars and working with figures. Right now—I may change my mind in six or seven months—I'm looking at dollar signs. I'm going to look at what holds for the future, like hospital work or accounting. But coal mining is my basic career.

The problem is jobs. Scab mines used to want inexperienced miners so they could teach them to work like they wanted them to work. But there are so many experienced miners out here hurting now, and willing to work, that they don't have to do it anymore.

I have thought about moving. Once before I went up to Illinois and put in an application. But S—— has been in her job for 11 years, and she has benefits. She doesn't want to move, and I can't blame her. It's like she said: Coal mining is not dependable.

In the coal industry I have no doubt that they're trying to break the union. My own feeling is that I'm not going underground to work for pennies. And I'm not going into an environment where I don't feel safe.

A friend of mine drove a truck for one of these little scab mines around here. He told the boss the clutch was messed up. He refused to drive it, so they fired him. And that was all because he had enough experience driving 18-wheelers that he knew it was unsafe. But they thought, why should we have to be out any money because someone knows something's wrong when we can hire somebody off the street who doesn't know there is a problem?

When I worked nonunion, they took advantage of people by saying, "We can hire you and we can fire you." In one sense I didn't feel any different than I felt at any union mine other than they couldn't fire me without a logical reason at a union mine. Whereas at the other mine, if you caused a commotion, they got rid of you. They got rid of two guys by saying they found cigarettes in their toolboxes even though neither one of them smoked.

I hope and pray to the good Lord that I never have to go to work in one of these little scab mines for eight or nine dollars an hour. I'm trying to get the opportunity through the government to get some education so I could make good money standing on top of the world instead of working underneath it.

*She sees a need for women's groups to focus on helping women get and keep high-wage industrial jobs.*

There is a definite need to help not only women miners but harassed women, or women who can't get the job they want because of discrimination.

My little sister would never have been a firefighter if it hadn't been for the Coal Employment Project.

Back around 1980 I was on the White Plains volunteer fire department. There were times, when the fire whistle would go off between eight in the morning and five at night, when I was the only firefighter in town. I was the only person who had a key to that building and the only person who could drive that truck.

I'd pull up to a fire, and my first priority was getting the water pumping. There would be five or ten men standing around. One person can't fight a fire, so I'd reel the hoses off and say, "This one's got pressure, pick it up and put it on." As far as being a professional firefighter, I wasn't [laughs]. Everybody knows our fire department couldn't save very many houses, because we didn't have the manpower.

My little sister called me when she applied for a job with the Madisonville Fire Department and she was being discriminated against. I told her to call the Coal Employment Project, and she took it from there. CEP confronted the Madisonville Fire Department with it. After they saw she was serious and they got a few letters, she was the first female firefighter in Madisonville. I look at it like this. The woman across the street may not be like me and know how to do things, but does she know she can find it out? Is the opportunity out there for her?

I don't feel the coal industry is over or done with. And I think women miners will still exist. I think women will always be in the male workforce. I don't see any change in women being harassed, because it has been everywhere in this world and it happens to secretaries and nurses, not just coal miners. You're gonna have your problems, and if you don't have something to fall back on, the majority of women might think, why should I fight? Why should I put myself through the mental anguish and stress? You take a 15-year-old girl, just out of high school, and she might be like me. She might have no choice but to go to work.

If the recession digs in real deep, women's rights could go backward. From what I've seen of the younger generation, I don't think they'd let it go that far back. They feel like their mama had to put up with this, but they don't have to. They were raised different than we were.

We fought a long and hard road to get what we've got. There may come a day where it happens again. There's going to be generations after ours, until the good Lord decides he's ready to take this earth and fold it in. Until that happens, we can't quit. We can't sit back. We can't take what we've done for granted.

# Carrying the Torch: Women in the Coal Employment Project and the United Mine Workers of America

By the late 1970s women miners had emerged as activists in two organizations: the United Mine Workers of America, the union to which most belonged; and the nonprofit Coal Employment Project. Formed in 1977, CEP began as a legal advocacy group that helped women obtain coal-mining jobs. Eventually the organization evolved into a national support network for women miners.

Since their entry into the UMWA, women have served as elected officers and committee members, convention delegates, strike leaders, and activists, and in a host of volunteer capacities. Women volunteered in large numbers in the 1982 UMWA presidential campaign of the challenger Richard Trumka. The following year Pennsylvania women lobbied International Executive Board members to gain the UMWA's official endorsement. In 1986 Joy Huitt became the first woman miner elected to district office when she challenged an incumbent slate in the union's District 22 in the western United States.

Women have also sought solidarity and support through the Coal Employment Project. CEP has sponsored national conferences every year since 1979, initiated legal actions on behalf of women miners, produced a lively newsletter, organized local support groups, and provided information on discrimination and other problems women face. CEP initiated a survey among its members to determine the effects of coal mining on pregnancy and launched the campaign for a family leave benefit in the UMWA contract and in federal law.

Through CEP, women miners have forged international links with women in Britain, Australia, Germany, Africa, and Canada. One of the most enduring alliances has been with the British women's group Women Against Pit Closures, formed out of the bitter 1984–85 strike. The relationship has been mutually supportive as women miners have begun to understand the broader issues raised by the global marketplace.

Despite layoffs that took their jobs, many women miners have remained active in the UMWA and CEP. Their commitment is rooted partly in their strong identification as woman miners and partly in a desire to keep working for economic justice in the ways imparted by mining work and union membership.

The interviews that follow reflect the experiences of women who have been activists for more than a decade. Bonnie Boyer was raised in the 1960s and decided at an early age to commit herself to fighting injustice. She found the underground mine and the UMWA a fertile ground. Joy Huitt entered the mines in her forties and rose through CEP to a position of leadership that prompted her campaign for a district union position. Kipp Dawson is a veteran political activist who helped both to build the union and CEP and to forge global ties. Through CEP and the UMWA, Carol Davis found her vocation as an organizer, a vocation that has carried over to other jobs. Sylvia Pye tells the story of the British group that has become an inspiration to women miners in the United States. Cosby Totten emerged from the mines as an activist and grassroots organizer in Appalachia who, like the others included in this final section, is looking ahead to gauge the impact of her efforts on future women workers.

# 19

# A LEGACY OF CHANGE
## Cosby Ann Totten

*"The women's occupation was the first act of civil disobedience of the Pittston strike. We called ourselves the Daughters of Mother Jones. I would have liked to have stayed and made them arrest us, but we brought media attention to the issues and we backed Pittston down."*

*In April 1989, 39 women—mostly miners' wives and daughters—sat nervously on the front porch of a house near Lebanon, Virginia, preparing for the first act of civil disobedience of the Pittston coal strike. One by one they numbered off, removed their identifications to preserve their anonymity, and named themselves the Daughters of Mother Jones, numbers 1 through 39. They drove the short distance to the Pittston regional office and quietly invaded the building.*

*In the lobby a secretary came out and stared at the group, dressed alike and wearing camouflage scarves as a sign of union solidarity. "Is this a sit-in?" she asked meekly. The women paused, and then a tall, blond woman, number 15, began singing "We Shall Not Be Moved," and the group joined in.*

*Women miners and other coalfield activists have looked to Cosby Totten for more than a decade for advice, humor, and support. As the current director of the Coal Employment Project and a laid-off miner, she has become a key figure in the women's network. From her home on a sweep of high meadow in Thompson's Valley, Virginia, Totten tries to keep the far-flung network of women miners and supporters up to date on current developments.*

*Working for social change, she has stayed close to her traditions. One of her ancestors was Sid Hatfield, a police chief who challenged coal companies in Matewan, West Virginia, in 1920 and was gunned down a year later. His death led to an uprising of around 10,000 coal miners who sought to liberate the nonunion coalfields of southern West Virginia.*

*Totten's father, a miner and farmer, taught her farm skills and challenged the family's eleven children to think about and discuss the broader issues of the day. Following his example, she has involved her six children in her organizing work*

Cosby Ann Totten.
Photo from Jean-Philippe

*and convened family "union meetings" around the supper table. Years after her layoff from the mine she keeps paying her union dues—to support the labor movement in building a better life for the next generation.*

*Our interviews hopscotched over more than a decade, taking place in 1981, 1983, 1992, and 1994. Our close contact in our work with CEP and the UMWA made these interviews the most difficult of all. Familiarity can make us take for granted the people on whom we rely. Our sessions often turned into discussions of labor issues or the host of projects on which we have worked together.*

I never saw myself going to work at the Bishop mine. Too many people had been killed there. Just before I graduated from high school there was two big blowups. The first one killed 31 people. I was a sophomore in high school, and my first cousin was in it, only he was more like a brother to me. His face was all swelled up. The methane gas did that. If he had stayed on the section and barricaded himself in, he would have made it, but they tried to escape.

In about another year, Bishop blew up again and killed 20-some. In between those two, the Amonate mine blew up right near us, and 11 people died. We lost a next-door neighbor in that one. Everybody who got killed lived within a small radius. It seemed like you didn't have time to get over one before the next one hit.

My dad worked at Bishop all his life, from the time it started in the thirties. He got laid off in the fifties but went back. He ran a crane outside. When I was about eight years old we went near the mine to trade cows. We were in the back of the truck, and I seen that driftmouth and the cars going in with men on it, and it scared the daylights out of me. Of course, back then about every day it seemed like they was bringing out somebody dead. I remember the funerals. Daddy's brother Lacey was the first man to get killed at Bishop.

*She can trace her family's roots back more than a century, to the Civil War and the Hatfield-McCoy feud. She says coal was an issue in that dispute.*

Mom's dad, Grandpa Elijah Vance, was a lawyer. He fought in the Civil War as a young man. My mom always sung that song "I Am a Rebel Soldier." He was a cavalryman, and he had a medal. He had been married three times before he married my grandmother. She was 22, and my mother said he was an old man when she and the other children were born.

Grandpa Vance came down through the Vances and the Hatfields, the ones that had the Hatfield and McCoy feud. One of the Vances left from Virginia and went to the Tug River in West Virginia. One of his daughters, Nancy Vance, married a Hatfield, and one of their sons was Devil Anse Hatfield. Mom would get mad when people said the Hatfields and McCoys got in a fuss over a pig. She said it was more a fight over coal rights when the coal companies came into West Virginia and Kentucky in the 1880s. She said the Hatfields didn't want to sell out, and the McCoys wanted to sell. But both had to sell for the companies to get the right-of-way across the Tug River. This was in the time before railroads were built.

We were also kin to Sid Hatfield, who fought for the union in the Matewan Massacre in the mine wars on the Tug River. That came after the feud, about 1920.

On my dad's side, my grandmother was the midwife up and down Dry Fork and Adria, Virginia. She took care of a lot of sickness, and she showed Mom how to take care of us. For colds Mom fixed up something out of wild plum tree bark. The women knew a lot about healing. I only went to an outside doctor twice, once for an ear infection when I was seven, and when I had appendicitis when I was about 14.

*During coal strikes, food supplies ran low.*

One strike lasted the winter and was sometime around 1949. We always killed hogs and had chickens, milk, eggs, and butter. If it was a good corn year, we ground our own corn meal. During the strike our food got down to where we ate corn mush for nearly every meal. We had a choice of whether to eat it with butter or syrup. But my parents never kept food stored up if other people were hungry.

Dad was always asking us what we would do in different situations. It was his way of teaching. He would get around the big homemade supper table—he made it—and we would have a discussion. He asked us, "What do you do when you see a picket line?" And the answer was clear: you turn around, you never cross it. Whether he attended his own union meetings or not, we always had them around the supper table.

We learned everything that way from him. I was the youngest girl, next to the youngest child. On Sundays everybody around here, all the aunts and uncles and children, would come to dinner. The number got up above 50 people. Dad would ask me how many were there, and that's how I learned to count.

On the farm at Adria we did most of the work by hand. When we came over to Thompson's Valley, Daddy bought a little John Deere tractor. He taught the girls how to run the farm equipment the same as the boys. We milked the cows and slopped the pigs and kept a truck patch of several acres. In the summer we would hit the hay field or the cornfield and work all day, then do the milking and slopping and feeding the chickens in the evening. Every season had its different jobs, with sheepshearing and canning and picking wild apples in the fall.

*Her parents taught the children domestic and farm skills and the value of building a family while developing an independent mind.*

Mom wanted us girls to grow up and raise a family and keep a clean house like she did. My mother was a real hard worker. And she was forevermore improving things, changing a recipe or sewing a dress better than the original pattern. She believed that everything should be in its right place, but it was hard because she had 11 kids of her own and took in any grandchildren who needed it. We put up just about everything we ever ate. They would go into a grocery store in the fall and come back with salt and sugar and flour. We had our own meat and canned the vegetables. All 11 of us were born at home. We didn't use doctors.

Dad wanted us to have an education and to go on and do something else with our lives. It didn't matter to him if we were boys or girls. We had a mind, and we should learn how to use it to think about issues in the world. I was going to be a doctor or a lawyer. I thought there was nothing I couldn't do. But when you're young, you don't think about the fact that there might not be any schools around a place like this to train you to do that.

My dad would have liked to have been a schoolteacher. Both of my parents weren't satisfied without a bunch of kids around. As we started growing up, they took in all the grandkids.

*She met her husband at a church meeting. She was a high school senior, and he was eight years older.*

I was raised to respect my elders, and Clarence was eight years older than me. His word was law. But I wasn't raised to let somebody else do all my thinking. After we were married he had six jobs the first year. We went to Maryland and then came back to Grundy, Virginia. After Clarence Junior and Rebecca were born, Clarence decided to build a wedge mill to make wooden wedges to sell to the mines. I helped him cut the trees, make the wedges, and take them around to the mines. But as soon as he got some money he would spend it on drinking and leave me and the kids high and dry.

From 1962 to 1974 I had six kids, and I had no idea about what was going on in the world. We had no phone and no TV. I never left my children except to go to the hospital to have another one. Clarence believed I was supposed to cook and clean and keep everything right and ready and take care of six kids, plus help him work on the wedge mill and cut timbers. He expected me to do all that, and then he would holler at me about why wasn't dinner ready. Oh, I was different then. He had me beat down.

I decided my kids wasn't going to be beat down like I was, and I started figuring out how I was going to get out. I had to bide my time, but I finally got my kids out of there and went home. It surprised me, because I'd took so much for so long. But when it comes to your kids, there is nothing that will make you fight harder.

Women have to stand up, not just for their kids but for themselves. When I heard about feminism I wasn't sure what a feminist was. It's like, what's a communist? There aren't any clear-cut answers. If being a feminist means burning your bra, I'm not one. But I have tried to shake a few of these women around here on the Equal Rights Amendment. They're not for it, but I can't judge them too hard, because I've been on the other side of it, too. Clarence beat me a couple of times, and when I took him to court the judge never asked him if he hit me. I have told these women that there is a law on the books in Virginia that if you have a dog, you have to give it a warm place to stay and feed it, but that there is no such law to protect a wife who is getting beaten.

I never thought I would be raising six kids alone, because we were raised up to believe there wasn't no such word as divorce. But sometimes it has to be done, and in my case it was necessary. If it had been just me involved, I might have tagged along and not said anything. But you can't raise six kids in a squabble.

About the time I was getting ready to leave Clarence, I knew they were having to hire women at the mine. I wanted to put in an application, but Clarence wouldn't let me. After I left him I took the kids and moved into the house on Dad's farm. Me and Clarence Junior fed the cows and took care of them, and in return we stayed in the house for more than a year while I got things settled. I went to work at the furniture factory, and it was the pits. I was the only woman in the carpentry part of it, but I couldn't get a raise. The highest I made was $2.45 an hour. If the kids had gotten sick, it would have been a financial disaster.

A woman I worked with told me about her sister going to work at Bishop mine, and she said, "Lord, she loves it." So I put in an application with Consolidation Coal Company. They called me in and told me I had to take the 80-hour course. I took it, working eight hours, and drove to go to school for four and a half hours at night.

My kids have always had a say in our decisions. I made up my mind that my children would know everything that was going on and have a vote. Clarence Junior was about 14 at the time, and he didn't want me to work underground, because I might get hurt or killed. But I told him I could get killed working at the furniture plant, too. I was working with the big band saw, and one day it broke and started flopping around in my face. If I had moved an inch, it could have killed me.

When I took the physical for Consol, the company told me I failed but wouldn't give me a good reason. When I filled out my application they asked for a family physician's name, and I put "none" because I didn't go to the doctor. I went to doctors all around, and nobody found anything wrong. I took all the doctors' letters to the office, and they didn't even go through them. They told me to start work. What happens is, since I had no medical records, I think they believed that if they told me I wasn't physically able that I would accept their opinion and give up.

*Totten went to work for Consolidation Coal Company in December 1976.*

I never had no wantings or ideas about going into a coal mine. But I always said I had six good reasons: Clarence Junior, Goldie, Vickie, Billy, Bobbie, and Rebecca. Can you think of any better ones?

In the mines the work came pretty easy. Maybe it was because I'd worked on the farm with Dad, who taught us to think for ourselves. If the tractor broke down, we had to tell him what we needed to fix it.

I was on the four-to-12 shift, and my youngest was four years old. I only had one vehicle. I would get up in the morning and straighten up and leave the kids a note telling them what to do that day, and they'd write me back. I'd only see them on weekends. Then I was on hoot owl shift, and that was terrible. When school was out I couldn't sleep with six kids in the house, but when school started back I couldn't get used to the quiet.

When I first went to work at Bishop, some of the men were still superstitious about women in the mines. There was a story about a redheaded woman and a couple of kids who went into that mine and got lost and supposedly starved to death. If you saw this redheaded "haint," you didn't have to work that day. It went way back and was part of the company's past practice. Of course, the company might have thought, if somebody saw a redheaded haint in there, they could be drunk or something.

The men tried to get me scared on the hoot owl shift, telling me somebody had seen her coming out from under the rocks. I said, "If y'all ever see that redheaded woman in here, it's a vision of things to come. We're in the mines, and we're gonna stay." This was about '79 or '80. The company would let them go home if they saw it. Most of the people were leery of the stories.

In 1976, when I went to Bishop, there were around 900 people on the payroll. At one time we had 25 women, but they started getting picked off one at a time. You'd hear that this woman wasn't coming back to work, and we'd try to find out why and find out that she missed two days without knowing that it could get you fired. That was really when the women started banding together. Consol was isolating the women and picking off the weakest to fire.

The first boss I had didn't want women or blacks to be in the mines. One black guy on the section had been in there nine months, and the boss wouldn't train him on anything. He would make us carry timbers out of the face in three-foot-high coal instead of letting us get a piece of equipment to carry it out.

Sexual harassment was never much of a problem for me. I had been raised up with boys, and it was easy to be around there. And I found out my big brother, who worked there, had told them to leave me alone. I got trained on equipment, too, because I had a black boss who understood how hard it was to get trained. He took a chance on me and put me on a shuttle car.

Before I went to Bishop I had never been in a union before. I appreciated the union. Without a union, companies would come in there and divide everybody up. The women would be fired, the blacks would get thrown out, and then they'd go after somebody with a long beard or something else the company didn't like. But the way things are now, ain't the company about got us where they want us? Got us passive? Can you see the unions holding it all together? Companies have the law on their side, and the federal government.

When the 1977 strike was called, we had special meetings at Bishop for people to sign up to picket. I raised my hand to volunteer, but they would never take me. I guess I was the only woman that ever volunteered to go, but they didn't want women out there where we might get hurt. But when you are in a strike you don't force yourself on them. You can get the whole local in trouble. I had found out enough about unions from my daddy's teachings that you do not force yourself in picket-line situations if you're not wanted. There was some rough stuff during that time.

*After the 1977–78 strike, and after seeing the effects of mining coal, she became active on safety issues in her local union.*

Three people got killed up there after I went to work. Two of them got killed by roof falls. There was a mechanic on our section. I had gone to school with him. He was working a Saturday shift, and he reached up under the canopy on the continuous miner. The lever was turned on, and the canopy came down and broke his neck. Another time a buggy got loose. Its brakes bled off, and there was a bunch of people in front of it working, and four of them got crippled up bad.

You never know what to expect underground. A piece of rock could fall out and kill you anytime. A lot of guys curse and talk bad and try to act like they're not worried. If I had my choice of who to be trapped with underground, it wouldn't be the big talkers, because they're the ones that fold first. The ones who just take it day to day would be the ones I'd put my confidence in.

When I was elected to the safety committee, I felt like I was doing something important. I think a lot of women take leadership roles in safety. I was used to taking care of kids. I'm not going to sit back and let the kids set fire to the house just because they're having fun. It's the same thing in the mines. You're not going to sit back and let some boss kill all of you just because he wants to run another buggy of coal.

*In 1979 the Coal Employment Project organized the first national conference for women miners. Totten reluctantly attended.*

My local president pushed me, saying I needed to go to this conference, but I didn't want to leave my kids. He called them when I wasn't there to tell them how important it was. At that first conference I got appointed to the board of directors, and we started working on getting family leave. At the time we argued that it couldn't be just maternity leave, because men needed it, too. We got in touch with [Colorado Representative] Pat Schroeder and [Pennsylvania Senator] Arlen Specter's wife and got the ball rolling.

Women miners led the fight for family leave, but it took from 1979 to

1992 to get it passed. We took it to the union's convention after going to all the local unions, and more resolutions were sent in on the issue than almost anything else. We kept pushing and worked with women's groups to get language in the bill that would include time off for taking care of seriously ill children.

My kids got involved in parental leave, which is what we called it before it became family leave. They wrote letters to congressmen and talked to other kids. Our local representative knew that these were future voters, so I think it influenced him. We have seven votes in our family.

Through CEP we learned a lot. The Coal Employment Project helped us break into the union. We needed help, because the UMWA was a men's union.

I went to China in 1980, and we went underground. It was weird. They had big seams of coal and would mine part of it and then leave the rest overhead, with no roof bolts and just a few spindly timbers. In China everybody worked. People in wheelchairs picked up trash in the streets with a pointed stick. The coal mines were more poverty-level. A whole family would live in one little room of stone houses they had in the mountains. There were women working outside cleaning up around the mine, but the Chinese told us women couldn't work in the mines because they would get lumbago.

*Cosby Totten was laid off, along with most of the mine's workforce, in 1982.*

I was 35 when I went into the mine, and I was hoping to work 20 years and retire with a pension and health card. But by 1982 we knew a layoff was coming, because the company was stockpiling coal. We just didn't know how drastic the layoff would be. By that time the workforce was down to about 450. And then they laid off all but 25.

At first being laid off was great. I crocheted afghans. But then it got boring. And the layoff was hard on the kids, because they had gotten so involved in my work with the union and with CEP. Their whole life after school was involved with what I was doing. I got an internship to do organizing work on parental leave. I didn't get depressed. It's hard to stay depressed with six children around.

It seems like there is a higher percentage of laid-off women who stay active in the union than men. Women put their time and energy into building our ideals and hoping for a better union to help everybody. With some men it seems like, if there isn't an instant return on their activity, they don't want to do it.

I was laid off in 1982, and I have stayed a dues-paying member of the United Mine Workers. I intend to stay a member, because I love my union. I stay active for my mental health, although sometimes being active makes me think I'm crazy. I stay involved because I want to make life better for my children. We need to work for good-paying jobs for them, even if we never get good-paying jobs again ourselves. Black people had more opportunities after the civil rights movement, and when their wages went up it helped women. If you fight to help the littlest one, it will help you in the long run. An injury to one is an injury to all. Our children need to understand those words.

*When the UMWA's national contract expired in February 1988 the union recruited Cosby to help organize in the campaign against Pittston Coal Company. Pittston had canceled health benefits to the families of retired and disabled miners and widows. She worked with two other UMWA women to set up a network of union "auxiliaries," or groups of family members, mostly wives.*

Women miners and the coal miners' wives mostly did not associate with each other. And if they did, it was never in the context of what the union was, or the job. And the men usually don't talk to their wives that much. They'll talk about their problems, and they'll gripe and growl, but they don't explain what the job is. So the wives never really have the opportunity to learn.

To learn what a union is, you need to go to the meetings and watch or help with the grievances and arbitrations. Wives weren't allowed to do that, although some of the [local union] officers' wives understood more because they did a lot of the work for their husbands.

But a lot of them were in the dark. One woman said, "I don't understand why my husband don't just sign that contract and get back to work." I had read about Martin Luther King, but most of the women did not understand civil disobedience. And up until then a lot of them didn't understand what caused roof falls or explosions. But being around retired miners on the picket lines, who would sit there and tell stories and explain why things happened, was an education for all of us.

With women miners, organizing was easier. We knew by experience that the company was our common enemy. Just trying to get hired taught you that. So it was easier for women miners to identify what was important than for the wives. And sometimes you have to keep an eye on the union.

The UMWA is mostly macho men. The women miners that got into the union hadn't been there that long, maybe ten years. What happens in the UMWA sometimes is that a few people will end up running a local, and it's not always the best people. And the auxiliary is not a dues-paying organization, and there was really no structure to go by. Even if we'd had a constitution and bylaws, it might have ended up being a whole lot like the union. But the union needs to be more democratic, with chances for people to have input.

The first I heard that there might be trouble at Pittston was when I heard that they might take the health care away from the pensioners, the disabled [miners], and the widows. My first reaction was, "They can't do that." [But] they did. Then you called me from D.C. and asked if I wanted to take a part-time job with the union, and I said yes. We were to organize the women—the wives of coal miners or any other women interested in promoting the union and the strike.

We tried to do it [by] going around to different locals, and we would ask the coal miners to bring their wives to the next meeting. Some of them didn't want their wives in. One man said he was the boss in his family, and the union was not going to tell him to put his wife on that picket line. They acted like being on the picket line was somehow degrading. But some of the men would degrade their wives in private, you know. Yet and still, most of the women that came out on that picket line turned out to be better wives and better homemakers, because they found out what their husbands had been doing all these years. The women started seeing that this was not just their husband's job. This was their livelihood.

If you stand out there by yourself on a picket line, it's just one person. But if you bring in your wife and your kids, your brothers-in-law, your sisters, your uncles, and your cousins, you've got 20 people instead of one. Let the company fight the whole family.

We had community marches in St. Paul, Clintwood, Haysi, Honaker, and Richlands. The women traveled back and forth. It was a way to link these little communities together. Before, when I thought of southwest Virginia I thought of everything west of Roanoke, but I didn't have any real feeling for it. The strike made southwest Virginia bigger and friendlier. Bigger, but yet and still closer. It opened it up. By the time the strike started we had been to all these little towns.

*After a network of local auxiliaries had been organized, the union assigned the strikers' wives the task of setting up an informational picket line in front of Pittston's offices in Lebanon, Virginia.*

We were there at least a year before the strike started. It started out that the women were going to take care of it, but the pensioners didn't have no place to go, so they started coming out. By the time the strike really got going good, Lebanon was more pro-union.

And there was Gail [Gentry]. He was the glue that helped to hold it all together. He was like our brother. He was out on that picket line through good weather and bad. And you know as long as Gail was around you would be protected. You wouldn't be hurt. He was just the type that would take care of all of us, even though he was in a wheelchair. He wrote fantastic letters to the editor and drove an hour each way to get to the picket line. He just said, "We've got to do this, because if we don't, what have we got?"

On a picket line where a strike is already going on, I don't think people have the time to bond and to get to feeling comfortable with each other. In front of the corporate building, there was no pressure. The scabs weren't going across and taking our jobs. Because it was relaxed, everybody could put what they felt into the discussion, and we could talk it over. If it seemed like a good idea, we would look at it and see if it would get us into trouble legally. And if it didn't, we'd get out there and do it.

Boy, we had some good times on that old picket line. We'd fix hamburgers and chicken, potatoes and corn on the cob, and tomatoes and onions from the garden. We had birthday cakes, and we had a Thanksgiving dinner and a Christmas party, and everybody came out, and the press would come. At Christmas we put up a little Christmas tree. I made everyone a little key chain with their pictures taken on the picket line. My family made ornaments out of yellow and blue beads, the colors of the UMWA, for people to put on their trees at home. I liked the idea of our unity spread all over.

There were some rough days, too. One time it started snowing and blowing, and we didn't have our tent. It was a miserable day. But we stayed, because that was what we were going to do, "come hell or high water," we said.

*After 14 months of working without a contract, UMWA members struck Pittston in April 1989. Union leaders promoted camouflage as a strike uniform.*

When the strike started [the international union] took over the picket line at the Pittston building. That hurt. That was a place where we had met for over a year, and we felt like it was our second home. The picket line was also a place where the women who had been active was used to having a say in things. That stopped, too. The feeling of really belonging and having a say wasn't there.

We had encouraged people to write songs, because that's part of it, getting feelings on paper so other people can share in it, too. But after the strike started it seemed like it went to professionals, which is not always that good. Maybe we don't talk proper English, and maybe we don't write proper English, but it's our English. It's our story, and people in other places could relate to our way of writing and talking, because they've got their own little ways of doing it, too.

Before that strike, if you'd asked me, "Are you going to wear camouflage?" I'd have said, "Why, hell no." I hated it, because I thought it was just this macho miliary style. But after a while people began to look alike, and that was the point, to camouflage them from the state police and from the security guards. We really did need a uniform, and camouflage was easy to get.

*On 18 April 1989, 39 women calling themselves the Daughters of Mother Jones occupied the Pittston offices for two days in the first act of civil disobedience of the strike. They attracted national media attention to the strike before ending their occupation peacefully.*

The women's occupation was the first civil disobedience act of the Pittston strike. It was scary, but it was a lot of fun. We felt like we were getting an opportunity to do something. My kids didn't know what was going on. I told Goldie that I wasn't going to be back for a couple of days, and for them not to worry until they got a phone call. She and a friend drove me down that morning. I just told her to pull off the road and let me out. She said, "Let you out on the road like this?" They thought I'd flipped my wig.

We were getting to the point that the men had worked for 14 months without a contract. And "unfair labor practices" don't really cover the humiliation and the hurt they caused to the miners. It was getting to wear on our nerves that this was going on and on, and not seeing an end. A lot of us knew each other from the picket line and had been through enough together so that if I said, "Well, I'm staying," I wouldn't be the only one.

We went in [the building], and a woman came out and asked if this was a sit-in. I could feel the tension building in everybody, and I started singing "We Shall Not Be Moved." We thought that within 15 minutes the state police would be there and read us our rights, or maybe just come tearing in to arrest us. As the day wore on, everybody got tired. We hadn't taken any cigarettes in with us. We didn't take no identification, no pocketbook, no chewing gum or candy.

*The use of numbers, not names, was another kind of camouflage for the Daughters.*

Some of the women didn't want to give their names for fear it might come back on their husbands on strike. It helped to give us more of a group feeling—that we're not individual, we are a group, and we're in this together.

We had television stations there, and newspapers. When a reporter would interview us, we would say, "I'm the Daughter of Mother Jones, number (so-and-so)." And the reporters would have to explain in their stories who Mother Jones was.

About dinnertime the widows and wives and some of the pensioners, and Gail in his wheelchair, came up the hill with pizzas and sodas. The guards had told them to stop, but they come right on through and into that building to give us dinner. It was an emotional few minutes for the people that was in there. [The union] negotiated to bring us blankets, quilts, and pillows to sleep on. [Pittston] said they could bring it to the door but that they could not come in. And the union kept sending food. Pittston let two women go out and get the stuff from two union men. The men would deliver cigarettes, aspirin, small items, and handwritten love notes addressed to the women by number. They brought us pizzas and chicken and hamburgers, and the next morning eggs and bacon and sausage and gravy.

Two of the women led gospel songs, and one gave prayer. It was so passionate, so heartfelt, the way those prayers came out of her. And people were making up new words to songs, and we were singing.

*The office was also being used as headquarters for the paramilitary-style security guard company Pittston had hired to police the strike.*

The guards kept prowling back and forth to make sure the rowdy criminals didn't tear things up. They turned the air conditioner off, and you couldn't hardly breathe. Then they turned it back on, and it was freezing. A company man came to lock the front door. He broke the key in one door, so they chained the two doors together and put a padlock on it. So we were really locked in. That sort of gave me a scary-like feeling, because they still had all those guards in there, and I didn't trust none of them. According to some of the reports, they had criminal records. That was one reason for staying together as a group and not doing nothing as an individual. It's pretty hard to pick on a whole group.

It started getting a little dark, and we could see the outlines of the men down on the road moving back and forth. The next thing we knew there was blue flashing lights and you couldn't see any headlights from traffic. Later we found out that they'd blocked off traffic. And a company helicopter came through about four o'clock in the morning. It come down so low that it was almost outside the window. That scared me, because they don't really have the lights to do low flying and there were power lines around.

If somebody had told me I would have done something like that a few years back, I would have told them they were crazy. After I left Clarence I would never have thought I would ever get the confidence to do this kind of stuff *[laughs]*.

We discussed leaving [the building], and we felt that we had achieved about all that we could have. I would have like to have stayed in there and made them arrest us, but we brought media attention to it. We backed Pittston down. We took it over and stayed for two days and a night. We left in a glorious fashion. We went down the hill, a-singing and a-whooping and hollering.

After it was all over [the strike leader] said that he had wanted the occupation carried out so that the women could show up the men. I didn't like that, because a wife does not want to be used to humiliate her husband or to whip him in line. I think a lot of women resented that. It seemed like most of the men in the decision-making part of the union would use the women when they needed them, and when they didn't they wanted them to get back out of the way. But if they called you at two o'clock in the morning, you were [supposed to be] Johnny-on-the-spot.

*The women's occupation was followed by mass picketing at selected mine sites where strikers and supporters sat in the road to block coal trucks. A center of civil disobedience was the picket line at the Moss 3 preparation plant near Cleveland, Virginia.*

From the very first day you went over [to Moss 3], it felt like home. You didn't feel ill at ease, because you walked into all that camouflage and you were one of them. Man, woman, or child, we had everybody. And the state police were there, the security guards were there, the scabs were going in and out. There were around 1,000 to 1,500 people on that picket line every day.

On the day of my first arrest there were a little over 500 people arrested. At first the police were going around the women and getting the men. We told [the strike leader] that he was going to have to let us in. Finally the troopers got all the women that was out in the road. We went to Blackford Prison. They took us through all those gates and all that barbed wire. They took me in there about ten or ten-thirty in the morning, and I got out about 12-thirty that night.

The second time I got arrested, ten of us spent the night in jail. There was 18 arrested, but one of them was released without bond. That left 17 of us. None of the women wanted to be bonded out, but they didn't have room for but ten. The women who had to leave were disappointed. In a way it would have been a reward. But I thought we were all in it together. I got to stay, but we were carrying out what the group started.

A lot of the kids wanted to get arrested. The kids had formed student auxiliaries and started singing and holding meetings and coming out to the picket lines. The union stopped them from getting arrested because they figured people would get mad if children started going to jail.

*Besides civil disobedience, the auxiliaries held jail vigils and cooked meals to supply strike support centers, including a camping area dubbed "Camp Solidarity." The abandoned camp was transformed as bunkhouses, a speaker's platform, and a kitchen were built to accommodate visitors, who numbered nearly 50,000 by the strike's end.*

I loved Camp Solidarity. To me, it was just another way of going home. There was a 22-foot-long by four-foot table, full of all kinds of food. I went over and took food, but you didn't have time to really sit down and eat. Everybody that came in stayed there in campers, and later they built barracks. We had rallies out there, and music was going every night.

All that cooking never got any real recognition. Women would cook and take it over [to Camp Solidarity]. A lot of times they were doing it on their own, expense-wise. Then the union brought volunteer chefs in, and the women didn't feel as comfortable. It wasn't the chefs' fault, but it changed things, because it wasn't the home cooking that we were used to, and the women felt like they were being kicked out.

*Gospel songs were common on the Pittston picket lines.*

When you drive through southwest Virginia, how many nightclubs and bars do you see? I'm not saying there's no alcoholics, but in southwest Virginia most of our social life is centered around church and family. And when you're part of a community that's church-oriented, the gospel songs are easier for you to sing than labor songs.

"Amazing Grace" was one song that everybody was familiar enough with to sing without the words. When the truck hit those people on the picket line, we started singing "Amazing Grace." When they brought over some man who'd hit one of the picketers, there was this sea of camouflage moving toward him, and we started singing "Amazing Grace." It calmed people down. If these people hadn't been brought up in the church, it wouldn't have happened that easily.

One thing that made people mad at Pittston was when [a company official] said that we used religion as an excuse to not work on Sunday and that we put religion on and took it off like a coat. He could not have been raised in southwest Virginia. People can leave the church, it can be 20 years, but their roots are still there.

Here in Thompson's Valley we've got three churches: Methodist, Church of God, and Presbyterian. They split off, but people don't just go to one church. They go to all three. It's just a part of who we all are. We are a product of our communities, and there is a church involved in everybody's life who is from this part of the country.

*In June 1989 a federal judge sent three of the union's strike leaders to prison in Roanoke, Virginia, on contempt charges stemming from repeated acts of civil disobedience. A car caravan of women soon followed to organize nearly a month of jail vigils.*

By this time I think the international was getting a little afraid of the Daughters of Mother Jones and the group. I think a lot of stuff went on to keep it from being strong. There was a lot of pressure. The women could not have carried out what they did without the additional women at Camp Solidarity, at Binns-Counts, the women who went to the courts and jails, and the women who would make phone calls at home.

There was no way to keep track of all the women, and that was a shame. I had a list of around 350, but I couldn't get home at ten or 11 at night and try to get back out there the next morning at eight o'clock and worry with trying to keep up with the names. I didn't have the time.

I think the women and the men had a different experience in the strike. The men picketed at their mine or joined mass pickets or went to the courthouses, but it wasn't the same thing. The women were closer and got to know each other better and traveled more. It was frustrating, because the women found a voice in the strike and then were kept from using it. If you're going to take part in a strike, you should have a voice.

Camp Solidarity brought women miners into southwest Virginia, and we all participated as union members, not women. That was a better feeling and made me feel like we were on equal footing. The Pittston strike brought unions together, and I would like to see every union come together to fight for better conditions. I say, if one group of workers are on strike, we should support them. If we have to, let's shut the country down. But human nature being what it is, people are greedy and some would tear down everything the unions have built over the last 30 years to get a week's work. The unions put a value on human life in this country. The government didn't do it. That doesn't mean I'm against our government, but I truly believe it was unions that put a value on human life.

*The issues she sees as important now are health care reform, economic justice, and the problem of increasing violence against women.*

We need to deal with the problem of low-income work. We need to get pay equity for everybody, including men. We need to get diesel equipment out of the mines. We all need health care. In one way health care without any money in our pockets won't help. But good wages without health care won't work either. We also need to work on the problem of violence against women, which is getting worse.

When we first got started with CEP it was easy to see that our enemy was the coal company. We figured that out real quick. Now it's harder to identify where our enemies are. Sometimes it seems like the union leadership comes out against us. The government surely doesn't care if you're a woman or a man if you're on strike. They'll go after you. But the question is, what are we going to organize around? It's a hard call to make, but we're going to keep working at it, putting one foot in front of the other.

Women went in the mines and showed the companies we could do the work, and we opened doors to truck driving and other jobs. It helped because women miners were so high-profile. I think the boundaries were pushed further out for men and women. There are more men now working as nurses and teachers. I like to work on issues that help everybody. If it's a women's issue, I believe it would help my daughters and my boys, if they get married. All for one and one for all—it still makes sense to me.

# 20

# ALWAYS A MINER
## Carol Davis

*"Every coal miner you see that's laid off says, 'We're going back tomorrow.' Coal mining is so hard to leave behind. People say coal dust gets in your veins, but I think it's more than that. Your life depends on that other person. . . . When you see them after a layoff it's like old home week. Hugs and kisses and all that. . . . I believe that's true for coal miners until the day you die."*

*It was during her second pregnancy working as a coal miner that Carol Davis decided to lobby for a federal bill and a union contract that provided a family leave benefit. Management had threatened her job, and she joined the Coal Employment Project's family leave campaign. Women miners drafted the contract language accepted by rank-and-file convention delegates in 1983; UMWA leaders then contacted sympathetic political leaders and met with national women's groups. Finally, in 1992, the Family and Medical Leave Act was passed.*

*Carol Davis is not the first activist in her family. Her grandmother, Juliana Perkins, disguised herself as a man to dig ditches to feed her family. She told her grandchildren the family stories of slavery in Alabama and stressed the importance of education to being free.*

*Like millions of people throughout the eastern coalfields, Carol's family left during the coal recession of the 1950s and 1960s to seek work in northern industrial centers. Carol moved back to western Pennsylvania from Cleveland in the late 1970s after hearing that companies were hiring women to mine coal. In the mine she found sisterhood as well as a family among her crew. She became a union activist and a leader in the CEP before a fire broke out underground, resulting in the mine's closure and the loss of more than 300 jobs.*

*Now Carol Davis works in the deli of a grocery chain and serves as the union*

Carol Davis with John, CarMelitta, and James

*steward to a younger workforce. Her contribution, she says, is in educating her children and other young people about union history and the principles on which the labor movement was founded.*

My grandmother took on a job as a ditchdigger when my mother and her brother and her sister were very young. It was in the late 1920s or early '30s. She disguised herself as a man because they weren't hiring any women. She dug ditches to lay pipes until she got sick one day and passed out. When they tried to give her air they found out her breasts were bound down, and they fired her.

Her name was Juliana Perkins. When she went as a man she called herself Julian. She was a butcher and a hairdresser and sort of a doctor. I remember one time my aunt told me a guy came to the house holding half his stomach. My grandmother sewed him back up and got him back to life somehow. She knew the roots and the plants to heal with.

My grandmother was the first freeborn in our family. Her mother was a slave in Alabama. She was set free, and my grandmother was born out of slavery. She would sit down and tell us stories. She said that we were to tell our kids, and they were to tell their kids, and on and on. That's why I know so much, because she would sit down and tell us how we got here.

I take a lot of strength from my grandmother, because she always told me there is nothing in life you cannot do. Do you remember Frances Perkins, when she went in the mines? Everybody thought it was my grandmother, because her last name was Perkins.

My grandmother was married twice. Her first husband—my mother's father—was a gambler and renegade-type person, and he must have done something against the law. They had the dogs after him. So he brought my grandmother and her three children to West Virginia.

He brought them up to a boardinghouse that my aunt's father owned. My grandmother got a job doing the laundry and cooking and cleaning, while my grandfather decided he was going to run around. So finally he had a meeting with my aunt's father and said, "You can have the wife and take care of the children—I'm gone."

He moved to New York, and my grandmother ended up marrying Aunt Rene's father and running the boardinghouse. She would make lunches and take them to the pit. Then they moved on up to Pennsylvania, and he got a job in the mines. Then he became real sick, and my grandmother took care of him until he died.

She went through a lot to take care of her family. My mother was the oldest, and she was born with a hole in her heart. She wasn't supposed to live past 18, so my grandmother took special care of her. My mother went to third grade in school, but she was very well read. Her big thing with us was that you have to get educated.

My mother got married when she was 12 years old because my grandmother was from the school that said, if a guy wanted to spend time with her daughter, he would have to marry her. Back then, if you were 16, you were an old maid. So she got married. My mother hadn't even started her menstrual periods. She didn't know nothing about it. This guy was a miner. He would come home from work and comb my mother's hair and cook and clean house. It was like he wanted a child.

My mother has been married several times, and I've never heard the word divorce. When they broke up she had to be 18 or 19. He was like 30 years old. He was a very nice man. I say that because he never forced himself on her. After that my mother got married, and she had my brother. My brother's father was a coal miner, and he got killed as a result of an injury in the mines.

My mother was the kind of person who liked to party and have a good time. Grandma would take care of the kids. If things weren't going good with my mother, my grandmother would come snatch us up and take us home. Or if my mother ran short, we'd have food out there. My grandmother was blind from a gasoline explosion, but you wouldn't believe that woman was blind when she took care of us.

We lived in Pittsburgh for a while, and my grandmother would come in from Cokeburg Junction. My mother told me that when I was born my grandmother started the day before and walked to our house.

Cokeburg and Cokeburg Junction are two different places. They're about two and a half miles apart. Aunt Rene was born in Cokeburg, but the town won't claim her because there's no black people in Cokeburg. Cokeburg Junction had about five families of blacks. It was a coal-mining community. Everybody struggled together. It was a real mixture of cultures. We learned some Polish and some Italian. That's where a lot of my values were instilled, because you didn't see color there. My white friends would be in my house. I remember getting up in the morning and running down to a neighbor's house for a big bowl of tea and homemade bread. Her daughter was my best friend. Every two weeks I'd get my hair washed and straightened, and she would come up and Aunt Rene would wash and straighten her hair with a hot comb you put on the stove.

The first house that we lived in in Cokeburg Junction was given to my grandmother by an old coal miner. Mike lived with another old guy who was a coal miner, and he gave the house and land to my grandmother after Joe died. He was just real kind.

Church was a very, very prominent thing in our life. I've been in a couple of different religions. I'm back to Baptist now. My grandmother was one of the founders of the Baptist church in Bentleyville. It was one of the old company houses. They moved it from Ellsworth and gave the blacks a plot of land for a church.

My grandmother brought a lot of her practices to the church to liven it up: singing gospel hymns and playing tambourines and a lot of clapping. She said church shouldn't be dead, because God is living. There was speaking in tongues, and you had to "tarry."

When you receive Jesus as your personal savior, the Holy Ghost is supposed to come in. My sister and me were tarrying, and you have to repeat Jesus' name while you are waiting for Him to come into your body. She was tarrying, and I was right next to her, and she threw up this white stuff, and they were all happy for her, thinking, oh, she was receiving the Holy Spirit.

I was tired of being up there on my knees, so I got a lot of spit in my mouth, and I tried it, but they said, "No, no, sister, you have to go back" [laughs].

At our church you had the pastor and the elders, who took turns preaching. Our pastor worked in a steel mill. The women were the mothers of the church. They were really strong. There was one woman preacher who was really dynamic.

My grandmother was very strict on education and on speaking properly. She told us, "They can take anything away from you but what you know." We would come home at night, and Aunt Rene had this Lincoln Home Library, and she would line us all up in a row. My aunt worked cleaning house and as a governess for rich people in Pittsburgh, and she would sit us down and teach us how to set the table properly and how to dress and how to act like a young lady.

When we went to school we'd look up at those little windows in the doors and see either my aunt's face or my grandmother's face. My aunt is well known in every school that my kids have ever been in, and if you are acting up, they will take you out of that room and take you in the bathroom, tear your butt up, and set you back in there. They do it to our kids in public school. And they are well liked and welcomed by the school.

When Grandma took lunches to the mine she stood away from the portal. Women weren't allowed around the mines. It was a big thing. We'd go to the post office, and my brother would walk across the tracks and through the mine yard, and I'd have to walk a half a mile around the shafts. But when I was little we had to go through the mine shop to get springwater, and there was an open shaft where they would drop the supply cars. My brothers and I would run down to where we could see the light and then run back up.

I heard stories about the company. My uncle would go to work, and he would work all day. They had to clean their own coal, and one of the men from the company would throw slate in it, and my uncle wouldn't get paid for that car of coal because it was dirty. He would come home with nothing, because everything went to the company store. My uncle would tell me about the treatment he got because he was a black man. To this day some of the men at the mine think he can't read or write because he'd be marking cars, and he'd see them write "Nigger," and he'd just go mark it and not say anything. And the guys would say, "He can't even read this. There's no sense in writing this stuff." It was his way of getting around it.

So there were stories like that, or people who got hurt from being on a picket line. My uncle was telling me that the company told you, you had to vote Republican or else. I said, "How did they know?" And he said, "Oh, they'd find out." The company controlled every aspect of their lives.

*The family structure changed when her grandmother died. Carol was then ten years old.*

I remember that day. She had sugar [diabetes]. She called us in. We were on our way to school. She called everybody in to say good-bye. She just peacefully slipped away. Of course, being ten years old, I was ticked off with her for leaving me, because she was supposed to teach me how to make biscuits. She was my biggest friend. After that my older brother raised us for a year.

There's seven of us, and only two of us have the same father. This man who was supposed to be my father lived in Pittsburgh, but he never came to see me. I don't really have any feelings for him at all. Maybe that's why I always wanted my kids to get to know their father and his side of the family.

The only male figure that I really cared about was a man my mother was married to named Harold Erby. He claimed us: me, my brother Layman, and my sister Julia. I was two or three. He knew he couldn't have any kids. When my mother would go away and stay for weeks and party and come home, he'd take her and put her to bed and never once raise his voice. And he took care of us. He had cancer of the throat, and he died. That's the man I called Dad, because he was just a nice, sweet, loving man. But she didn't like that type.

*In the early 1960s her family moved to Akron, Ohio, after layoffs hit the Ellsworth mine.*

My uncle had a job at Chrysler and got a job for the man my mother was with. I was in ninth grade, and my brother was in seventh grade. It was around 1963, around the beginning of the riots. We never experienced racism until we moved to Ohio. It was devastating. My brother and I weren't used to seeing so many black people. It was really, really scary. It sounds strange. They knew that we were different by the way we talked. They said we talked so proper. They said we talked like white niggers. My brother and I would cling to the people we were more comfortable with, white people. We were around them a lot more than we should have been, I guess.

My brother was a great athlete, and I remember him standing out in front of the gym crying because he was afraid to go try out because of the black boys. There were many days we got chased home. We got to the place where we had battle stations, where we had rocks and bottles.

My brother and I didn't understand what the problem was with racism. You had to prove that you could be just as black as they were. There was a riot in the school, and the teachers and the white kids had locked themselves in a room, and the teacher looked around and there was my brother under a desk [laughs]. The teacher said, "You're in the wrong place, aren't you?" And he said, "Oh, no. Those people are crazy!"

It's a strange thing. It's hard to talk with black people about the experience, because during the riots we were supposed to be totally against whites. When we went to grade school there was only three blacks in the whole school, me and my sister and my brother, out of about two or three hundred kids, but you didn't feel like you were singled out.

Our grandmother and our aunt taught us that people are taught a certain way and that they don't know. We spent most of our time educating the people around us that we should be all one. That's the way I raised my kids, and that's the way it's always been for us, that everyone has always been welcome.

My grandmother said you should never compromise what you believe, no matter who you're with. I didn't strive to be accepted by black people. It just sort of came on. They knew me enough to say, "She's like this with everyone." I've got friends right now who are strong militants, but I also bring my white friends around them, and they've learned a lot.

*When she was in tenth grade Carol became pregnant with her daughter Michele.*

I missed eleventh grade, but they put on my record that I was physically unable to attend, because I was such a nice girl. They threw you out of school back then if you got pregnant. My friends couldn't talk to me anymore. It was isolating. In my family it had been a part of our life, but I didn't tell my mother. I was trying to hide it. I was seven months pregnant, and he told my mother. She said if the man she was living with said I had to leave, I had to go.

I said, "I'm your child. You're going to throw me out in the street?" This man had been trying to molest me for five years. I thought, she doesn't care about me. I stayed at my friend's house overnight, and there was a big snowstorm, and we came home the next day. I said, "You can come in and meet my family." And after my friend left this guy starts on my mom about us being together. He said, "She's going to be just like you, seven kids and seven dads."

So my mother came with this broom, and she beat me until the handle was gone. I took that, and then he tried to mess with me. The day my aunt died he said, "Either put out or get out." I left. He told my mother the reason he put me out was that I wouldn't fix him nothing to eat. Thank God Aunt Rene was there. She said, "That's not Carol. She's a very obedient child."

*She started working after Michele's birth.*

I was washing dishes at a restaurant and going to high school while my mother took care of Michele. I walked from one end of the town to the other to get to my job because I couldn't afford to ride the bus. It was my first job, but I had to pay my mother for taking care of Michele and for staying with her. And I had to buy our food and clothing while I was finishing school. I figured, if I was doing all that, I could make it on my own. So I moved out.

I worked three jobs and never went back home. I was pulling sheets in a laundry and peeling potatoes, standing in water for a dollar and a quarter an hour. I cut myself down to two jobs because I wanted to go to Akron University. I took her to college with me. It worked out well for Michele, because she started reading when she was four. But I just couldn't keep it up.

Our first apartment was the pits. One room with a shared bath, and that was it. The National Guard was parked right out in front of our building. It was like a war zone. Then we moved to a friend's apartment building, and that was really terrible. The gas was turned off on us, the water was turned off, and me and Michele was going to the gas station to wash up and get bottles of water.

It got so bad that I sent Michele out to live with my aunt for six months while I found a place. I called the board of health, and they put me in a project. We were moving up. But times were hard for me and Michele. She went through them with me. She turned out to be my best friend. For 14 years it was her and me against the world.

I guess it was inevitable for us to come back home. We were in Ohio for about ten years. Aunt Rene called me and said they'd hired a woman in a mine. She said, "There's good money for you and Michele." I told her I would try anything once. I put my application in, and my uncle said, "I don't hold to women working there one bit. Don't ask me for any help." He was working at Ellsworth.

I went down to the mine office every Thursday for a year. Finally they called me, and when they did my uncle took me and bought my work clothes, and then he went to the mine office with me to make sure my safety equipment worked. He said, "I've got two things to tell you. Never run from a fire, and go to those union meetings!" In a coal-mining town, union was very, very serious. It was next to church. We were told, go to that union meeting, go to church.

That was how I got back home. It was like it was meant for me to be back here, and to be in the mines. I chose Marianna simply because there was none of my family there. They were proud of me, but it would have been hard on my uncle for me to work there.

I went in the mine in 1976. I was 28. There were already a few women there, and I knew two of them. So I did get a little coaching. My first day I remember standing there, and the men came off the shift, and they were saying, "Look what they hired now." I thought, oh God, what am I going to do? Even when I took the 40-hour training the guy who was training me said, "I'm gonna show you how they pat you down." And he starts feeling all over me. I said, "You don't have to fondle me." He stepped back. But when I first started at Marianna it wasn't much of the hands-on type of harassment, it was, "Let's see how dumb she is."

The very first day we went in the guys used every cuss word out of the book. And I just kept on with what I was doing. The next day the regular boss came back, and he gave a talk in the dinner hole. The boss said, "There's a lady on the crew, and you have to watch your mouth." I'm thinking to myself, this ain't good for me. So I said, "Can I say something? I didn't come here to change anything you say or do. As long as you don't direct it at me, I'm fine. I'm in your world." After that it was basically good for me.

But there were times I was pushed to the limit. Some old guy would rub up against me, saying, "It's good for my arthritis." Then you feel that you have to take this. It's your job, and as long as it doesn't get real bad you want to get on with your job, thinking, I'll pay my dues and they'll leave me alone. I would say to them, "Would you want that done to your wife or your sister?" I'd say, "I know your wife. Let me tell her about your cure for arthritis." Then he'd back off.

I never took anyone to the office on sexual harassment, because I felt there were ways that you could deal with it. I remember them having *Playboy* pictures. Our dispatcher's shanty had them from the beginning. They were like collectors' items. They had a centerfold up at the dinner hole one day, and I looked around and said, "Where's mine?" [laughs]. They brought me in one the next day. It was embarrassing to me, but it was my fault because I opened my mouth.

I never had some of the real bad things the other women had. I know a lot of women from my mine who took it, and I told them, "You don't have to take it." There's one guy that I hit with a shovel because he got out of hand, and I figured this was the way to handle him.

At my mine I decided to be the welcoming committee for new women, because I didn't want no other women to go through this. When the next woman came in I introduced myself and pointed out the officers and told them when the union meetings were. I wanted to make them feel at ease because they were so out of place. Then we started going out once a month for breakfast to keep banded together. It was before the Coal Employment Project.

They liked to tell black jokes. All I'd take was two jokes a day, and I would tell two jokes a day [about white men]. It got to where I got along with everyone, and I think because I really snipped it off. I had a good time with them. On the top of the ground there would be times when they'd come to my rescue. We'd go out drinking sometimes after the four-to-12 shift, and if some guy would come up to me, the guys would tell them, "That's my sister. If you've got something to say about it, let's go outside."

At first I was viewed as a woman. A few men brought out the fact that I was black. One man asked me if I prayed to a black God. But then, as the years went on, I was viewed as a buddy. It evolved after I showed them I was going to work as hard as they did. Now some guys I could talk to, they wouldn't even refer to me as a female.

*Only a handful of blacks worked at the mine by the 1980s, although black miners had been founders of the local union.*

At one time there were up to ten blacks out of more than 600 workers. At the mine most of the workers came from someplace else than Marianna. Some had retired who were brought in as scabs during the forming of the UMW. The numbers of blacks from this area really dropped off. It seems like when the pay got higher black miners did not get hired.

Years ago the very first president of our local was black. No one mentions that. And there was a black woman who came from Washington, Pennsylvania, who was a psychic. She told the men they should accept the union. This is in the archives. I started digging into the history of Marianna because I wanted to know what happened to blacks. I wanted to know, how did we die out?

Race was played low-key at Marianna. We didn't have things like cross-burnings. A couple of times the dumpers said they'd erased the word *nigger*. A group of men brought in the NAACP. They met with a group of us because another mine was having problems. Things blew up there, whereas they did subtle things to us, like talking you down. I usually considered it was because I was a woman. But there weren't that many blacks at the local, and we weren't going to stand together. So mostly I just let it slide off.

I was afraid of all men before I went into the mines. I'd had relationships with men, but I was always shy and in a corner. My younger brother and I were close, but I was really afraid of any man who was in authority. My uncle worked in the mines, and he was a drinker. And men who were with my mother were real abusive. She got beat up all the time. That's how she died.

She died in July, and I got hired in the mine the next February. I had planned on taking care of her after the beating, but she died before I could get to her. She would pick these abusive men, and my brothers would have to beat them off. My uncle would say, "Well, she likes that." But I think she probably just wanted someone to love her.

She was the old type who believed the woman was supposed to listen to what the man said. I was talking to her afterward—she died several weeks after the beating—and she told me to be sure and pay his electric and gas bills. I said, "No, I'm not doing that for him." She said, "You'll never get a man then." I said, "If it takes that, I don't want one."

*She overcame her fear of men by working with them underground.*

Working alongside of them, I was working hard to prove I could do the work. I did some crazy stuff, taking on a lot of heavy work to prove I could keep up with them. I had low self-esteem for years, and they would tell me, "You're a beautiful person." And I would say, "No, I'm just a fat ugly person." And they'd say, "No, you're not. Don't say that. You're beautiful to us." So it built me up.

In third grade I used to have big boobs, and I walked hunched over. My aunt made me walk up straight and said, "Be proud of yourself." Underground I guess I was just with the right group of guys. They built up my confidence. We've had plenty of therapy sessions down there, especially when you're in a one-on-one thing with someone.

If you see someone looking depressed, you can say, "What's the problem?" and they'll open up. It's a twofold thing. That's what I like about mining. It's not the work. The money's good. But it's that camaraderie. It's like a family. Even the wife can't get that close, because there's some things they would tell me that they wouldn't tell their wives.

Being underground is a stress releaser. You can go in there and be a regular bitch! And they'll laugh at you and say, "Calm down. What's the problem?" And you come out on top of the ground and it's okay because you don't have that big thing that you have to carry around with you. You just let it go. I've been known to go way back in the returns and scream to the top of my lungs. To me, it's better than therapy.

At first the wives were standoffish, so I made it a point to go talk to them. It was important to me to get the wives to know me, because I lived in that community. I would go to see the wives, and I'd explain to them certain things about the contract, because the men don't look at nothing but their time off and the money they get. I would explain their medical benefits to them. Then it got to where they'd talk to me about their husbands, and I'd be in the middle, holding the families together. I became a priest [laughs]. It was hard. It's hard now for any man to pull something on me, because I know every excuse. Some of the wives and I got to be real close friends. In fact, they would take me out on my birthday.

I think one of the best things that happened was that this guy told me he was having some sexual problems with his wife, and I gave him some advice. And he comes back in the mine and says, "You saved my marriage. Twenty years, and it's like a new marriage. Of course, I told her I read it out of a book."

*In 1979 she became pregnant. It was the company's first experience in the division with a pregnant miner, and little was known about the effects of underground mining on pregnancy.*

Michele was 13 at the time. I was excited about it, because I wanted other children. I told the superintendent, and he was excited, because he and his wife had been trying to have children. He said, "Work as long as you can, and take off as long as you need after you have the child."

When I was three months along my doctor did a sonogram, and I found out they were twins. She wanted me to take it easy. I took off after four months, and then breast-fed, so I stayed off an extra six or seven months after they were born.

When I was off with the twins I had no guidance about what to do. I could have collected unemployment until the day I delivered. Right after that I got pregnant again with CarMelitta. I had her in 1980. It was like, I'm pregnant again. When I was pregnant with CarMelitta I didn't know what to do. I thought about abortion. I had a good friend who said, "You're going to have this baby." She had seven kids. We had made a pact that if anything happened to either one of us, we would raise each other's kids.

The second pregnancy was completely different. We had a new superintendent. I told him, but I kept on working because I needed the money. I worked underground eight and a half months into the pregnancy.

I didn't get big. The girls didn't even know I was pregnant. When I left the mine to have the baby at eight and a half months along I told them I'd see them in a few weeks. They said, "Why?" I said, "I'm taking off to have this baby." They said, "What baby?" I don't get a big stomach. I get big everywhere else.

I had a really good boss on the section who would tell me to sit down and put my feet up. His wife had five kids, and he knew how it was. I was on buggy and doing brattice work. I maintained my same job. My shift foreman finally came to me and said he didn't want to see me back until I had the baby because they were getting nervous. I called in and told the company. A few days later I heard through the grapevine that I was going to be fired.

I went to the mine, and the superintendent said to me, "Pregnancy is not an illness." I said, "Fine, I'll come back to work, but you'll take responsibility for my unborn child, myself, and probably three or four of my buddies who are going to be running around not knowing what to do when I go into labor." He backed off. I left and came back to work six weeks after I had her. I was breast-feeding, so I had to pump it and freeze for the time I was at work.

CarMelitta was the easiest pregnancy I had, maybe because my body was built up from working. With the boys, I didn't work long enough to know how it was affecting me. With CarMelitta, it seemed like she lay dormant all the time I was underground and started moving when I got home. She's a very strong child now. I don't know if it has anything to do with me working underground. The work wasn't hard for me, and I had a good boss, which is unusual.

The guys understood, too. They were bringing clean white cloths underground, and they said they would have the baby's hardhat ready for it. When I came back they had a "welcome home" party for me underground and gave me a cupcake.

I was sort of a test case. Denise [a coworker] got pregnant a few months after I had CarMelitta. It was easy, because I had everything worked out for her. I told her she could get unemployment and union benefits and everything. She had no problems. It takes one person to work out the bugs.

*She educated others about their union rights.*

We got the union on a silver platter. We got it handed to us like it was our God-given right when people died for it. I told the guys, "You're gonna have to be busted down to ground zero before you find the significance of the union." That's why you see the pensioners so active. They're willing to do whatever they can, because they've been there. But no, it was like, damn the union, with only five or ten people at union meetings. There wasn't many women going down there. There was one or two of us, and then there was just me. I remember when my uncle would go and there would a big crowd at the meetings at Ellsworth.

For women to get involved in the union means a lot of stress. It is a struggle. There were meetings and they would not tell me. I had to find out myself. Then I'd show up, and they'd say, "How did she find out?" I think men are still up there, number one. The first time I went to the international headquarters when [Arnold] Miller was [president], I raised my hand to ask a question, and he asked me, did my husband work in the mine. I told him, "Wake up. There are women in the mine, and I'm one of them." It really turned me off. I said, "If you don't even know that you have females working in your organization, what good are you to us?"

My first position at Marianna Local 2874 was sort of thrust upon me, because they wanted to get rid of a man they didn't want. So they ran me for mine committee. I'd only been in the mine for a year and six months. I was just a local union woman attending the union meetings, so they figured they'd throw me in there, because they thought I'd just follow orders and not know what to do. They ran me, and I won overwhelmingly. I made a commitment to learn more about the contract and grievances if I was going to defend my union brothers and sisters.

Later on I decided to run for office myself. They nominated me for vice president, and it was a joke. You had three days to accept or decline. I declined. Then the guy running for it said, "I'd bet you ten dollars you couldn't get five votes." So I said, "I'm running." Me and my kids did a big campaign thing. It was the first time they ever had a campaign in our local. I had fliers on everyone's car. I beat him by a lot.

When I first ran for office, I had the worst time with the black guys. They called me "stupid black bitch." They said, "Why do you want to be used by these people?" I said, "Well, if I get in, I can do something for us." But that's not the way they saw it. On one hand, our culture keeps black women down. But someone had to get out there. And then some of them were glad I got involved. I couldn't get my brother to attend union meetings, but he was happy I was in there.

Then one of the guys resigned from the safety committee. The committee goes by set rules—the federal and state laws and the company's mining plans. It was easier for me, and it made me more safety-conscious, too. I think statistics will show that when women got in the mine things got a little bit safer. I think we have this maternal instinct to make sure everybody is doing everything right so that they can come back and do it tomorrow. And being new, we were afraid to break the rules. They used to have one woman on every crew, and the women said, "Hey, you're not supposed to be doing that."

Safety committee became the love of my life. We had pop inspections. The company was afraid of us. One of the bosses said, "Carol's like horseshit. She's everywhere." When I became chair of the safety committee, I put in a lot of extra time that I didn't get paid for. I had only been in the mine three or four years, so I'd go in to learn the mine. The superintendent would let me go in anytime I wanted to.

I went back there, and you could see the different type of mining. They used posts, and you had to walk bent over. I'd go back there and sit, and it was so peaceful. Sometimes I'd be the only person in there. I thought it was my responsibility to know every inch of that mine. I spent a lot of time there. My family didn't like that.

The first union convention I went to was the District 5 convention, and that was a frightening thing. I was the only woman out of maybe 100 men. The guys had voted me to go to recommend things for the contract. After the meeting we all caucused from different locals, and everybody had guns. It was frightening. It was all secret. They locked me in my room at night. I thought, what the hell is going on?

It was during the time of [Arnold] Miller, around '78 or '79. I asked one guy, and he said, "None of your business. You don't have to worry about this."

*She was elected to represent her local at several international conventions.*

As I got involved it got better. The convention I did not like was the [1986] special convention about the selective strike fund. I supported it, but I am a firm believer in doing what your local tells you to do. At one point the international union was going to assess 1.5 percent and cap it at $100 million. My local said, "Don't go any higher than 1 percent." So when we got to the convention, the opening prayer was, "Please let the delegates accept this package deal." It turned me off completely! I thought, they're going to shove this down our throats. Normally you vote on one thing at a time.

The international had their people lined up at the microphones. If somebody else got up there, the president would recognize another mike, and that guy would recommend debate cease. We could see it was a setup. A lot of people said, "Screw it, it's already a done deal."

*She also became active in the Coal Employment Project.*

CEP was in existence four years before I knew about it. [A local officer] had gone to a conference, and he came back and said I needed to get involved because the next CEP conference was going to be in Pennsylvania. There was going to be a meeting at District 5. I said, "I don't want to be involved in anything that's going to break us away from the union."

Another organization started up about the same time, some sort of democracy union. And they were trying to pull in the women miners. They pinpointed me to tell me how I was oppressed and said I needed to speak to them, because I was an officer in the local. They were pulling away from the UMWA. I said, "I want it all to be one and the same union." It was hard enough for women to get in for us then to pull away. But a local official coerced me into going. At that time we had been laid off the first time for six months. I took a bunch of the girls with me and went up to District 5.

The Coal Employment Project let us know that we weren't alone, that there were hundreds of women out there having the same problems. When you go in to fight these guys, you might be physically alone, but you have these hundreds of women behind you. That turned my life around. I thought, I don't have to be afraid, because I've got these women. I got really involved in the 1983 conference, which was the first conference to be endorsed by the union. Richard Trumka had just been elected, and all the women had campaigned for him. He was our keynote speaker.

When Richard first got in, I thought he was this bright young man who was real educated and said all the right things. I thought we would no longer be considered dumb coal miners. But things have changed, and I don't have the faith I used to have.

I'm worried the union movement is going to be lily-white. The only way to reform the union would be through another battle. Someone would have to be willing to sacrifice and stand up. And then you'd have to have a group behind you. Most people feel like they are alone. That's the importance of CEP. That's why CEP should be there, forever.

CEP had them nervous years back. The international was really nervous when Joy [Huitt] got elected [to District 22 office]. I felt that they thought the women were taking over. A couple of men in leadership at our local thought they created me. But they wanted me to be that dog in the back of the car that just nods its head. When I challenged them they got scared. We had them on the run back then and didn't even know it. We lost our hold.

In District 5 at one time Margi Mayernik wanted to get together a woman's slate for District 5. The leadership got nervous, because Margi was practically running the district. I should have listened to her. But I had low self-esteem. I think I'm not good enough to take that next step.

But it's time to do something. I think we have the women who could do it. My aunt used to say, "Do something, even if it's wrong!" To let them know that we're still here, and we're not going to take any crap, regardless of whether we win or lose.

In any industry that has nontraditional jobs for women, you should be able to sit down and talk to your own sex about what's going on. Sexual harassment was really big at that time, and women needed to know they had a recourse. The UMW don't know how to face a woman being harassed. They turn their backs. They say, "That's a union brother you're talking about." So women had to take measures into their own hands. Most times you had to go through an agency outside of the UMW, which we shouldn't have to do after all these years.

When I got into the union I took an oath to uphold my union brothers and sisters. That means something to me, and it will mean something to me until I die. The men take the oath to help a sister in need, but when the sister is in need they won't help you.

*Through CEP women miners developed a national organizing campaign for parental leave.*

I got involved with the parental leave campaign because of all the problems I went through in my pregnancy. We also heard women at one company had to sign a paper saying they had to notify management the moment they knew they were pregnant. You shouldn't have to do that.

CEP called a meeting to discuss issues in the upcoming contract. I started wondering how many other people were being threatened with being fired over a medical leave. There were a lot of guys whose wives had left them with children. At the meeting we were coming up with things that we might want in the contract. First we said we wanted maternity leave. Then a few of us said, "We don't want the guys to think we're fighting against them. Let's let the men have a leave, too." So it was maternity-paternity leave. We worked on it for hours. And we drew up what we thought would be good.

We got a meeting with [UMWA Vice President] Cecil Roberts. He told us the problems his wife had had during childbirth and the hard time he had getting off work to be with her. It snowballed from there. He told us how many people we needed to send in resolutions before it would be honored on the convention floor. All of us were from different districts, and we sat down and came up with a game plan.

We decided to hit all our districts and all our locals to see if we could get resolutions in. It was a contract year, too, which was great. We had the highest number of resolutions ever sent in. They flooded the international. The union had no choice but to address it. We got a memo of understanding in the contract. Then they set up a committee to study it, which I had the good fortune to sit on.

[The parental leave campaign] taught me that the average person can change the course of what happens in the country. Getting the family leave bill passed took years, and we went through it year after year after year, but we hung in there, and finally it was signed. What disappoints me is that women miners never got recognition for starting it. But we know. Every time somebody takes a family leave, I know we started this bandwagon rolling.

*Her children were affected by her activism on social issues.*

When you live with a union person you pick it up. CarMelitta is into saving the environment. John wrote a letter to President Reagan about the ozone. Litta has been known to set up boycotts at the store about the grapes and the farmworkers. She would station herself by the grapes and tell people what was going on. James is quiet but is for American-made things. John and James had been to picket lines. They had sat in front of trucks and enjoyed that.

This year, during the coal strike, one of the guys took [CarMelitta] and some other kids to a rally. She was so excited, and she wanted to stay. I hope they continue to have these ideas.

Michele is my political person. At the age of six, when she was watching the Democratic convention, she called to donate her five dollars in the bank.

*On March 7, 1988, a fire broke out in the Marianna mine. No one was injured, but the mine eventually closed, leaving several hundred miners without jobs.*

Two weeks before the fire the federal agency said everyone had to practice with the SCSR [self-contained self-rescuer] and put it on before they went underground. As chair of the safety committee, I had to be there to physically witness it. A lot of the old-timers said, "Carol, just sign my name." I refused. We argued back and forth, and they finally put it on. The night of the fire, a lot of them thanked me for badgering them so much.

No one ever would have dreamed this would happen. It was like a nightmare. There was a big coal spill on the belt, and it caused the belt to drag down. The friction from the belt caused the fire, but there were a lot of things leading up to the fire. The company responded to a belt flooding problem in a way that limited use of the fire suppression system.

When the belt caught on fire the dispatcher called the guy on the dumper and said, "You get those people off the section, and get them the hell out of there." The guy said, "By whose authority?" He knew the place was burning. So the dispatcher said, "By my f—— authority. Get them out."

And they got everyone out. It was a miracle. There was a full shift in there of 60 or 70 people. Usually in a fire that intense someone dies. A good friend of mine was in there, and she said some of the [cinder-block] stoppings glowed cherry-red with the heat. Some of the overcasts [overhead metal airways] were melting and falling around them. Some bosses got lost in the smoke in the return entries. Everyone was disoriented, but the ones who lost their heads had somebody with them who guided them out. There was one guy who had a bad heart, and they carried him out. Some guys helped other people put their self-rescuers on. The strange thing about it was the people in the mine who were the most levelheaded were known as the craziest people in the mine.

Before MSHA came in, our mine foreman had a command center set up. He knew every inch of the mine. He was telling them how to fight it and what was going on. When MSHA took over, it started going downhill. They were sending people through the wrong door to take air readings. The foreman knew they weren't in the right place. Finally he went under, but the fire got so bad it pushed us all out of the mine. Then MSHA proceeded to drill holes in the top, and they were dumping some kind of foam to smother the fire. It didn't work, because there was too much air.

I didn't see my family for a month. They had set up a command post in Marianna at the carpenters' shanty. As safety committee chair, I felt I could not leave there. I felt responsible. It burned for a year before it was out, and we lost the mine.

It was like your whole life was pulled out from under you. The first year our local really banded together. We did food banks, we did entertainments. We were getting unemployment. They had job postings and a job fair set up for us. There was a lot of grabbing at straws. We had people signing up for the union's educational fund. But a lot of people didn't take advantage of it, because we thought we were going back to work. I thought like that. After six years people still think they'll get their jobs back. Everyone's holding onto hope that the mine will open up again. That's why every coal miner you see that's laid off says, "We're going back tomorrow."

Coal mining is so hard to leave behind. People say coal dust gets in your veins, but I think it's more than that. Your life depends on that other person. They don't have time to think if you're a woman, man, black or white, because when that roof's falling down it's falling down on everybody. And when you see them after a layoff, it's like old home week. Hugs and kisses and all that.

And I believe that's true for coal miners until the day you die. In me there is yet a glimmer of, someday I'm going back to Marianna. There's all this coal left back there, I'm going back home. That's my home.

*The mine did not reopen after the fire. In 1992 Carol went to work slicing meat in the deli department in a local grocery store. She quickly became active in the local union of the United Food and Commercial Workers.*

The shop steward came to me after I'd been there about three weeks and said she was going to step down and wanted me to take over. I hadn't been there that long. But since I had been in the UMW she thought I could handle it. And at that time I had no phone, no car. But they helped me out.

They only had union meetings during contract time. I am trying to change that. These kids don't know anything about unions. They've had more grievances since I've been there than they've had in 12 years. I felt sure the company would let me go within 30 days. On the written interview I wrote about my union experience.

I started out at this job at $4.25 an hour. Even now when I get a check I think, I made this in one day in the mines. At first it bothered me. I had to work and still get food stamps.

Once I asked for the day off for the miners' holiday to celebrate the eight-hour day. The deli manager asked why, and I told her. She said, "You're no longer a miner," and wouldn't give me time off. So I wore my camouflage [a sign of UMWA solidarity] to celebrate anyway. I was serving meat in my camo. The store owner looked at me, and I explained the holiday to him. He said, "Oh, I thought you were going to war." I said, "Well, in a way, I am." I felt sure they were going to send me home.

*Her coal-mining experience continues to be a source of inner strength.*

It gave me so much confidence. Working with the guys let me know I could accomplish the same things they could. And they found out I could accomplish the same things, too. Being in the union gave me a place I really belonged. I will always be a union person, whether I'm a miner or not.

I think things happen for a reason. I think the reason I was put at [my new job] was the young people who don't know about unions. Maybe I can educate them and show them unions are a better way to go. If it wasn't for me being with the miners, I don't think I would have ever realized my place. That's my place—to educate people about the union and to keep it going.

*She believes women miners' contribution may be downplayed by the industry and the union, but that their legacy will be passed on to their children.*

I think it was important. I hope that people remember, because some men would rather sweep it under the rug than give us any credit. We had a subtle impact. If anything, I think it's straightened men up for a while. I think we were scary to them for a while, but with us laid off now, we're not that much of a threat. Women who are in there now are a little afraid, because there's not so many of us to back them up. The union did a number on some women who got higher up. They put us back where they wanted us.

We don't have any young women coming into the coal industry. We needed them to take our place, to remind them that we're here. We could have made a better union. But a lot of women miners are raising union kids. Most of the guys didn't think that was important for the family. The men treated their union membership like a secret society. You wouldn't hear much of what went on at union meetings. But we had to sometimes take our kids to union meetings, because we didn't have child care. You didn't see many kids on picket lines when there was just men in there. They were trying to protect their families from that. So maybe, inadvertently, we're getting back on track. But it's taking a very long time.

# 21

# FOR THE FUTURE
## Joy Huitt

*"Being the first is always the hardest. But in the long run you have to say it's worth it to lay the groundwork for the future. If I had to do it over again, I would do it. You just can't let people beat you down, even if it takes a heavy toll."*

*Joy Huitt started working at a Utah mine after toiling at a low-wage job at a nursing home where she was denied time off for a family vacation. Her husband, a miner, supported her decision, and they found their marriage strengthened by sharing the same occupation. Huitt campaigned for a district union office as an independent against an incumbent slate and in 1986 became the first woman in the history of the UMWA to be elected to district office.*

*The flush of victory soon paled as Huitt faced a battle larger than the campaign: dealing with internal union politics. The problems proved devastating. After losing her reelection bid by a narrow margin, she filed a compensation case based on the stress caused by alleged harassment by other union officials.*

My father and grandfather were coal miners. My grandfather worked in the Alabama coalfields, where I was born. When the coal industry went down before the Second World War, my dad hopped a freight and went to Illinois because he heard they needed coal miners. There wasn't any work in Illinois, but some coal company was paying train fare on to Utah.

Most of my early memories are of East Carbon, Utah. We had a lot of Hispanics and Slavics, proud people who held to their traditions. We grew up with stories about the union just like we grew up with Mexican food. Like a lot of people, we had a picture of John L. Lewis with a picture of Franklin D. Roosevelt in the house. I heard the labor history stories from my grandfather about how people had been thrown out of company houses during strikes and forced to live in tents through the winter. We heard about the strikebreakers and all the sacrifices our grandparents made so that we

Joy Huitt

would have more security. I tell my children these stories. You've got to know what has been planted for you, and that you have a responsibility. You can't take lightly the sacrifices that were made for you.

When we were growing up Labor Day was the biggest event of the year. If there was a baseball team formed, it was a UMWA team. And at Christmas the union had a Santa Claus that would give children sacks of candy. The whole community was centered around the UMWA.

My father was a great union man. When we were young he'd tell us that when we grew up and married coal miners we would understand why the union was so important. I did marry a coal miner, and then I became one, and I know what he meant. Our good living and our retirement benefits weren't just handed to us. [My husband] Red has worked 15 years at the Kaiser mine, and we raised six children on wages negotiated through UMWA contracts.

*She entered the mines in 1978, at age 43.*

I'd been working as a licensed practical nurse in a nursing home. Conditions were terrible. You had no time off. We worked rotating shifts, and we were on duty a lot of weekends and federal holidays. Management kept us shorthanded because they knew we were too conscientious to neglect those old people. I started out at five dollars an hour, and after four years my wages were only up to $5.80. In that time I never got a two-week vacation, and most of the time our family didn't even have weekends together.

That really got to me. I had noticed other women going to work at the mines. But I didn't know how to bring it up with Red. Luckily he had worked with the first woman miner to get hired in Utah, and she was such a good worker that he couldn't say enough good things about her. I thought, if she can do it, I can. And I knew life would be a lot easier with six kids if we had two paydays like Red's coming in.

When I did talk to him, he wasn't too sure about it, because I'd never done manual labor. I'd been a waitress, a cook, and a nurse. You can argue that those are hard physical jobs, but it's not like working on equipment or doing farm labor, which a lot of women miners have done all their lives.

But I still thought I could do it, and when it came time for our summer vacation and the nursing home refused to let me off for two weeks, I quit. Before I went into the mines, though, I sat down with each of my kids individually and told them that my working was no different from their dad working in the mines. They were surprised but supportive. Being teenagers, they didn't need me around the house, and they were happy about the extra paycheck. But it scared them a little. It's more threatening to children to think of something happening to their mother.

When I took the job I was 43 years old. I was one of the older women. There are a lot of younger ones now, because they see it can be done and they can make a good living. When I went in at Price River Coal there were four women working there out of about 160 workers. People couldn't understand why I wanted to do it. The bias against women miners was everywhere then. The public feeling was that you wouldn't go in a coal mine if you were a lady.

Like every new miner, I was put to work shoveling loose coal around the conveyor beltline eight hours a day. I hadn't held a shovel more than a few times in my life. I got blisters and backaches and bladder infections, but it didn't take long for my body to toughen up.

*Her mining job strengthened her marriage.*

With both of us in the mines, we had an important part of our lives in common. We had the children, of course, and our day-to-day life together, but you get closer when you share work like coal mining. It gave Red a new respect for me. And it gave me a much deeper understanding of what his years in the mines have meant to him.

After I started in the mine Red was even more gentle with me. He tried so hard to make things nice for me at home, because he knew what I'd been through during the day. We didn't have that much time together. I worked graveyard shift, and he was on straight afternoons. But we had mornings and weekends, and it was what you would call quality time.

Coal miners are a tight-knit group of people. You work so closely together, and you realize that your life might be in the hands of your coworkers in the blink of an eye. As a whole, the mine functions as a very close community. Even though people are working throughout the mine and aren't in direct communication, news travels fast.

On a crew it's like the group becomes a single person. Some wives have trouble with their husbands' developing such a close working relationship with women. It's not sexual, but a type of friendship that they can't share. Before women went underground that bond was only formed with men. When it exists with women it becomes a different matter to the wives, and harder for them to deal with.

In the beginning women were not encouraged by the local officers to go to the local meetings. They let us know they wanted us to pay our union dues and shut up. But women miners are not the type to be told to keep their mouths shut. Some of us thought, if the men don't want us to come to their meetings, we had better go find out why. We had seven women working at Price River, and we always had two to four women at union meetings. The local's attitude changed then. They saw that we were willing to work. But it's a slow process. We were just starting to make progress when the layoffs started.

I think women have had a real positive effect on the union, considering our numbers. We never made it over 2 percent. The union saw how successfully we organized around the parental leave issue, and they respected us.

*Sometimes she worked alone on midnight shift as a "fireboss"—traveling the mine to check for methane gas around the abandoned sections.*

The sense of danger is always with you, especially if you're alone. I'd be there alone for hours. It's strange, and kind of pleasant. You have a lot of time to think. Sometimes it gets eerie. You will see a reflection in the groundwater and think it's something else. One time I climbed a ladder and there was a big rat staring me in the face. Or you'd open a man-door and bats would fly right at your caplight.

Around the old works where I firebossed, I noticed that the roadways were staggered. The guys said that in the old days they built them that way because the horses pulling carts of coal would go too fast if the roadways were straight.

You get frightened at times. One time I was alone, and the roof started falling in behind me. It sounded like a loud shot, and a big section fell. The concussion hit me so hard it nearly knocked me down.

It is unfortunate that so many miners don't understand the safety protections they have under federal law. That is why I got on the union's safety committee at our mine when a vacancy came up. It takes a lot of time, but it's worth it, because management is always pushing production and a lot of people don't know better than to take the risk. The union has to take the burden of teaching and enforcing mine safety, because the company's sure not going to do it, and neither is the government, which supports big business. We've seen the tragedy that kind of attitude can bring.

*In December 1984, 27 people died in a mine fire at the Wilberg mine near Joy Huitt's home in Utah.*

The whole world stopped when Wilberg happened. I was in Salt Lake City, and my son heard it on the news. I panicked. I knew people who worked there, and I started frantically trying to remember who worked on what shift. Two of the Wilberg women were very good friends. They were in our women's support group. I couldn't get hold of either of them, and the television was giving such spotty information that I jumped in the car and drove three hours home. When I got there I found out that one woman was trapped, and then I was really frightened. Not that it was any different for a woman to be in there, but some were good friends. I finally found out that it was Nannette Wheeler who was trapped with the others. I had met Nannette, but I didn't know her well. She was a kind, quiet person.

You feel so helpless, so lost, in that situation. You don't know what to do, but you have to do something. Miners in East Carbon have always congregated at the post office, so that's where we went first. Then we went to the union hall, and then to somebody's house. But you want to stay together, with other coal miners. There was a big snowstorm that night, and it was hard getting up to the mine. As the hours went by we knew they probably couldn't make it. But you never give up hope. When word came that rescue teams had reached the first bodies, we knew the others couldn't have lived. But we still waited and prayed. They were very young.

Our women's support group immediately went into action to help by providing food to the rescue teams and arranging transportation for the families flying into Salt Lake. We were on call 24 hours a day. We wanted to be there when the families needed us, because we knew how hard it was for them trying to get down to Wilberg during Christmas week. After we put out word in the community that hot food was needed for the rescue teams, the retired miners' wives sent up homemade tamales and roasts. The women of the Mormon church really responded.

The church has such a strong influence in this area, on both the people and the coal industry. The Mormon elders don't believe that women should work outside the home. Its anti-union position is something the church elders don't mind voicing. They feel that workers don't need a union, that you should be grateful to the company for giving you a job. . . The families [were made to believe] that the Wilberg disaster was an act of God.

I don't think it had anything to do with God. It was an act of greed that 27 people were killed while the company was pushing for a world production record. The mine was designed for maximum production. They were using a system that, under the best of conditions, restricted the possible escape routes. The escape routes that did exist had been blocked by roof falls, and the company obtained a waiver to go on producing coal. With those hazards, the company decided to go for a world record and sent a lot more people down there than they would have normally. There was one way in and one way out, and when that entry was blocked by smoke it was just a matter of time.

*She was laid off in 1984 after five years underground.*

When I got laid off I was shocked. I thought I'd be called back. Everybody does, I think. But when it doesn't happen, you get a devastated feeling after a while. My husband was still working, so it could have been much worse. But you feel empty, like you're not contributing.

I went back into nursing for a real short time, making six dollars an hour. It was difficult going back. Once you've been a miner and had a taste of being in a union and making good money, it's very hard to accept what you had before. The women knew that I was pro-union because I had worked in the mine, and they'd say, "What we need is a union." And I said, "Yeah, but you've got to pay a price for it. Nobody's going to walk in here and hand you a union." And they weren't really willing to pay that price, partly because women are conditioned to accept conditions instead of standing up and fighting them. I couldn't handle that situation, so I was pleased when I got the opportunity to work as regional coordinator for the Coal Employment Project.

Through my association with CEP, I've met nonunion women, and it seems like they work under great fear. They're not really free to communicate. They have to be so careful about what they do so that they're not implicated for trying to organize a union. The companies often rotate the women on different shifts to keep them separated. They have no job protection. They can't relax. There is a lot of turnover. You make money, but you don't have any rights.

If we had a problem with sexual harassment, we'd go to someone in our local, or to the district or international union. There are channels, and eventually someone is going to hear you. If you try that in a nonunion mine, you don't have a job the next day.

*In 1985 she traveled as a representative of the Coal Employment Project to visit mines in India, Great Britain, and Ukraine.*

India was a devastating experience. Conditions were horrible. Women and children worked like animals, handloading coal and pushing the loaded carts to dump into bins. You can't breathe because of the dust, and I saw no real roof supports. Some of the Indian women were so poor that they wrapped themselves in rags to work because they only owned one set of clothes. They live in shacks with no heat. The mine we went into was supposed to be one of their better ones. If that's true, I cannot imagine what the fatality rate must be like.

The mines in England were a lot like those here, except that they are more advanced in their technology. What surprised me most on the trip were the mines in Russia. Conditions were better than in any coal mine I had ever seen. Everything was extremely clean. In some mines they were using video monitors, the first I had seen. We were in the Ukraine, and I know it wasn't one of their showcase mines, because they didn't want us to tour it.

In Russia they have established a national institute for mine safety where young men work full-time to improve their training in mine rescue techniques. In this country our mine rescue teams can only train in their spare time because they're also working in the mine. In Russia they are willing to use mine safety technology because they are so desperate for coal. If we were less dependent on foreign oil, I think we would be running safer mines.

There are no women working underground now in England or Ukraine. In England women and children were taken out of the mines in the 1800s because they were being used as beasts of burden. Women worked in Ukraine during the Second World War, but they lost their jobs after it ended, and now they only work outside, in coal preparation plants. A Ukrainian man explained to me that working in the mines made women sterile and that's why they had to leave. I told him I'd like to introduce him to a few pregnant coal miners in America who could disprove that myth!

I will never forget the toast given later by one man who presented me with a set of babushka dolls. He said, "May your grandsons grow up to be strong trade union men. And may your granddaughters grow up to be women of peace, as women of the world have always been."

*She returned to the United States and helped organize an annual national conference of women miners in Utah.*

It was exciting for us to see women in this area finally getting some recognition from the community. It was a major news event here, with all these women coming in from all over the United States. Some came from Canada. It changed a lot of attitudes, and I think it opened the eyes of some nonunion women who came and saw how the UMWA backed its women members.

When the Coal Employment Project first came to Utah to help set up a women miners' support group, 100 women showed up. I think so many showed up because we are so hungry for stimulation out here. We set up the group, and that helped us lay the groundwork for organizing the conference.

*Chairing the conference gave impetus to Joy Huitt's decision to campaign for UMWA district office in 1986. The following comments were made during an interview in the midst of the campaign.*

As far as I know, I'm the first woman to run for district office in the union. If I win, it will definitely be a first. Right now I'm leading my opponent, which is pretty good, because I'm running independently against a popular slate and none of the other slate candidates has any real opposition. So they're aiming all their guns at me. You never know what's going to happen in a campaign. But I believe I am going to win. You have to believe in yourself, or you could never ask other people to believe in you.

In the West, [political] slates don't carry as much weight as they do back East. That's helped me a lot, and I think it's surprised people in the East that I could do as well as I've done as an independent. Out here getting votes depends on how well you're known in the local unions, so you have to get out to meetings. That's not easy, because our district covers four states from Wyoming to Arizona, and the election is in February. Even if you've got the gas money to drive hundreds of miles, you never know what driving conditions will be. It can get rough in winter.

I'm running for the office of district secretary-treasurer. My duties would include handling money, keeping records, and helping pensioners with their problems. Our district represents a diversity of members—Navajos, Mormons, Mexicans, and miners from the East who migrated here during the coal boom. You need to learn about a lot of different cultures if you're serious about representing their interests. It's been rewarding, but it takes effort.

Many of the Navajos don't have telephones on the reservation, so you have to work harder to get in touch with them.

The Navajo people pass their property and money down through the mother's side of the family, so the women have a lot of influence. The women miners gave me a lot of help down there. Women usually don't have such a strong influence. On the reservation especially the women who were aggressive enough to get a job in the mines were listened to quite a bit.

I've often wondered why women haven't become candidates in these elections before. I don't think they lack the desire for elected office, but they hesitate to commit themselves to a campaign because of the time involved and the money. It's a tremendous expense. And when you're unemployed like a lot of us, it's money you don't have.

But I've been lucky. I've been laid off for two years, but my husband works at another mine, and he's behind me 100 percent. With his support, I'm running with a full head of steam. He is my campaign manager. We've been married 30 years and have six children and six grandchildren, and we've had a very stable family life. He has encouraged me in everything I've done, and without that, I don't think I could take on something like this.

He has a feel for how the men think. The women miners have been instrumental in giving support and ideas and in traveling to different areas and testing the waters. All my kids are involved. My son-in-law is a miner, and he was one time the president of a local, and my brothers-in-law were all involved in different locals, and I had a lot of friends from different areas.

*Huitt found support from union members who wanted reform in the district.*

I had grown up in the union, and I knew what the UMWA had given to my family over several generations. I wanted to give something in return, because it's been such a part of my life. So I came by it naturally, the desire to build the union. It takes more than desire, of course. It takes confidence and the skills to do the job. I don't think I would have been prepared to tackle this campaign without the support and training I've had from the Coal Employment Project. Through CEP I've learned administrative and accounting skills, helped put on a national conference, and traveled overseas.

CEP has worked to provide opportunities like that to women miners. Without our organization, women would have had a lot further to go in becoming a recognized force in the union and in the coal industry as a whole. Having the women miners' conference in Utah was a great help in the campaign, because as chairperson of the conference I got a lot of exposure. And the union's participation showed the men that the UMWA was behind its women, and that helps any woman who decides to run for office.

I had a base in Wyoming through my son-in-law, who lived there and was acquainted with leaders of some of the locals. They were happy to see a woman take on the responsibility of the job. They felt an outsider could pick up on things that were amiss. They felt there were a lot of things that were wrong in the district. They felt if I was willing to be the first woman to run, they thought I would be willing to push for needed changes.

When I think about it, every step has been exciting. Going underground changed my personality because it broadened my horizons. I found that I could do things I would never have dreamed of. If I'd stayed in nursing, I would never have thought of running for political office. I would never have organized a national conference. I would never have been involved with the feminist movement. I would never have gone overseas. My life has been enriched, and I am very thankful.

*She won the election but found that the battle had only begun. In a 1994 interview she was disillusioned.*

In the beginning the membership was very supportive, but not my fellow officers. They let me know it the night of the election when they refused to shake my hand. I felt they undermined me constantly during the four years I was in office. If there was an important meeting, I would not be sent. But if it was not important, or a lot of work and not a lot of prestige, they would send me. There were doubts they planted in people's minds about my ability. They wouldn't say things directly. They would insinuate things about me, as if they should investigate the books.

One week after the election a secretary retired and the bookkeeper took her place. The district expected me to do the extra work. Along with that, they made the workload so heavy there was no way to keep up. Closing out the books and sending the money back to the locals is a full-time bookkeeping job. I thought I should do the job as a way to save the district some money. But they expected me to continue to have the workload that the other secretary-treasurers had carried in the past, and it became overwhelming.

Then the district remodeled the office, and they placed my office in front, to deal with any walk-in problems. The other offices were down the hall, with a back door locked from the inside. If problems came up, the other officers could duck out the back. Handling these complaints put my other work on hold, which increased the pressure. I came in and worked on my own time, and I was refused the use of a secretary to help with this additional work.

I wanted to update our records and put them on computer instead of the index cards we were using. They hired a computer programmer, who said it would take five years to put these records in. He put only part of them in, which meant I had to work from two sets of records, which made things worse. Gradually these things chipped away at my self-confidence. By the end of my term I was losing sleep and having trouble functioning.

One time I was directly threatened by a union official, who told me he and some others would destroy me. I began to be fearful of traveling the long stretches of road to Wyoming and Arizona by myself. The international union audited the district books for the first time in years, and I overheard comments to the effect that officers were looking for any discrepancies they could pin on me.

I was also set up at local union meetings. I would be sent to a hostile local. They were angry because they had not received their money, and the district had blamed me. I showed the local union proof that the company had not sent in their dues money, but it was hard for me to go to that local after that.

Then the district president humiliated me again when I had an invitation to be interviewed by the British Broadcasting Corporation (BBC) on behalf of women miners and the union. At the last minute he canceled the interview because he felt I was not capable of representing the union. They didn't even notify BBC, which made me look incompetent.

At the same time all this was going on, Kaiser Coal shut down. My husband was local union president, and we lived in the community with all the miners from that mine. People were calling us night and day with their problems. We formed a women's auxiliary to help, but I received no support for that from the district, since the district officers weren't interested in women getting involved.

Toward the end of my term I was extremely depressed, and it was difficult to concentrate. As reelection time approached, the problems mounted. Nothing was the same as the earlier election. In order to run you need a letter from your local union stating you are a member in good standing. District officers are exempted from having to attend local meetings, but the local officers who were in tight with the district refused to sign my letter. I won this battle after a hard fight and then was not informed about the meeting when local presidents would declare who they supported. One local officer told me about the meeting. When I got there a district official said they didn't want me there and would not support my reelection.

*In 1990 she lost the election when a tie vote was broken with challenged ballots that favored her opponent. After intensive therapy over several years, she filed a compensation claim against UMWA District 22 for stress-related illness.*

From the minute I took office I felt that the people who were afraid they were going to lose power immediately started undermining me. They wanted to tear me down. Maybe you have to be stronger than I was. I'm not trying to discourage women from running for office. It's important that they do run. But people have to really count the costs before they decide to go into an elected position. Unfortunately for us, the people who are in power are often more concerned about themselves than they are about the union as a whole.

And I had no real political base. There was my family and the women miners and a few other supporters. The miners wanted a change so desperately in that office they voted for me. But as soon as the officials started undermining me, that support backed off real quickly.

I am a very strong union person, and I realize it's the only way to stand up. Then I see that it's not what you think it is. The inner conflict was the most devastating thing. The union needs to go back to serving the common good. Our union is in for a big fall, maybe a complete fall. It breaks my heart to see what is happening with the United Mine Workers, because I have always been so proud of the fact that our union was clean and free of crime. But what they consider the little things, like destroying a person's life, doesn't seem important to them.

*Despite the political problems, Joy Huitt believes her experience had some positive results.*

There are a lot of good things that happened. There was enough support for a woman to be elected. I paid a high price in being devastated mentally. I became very disillusioned about the union. I lost a couple of years out of my life, where I had to regain my balance. I have a stronger relationship with God, and with my husband, because I had to depend on them.

But for a person who's going into it, I would say, yes, it's worth it. You have to do it. It has to be done. Being the first is always the hardest. But in the long run you have to say it's worth it to lay the groundwork for the future. If I had to do it over again, I would do it. You just can't let people beat you down, even if it takes a heavy toll.

# 22

## THE WORK OF HOPE
## Kipp Dawson

*"Even though it's a world unto itself, the mine is a world that's very much affected by what's going on aboveground—for example, by economic and political changes."*

*A woman of high energy and with an expansive vision of social justice, Kipp Dawson worked more than a decade in a Pennsylvania mine. She dug coal, but she also mined insights about the work and the history of women in the United Mine Workers of America. For women miners active in the Coal Employment Project, petite, curly-haired Dawson has been an enduring presence and a buoyant source of support and activism.*

*She was born into a family that took political risks and paid the price: her maternal grandfather was lynched for his activities in Erie, Pennsylvania. The family's political involvement was balanced by its love of music, literature, and history, which Dawson shares.*

*After her layoff Kipp Dawson intensified her research into the history of women in the UMWA, went back to college, and now works as a day-care teacher. She lives with her partner and two daughters in Pittsburgh, where she gardens and has begun probing the political history of her immigrant grandfather.*

My family's history is relevant to everything that I do and that I think about. My family is Jewish and working-class. My grandparents immigrated to the United States from Poland and Russia, after the 1905 revolution, when Russia was crushed. They were politically active on my mother's side. My mother's mother and her husband moved to Erie, Pennsylvania, after a long [trip] to the United States, where they were very active in the Jewish community and in organizations that supported the then-new Russian revolution after the First World War. They were also very involved in organizing support for miners who were on strike in the early 1920s.

Kipp Dawson

At that time there was a real right-wing, xenophobic, anti-Semitic, racist wave of terrorism in the United States. The Ku Klux Klan was in its heyday all over the country, not just in the South. In the area in which they were living after World War I there was a real crackdown on people who had opposed the war, and my grandparents were involved in activities that were not looked on positively by the United States government. The U.S. attorney general, a man named Palmer, organized massive roundups of immigrants and deportations. In the area in which my grandparents were living, the League of Women Voters was involved in organizing anti-immigrant activities. Lynching wasn't just happening to black people. In 1924 my grandmother's husband was lynched for his political activities.

The ideals for which my grandmother and her husband lived have been alive in our family ever since and have been part of my heritage, my goals: being a part of society's attempts to make life better for common people, and having a particular respect for ordinary working-class people all over the world, a sense of common humanity, and an optimism that the human race can move forward and that people don't have to be subject to the whims of wealthy people, who make life miserable for the rest of us.

I grew up in a household where those ideas were very, very much alive even though the social atmosphere when I was a child was not conducive to that. I was born in 1945, between V-E and V-J Days. I was not a postwar baby. My mother followed my father around. She was very involved in the war effort herself. He was in the army in California, and she drove jeeps for the army and followed him. In an attempt to get pregnant, she says. She was trying to have me. And she was successful.

But my mother also was a Rosie the Riveter during the war. She worked as a machinist. She was very proud of that kind of work and in fact stayed in it when the war was over, even as all of her successive children were born. She was also, like my father, very involved in her union. She was part of the steelworkers' organizing committee. Through that experience she really learned a great deal of respect for the United Mine Workers. I grew up in a house where John L. Lewis's picture was around, and UMWA was heroic, and miners were considered the salt of the earth and not just coal miners.

My mother was in the longshoremen's union most of the time when I was growing up. She worked in a warehouse making Halo shampoo for Colgate Palmolive, and she was a shop steward and a delegate to union conventions. I grew up attending union meetings with her and being on picket lines with her at six years old. I was the oldest, at that time, of four. About a year later my mother married my stepfather, who was also involved in her union. He was black. This was in 1952, when such things did not happen. They courageously set out to build a family in an atmosphere that was not very conducive to that.

Simultaneously my mother was also active as a socialist. At that time she was a member of the Communist Party. Again, in 1952 that was a rough thing to be. We were also quite poor. We lived in a housing project. But we never thought of ourselves as poor. That was the great thing about those housing projects. Everybody had the same income that we did, and so we were part of a huge family, largely black immigrants from the South who had come to work in the shipyards in the San Francisco Bay Area during the war and who, like us, found no place to go for housing after the war. So we stayed in these projects until '56.

So we had a very iconoclastic family, but a family that was full of warmth and enthusiasm for life despite incredible obstacles. The FBI came around harassing my mother all the time when I was growing up. They harassed our neighbors. My mother was the kind of person who inspired confidence and loyalty in her friends, and her apolitical friends and neighbors stood up for her to the FBI. And yet we learned that we had to be discreet about our politics, what we believed in, and even our family makeup at times, because to be otherwise would possibly cost my mother the custody of her children.

One of my biggest memories of growing up was in 1953 when Julius and Ethel Rosenberg were executed, because I thought of my mother and Ethel Rosenberg as being so similar—brave women who stood up for what they believed in and got executed for it. I was really frightened that if she could go, so could my mother. My mother was a little frightened of that, too. But despite all that, my mother (who was a really tiny woman, smaller than I) and her mother (who was also tiny, and whose husband had been lynched for what he believed in)—they were my examples of pride and confidence and standing up for what you believed in. Strong women who believed to the depths of their soul that the human race was good, that if people had a chance to live their lives in a way in which their inner traits could come out, they would show themselves to be remarkable, wonderful. And that it was our *honor* to be a part of whatever we could to make the world the kind of place where that kind of human potential could come to the fore.

Our house was always full of music, different kinds of music. My mother loved classical music, my stepfather loved jazz. Folk music was something else my mother liked. We all listened to it and sang together. The house was full of books that you could [read to] escape into a world you would never know anything about and in which your imagination could soar.

My mother never finished high school, but she is one of the most educated people I've ever known. She loves to read and talk to people and explore things. So that was encouraged in all of us kids.

The thing that began to break the ice for us when I was growing up was the civil rights movement. In the late 1950s, when I was in junior high school and high school, black students in the South had the bravery to sit down at lunch counters at Woolworth's and Kress's, which had been off-limits to blacks. They got arrested, and they made national headlines, and all of a sudden the world opened up for all of us, and it became really clear to me that things could change, and these people were showing us how to do it.

My mother and I—and some of the other kids, too—started to walk the picket lines at Berkeley, California, around the Kress and Woolworth's stores, in solidarity with the students in the South. I began to feel my own wings there. I was 14 in 1959.

When I went to high school that year some friends and I started the Civil Rights Club. It was called the Anti-Anti Club at first, and then we changed the name to Students for Equality.

We were just soaring. The McCarthy thing was crumbling, and we could say what we believed in. Three of us—two young black women who were my best friends and me—formed the core of the Civil Rights Club. And it really took off in Berkeley High School. We had demonstrations, and I won a "good citizenship" award and a trip to Washington, D.C., for starting the club. It was a great experience, and it was something my family did together.

The same year, 1959, something else very significant happened. The Cuban revolution happened. At first even the United States government saw the overthrow of Batista in a positive way. This was a very poor and backward country where people had been poor and downtrodden for so long, and where they had stood up and overthrown a dictator and taken things in their own hands. And soon, in the name of socialism—which was the hidden, secret word we weren't allowed to say—they became an inspiration to people all over the world. And part of the civil rights movement looked to Cuba for inspiration.

To me, all the old barriers were breaking down, and I began to feel that all of us poor people and working-class people were going to be able to do what we'd only dared dream about before. And my mother and grandmother encouraged that in me, too, although my grandmother was more frightened, because she had been more affected by the repression she'd lived through.

Those were my roots. And at the age of 16, I graduated from high school and went to San Francisco State College, where I studied for a couple of years. That was before it became a radical center. Then I got involved in the free speech movement that was built on the civil rights movement.

Then came the Vietnam War. I was extremely involved in the antiwar movement from the very beginning, in Berkeley. By this time I had also joined the Young Socialists' Alliance, which seemed to me to be an organization that aspired to the same things that I did. It was active in the civil rights movement and the anti-Vietnam War movement, but it also had a long-term perspective of working-class people taking over and running the world in a humane way. And that did, and still does, make sense to me as an alternative for moving the human race out of the morass that we're in now.

So those were exciting times for me, and it was with that kind of spirit and that kind of optimism that I moved to New York to be on the national staff of the antiwar movement. And we won. The antiwar movement won. And that invigorated me to keep moving forward. For me, moving forward meant, in a way, coming home to my class, to my kind of life.

I moved to Pittsburgh to try and become a steelworker and ended up instead, to my great luck, as a coal miner. I got hired in 1979, among the last of the affirmative action hirees in the mine that I worked in. I've been there ever since. I've always thought of my work in my union and as a coal miner and in the Coal Employment Project as a continuation of everything else that has ever been important to me, as an optimistic struggle for a future of the human race.

I was also incredibly nurtured by the women's liberation movement, which I was involved with also from the beginning in New York City. I was on the steering committee that organized a massive 1970 march in New York, which is considered the beginning of the public part of the movement's second wave. We met at Betty Friedan's house to organize that march.

I've been lucky to be in a lot of good places at the right time. I was also in New York when the Stonewall riot happened, when police raided a gay bar in [Greenwich] Village in 1969 and gay people stood up to the police and the gay pride movement began. I helped organize the first anniversary march that now is Gay Pride Day in New York, the last Sunday in June.

All of this I brought with me into the mines, and I found women who had come [down] many different paths, but women who shared a sense that we can do anything we put our minds to, against all odds, and who brought that kind of confidence into the mines, and with whom I was able to share my past without it seeming to be a real strange thing. That happened over the years, not the first day. In turn, I get from them their individual struggles in a world that tells them they're worthless, that they should limit their horizons. But [they] refuse to have their horizons limited.

Even though the mine is a world unto itself, it's a world that's very much affected by what's going on aboveground—for example, by economic and political changes. When I first went into the mine, I was still, like everyone else, riding high on the crest of the Miners For Democracy [reform] movement, and there was an atmosphere of strength and pride and courage among the union people—who had been through a fight to take over their own union and had won it—that I felt really inspired by, even in '79.

In 1979 miners had just won the '77–'78 strike, which had reinforced the idea that, even though the leadership had wanted them to settle for earlier versions of contracts they didn't want, miners decided themselves that they were going to do what they needed to do. Jimmy Carter tried to invoke [the] Taft-Hartley [Act] against the miners, and that didn't stop them either. They stood up to the United States government, their own leadership, the BCOA [Bituminous Coal Operators Association], and they won.

The mine itself is a whole other thing. Then and now, it fascinated me. Like a lot of other women, although I like working with other people, I especially covet the time when I'm by myself in there, too. There's a kind of awe and wonder that comes for me from being among life that existed so many millions of years ago, and that you can sometimes see the outlines of in the fossils that hang over your head. And the silence, when there's no equipment running. Sometimes in the darkness when I'm by myself, like when I'm trying to pee, I turn my light off. And you can really feel much more of a connection with the life that came before us, and that which is to come, than in the hubbub of our everyday life. I think I got a lot of strength from being where I was, a kind of inner peace to help me get through the other stuff that would go on in the mines. I still covet those times. But the production schedule is so hectic now that there's not much chance. But that is a part of it, and I can really understand the words of the mining songs, of how mining gets into the blood.

At the mine where I worked women had been there for four years before I was hired, but were still really not accepted. Looking back, I see that the company was on a real campaign to convince us women they had to hire that we really didn't belong there and that we didn't want to be there. They did things to us that, to me, I didn't realize at the time, were special negative treatment. But most of the women had fought so hard for these jobs that we weren't going to be scared off by their tactics.

For example, one of the things they did with me early on was to put me shoveling and building cribs with five or six black men. They thought that was going to scare me. That day turned out to be one of the best days I had. We sat around and talked about our families and paths, and I talked about the civil rights movement. It was a great day, and I felt a common bond with these guys who I was supposed to be scared of. But there were much worse things that happened, too. Like giving us work that was more physically strenuous than they would ever give men, and then have me work for months with a bunch of men that thought women had no place in the mine and were determined to make me feel inadequate. They are now, mostly, my friends. Also, much more explicit and graphic experiences of sexual harassment, primarily for me from bosses, that were not part of any conscious plan by the company to drive us out but which were part of a real atmosphere of prejudice against women miners, which still exists to some extent.

There was another very, very important thing about the mine that I worked in. The women who had started four years earlier had faced much worse situations and had really, literally, paved the way by the time I was hired. Most of those women described themselves as "twofers," which meant they fulfilled two affirmative action needs for the company: they were black and they were women. And they stuck it out through incredible stuff. And when we came, a group of white women, they greeted us as sisters and comrades in arms and as people they were going to help.

I was hired as part of an affirmative action plan. I was hired along with 23 others, and in that group there were only the following categories: experienced miners—among whom one was a woman—veterans, blacks, and women. That's all. They didn't hire any inexperienced men who weren't black or veterans. We were all fulfilling quotas, except for the experienced miners. That was because people had fought for, and won, an affirmative action program.

I have to give a lot of credit, at Bethlehem, to the black steelworkers who brought suit against the company years before that, to win the consent decree in 1974. I believe that I was hired because the company believed that I was too small to make it, because the women who were hired at that time were either very overweight, old, or very small. It just seemed to me that unless no one else was applying for these jobs—and I know that wasn't true—they were looking to fulfill their quotas and then get rid of us.

It didn't work. We all made it. None of us changed. The little ones didn't get any bigger, and the overweight ones are still overweight, and we're all working.

One woman was a single parent raising two kids by herself. She had been through all the labels, since she was a black woman who was gutsy and who laughed. They decided she was a whore. She befriended me immediately; hers was the first female voice I ever heard underground. When I first ran over to her and said, "It's great to see you," she laughed and said, "There are a lot of us here. You're going to be okay." From that time on she has been a real source of inspiration and strength to me.

About half the women who were there when I started are gone now. Some of them were injured. One quit because she hit the lottery. Others have been laid off. There were probably a dozen women when I started, and the workforce then in our mine was about 400. Then in '84 our mine was combined with the one next door, where there were a lot of women working. But most of the women over there were laid off. They were given lower seniority than we were. Now a number of them have been called back. We now have around 20 women in a workforce that's still around 400.

For me, there were a number of things to adjust to. I had done a lot of manual work before I went into the mines. I had worked as a printer and ran a big printing press where I carried around 50-pound cans of ink as a matter of course and dealt with large quantities of paper and big machines. But in the mine I was with a bunch of people who had had life experiences that I never had had before. I realized that I had a lot to learn, and it was very positive. Especially from the old-timers. It was wonderful. Instead of being an organizer and leader, I found myself being a student, and I loved that. I got books in the library on archeology to look up the fossils.

Then there is the opposite of that learning experience, what I call the "brain death" aspect of the work, the way that the company seemed to go out of its way to discourage any creative thinking on the part of any of the workers, to discourage initiative so that we just did our jobs, and especially with women, to not offer suggestions about how things could be done better. And the monotony, the monotony. . . . When I first started we used to bring newspapers in to read during our breaks, and people would bring different things to read and share at the dinner hole, and then all of a sudden it stopped. No reading material was allowed in the mine, supposedly because it was a fire hazard. That was a big joke. There is so much other combustible material in the mine. The dinner holes were done away with. Shift changes were reorganized so we didn't have contact with people coming in and out of the mine. Everything seemed to be done to discourage social discourse. This started in the early eighties and has continued. It has coincided with the decline in strength of the union and the attacks on unions after PATCO [the union of air traffic controllers who struck unsuccessfully in the early 1980s]. Even though UMWA did stand up to the challenges to the contract in '81, it wasn't the same union as it was in '77–'78. There wasn't the same organization in the coalfields to push for a stronger union or for community involvement in the union's fight. Even now, looking back, we've gotten much weaker still.

Speedups were part of that, too. We were encouraged to work through dinner, to not take breaks, to keep production going, to work in conditions which, when I started, old miners would yell at young miners for even thinking of working in. Dust and bad roof and water especially. If the old-timers could see what we are doing now, they would be appalled and angry. So everything went downhill together, and this "brain death" thing that is so deadly got to be more of a factor in our lives.

Added to that, for a lot of us, was working rotating shifts, where our minds and our bodies never could adjust to any routine. We're on permanent jet lag. And that was very frustrating to me. I had been used to being very intellectually and socially active, and I kept up a lot, but it's very hard to find time to read and to keep an alert enough mind when you're on rotating shifts. It's very hard to organize a social life when your shifts are changing. And then, at the drop of a hat, the company can tell you to come in on a different shift the following week, with a different crew. That was hard to adjust to.

Someone who was doing some labor history research came across a quote from Andrew Carnegie, I think it was, about why they were organizing the steel mills on rotating shifts. He said straight out that if you get people confused enough about when they're asleep and when they're awake, it's harder for them to do unpatriotic stuff like organizing unions.

My local was very involved with Miners For Democracy and in organizing the insurgency around the '77–'78 strike and in the rejection in '81 of the contract and the movement that eventually that helped Trumka defeat [UMW president] Sam Church. There were differences in the locals, but my local has never voted the contract up. They're proud of that. I'm not so sure that's something to be considered a real measure of militancy, but they were involved in the antinuclear movement. Women miners were sent to CEP conferences as delegates from the local from the very beginning. Today, though, the local is as weak as any other local, and that's frustrating, too.

In 1984, the company merged two mines with two UMWA local unions, which set the members of each local against each other over the issue of seniority. There was no leadership in the local or district or international to explore those issues. We had the larger local, and I was one of a few in our local who argued against us getting more seniority than the other local. I was considered nuts. People would say things to me like, "If you want to volunteer your check number to someone, I'll take it." This was in '84, and unfortunately, no one was able to rise above their own panic-stricken need for their own job, to be able to see that we were being pitted against one another and we were being weakened. We've still got a lot of scars from that. I go to work every day beside guys who had worked in the mine beside ours, who have worked for 20 years, who have worse jobs than mine, and I've worked just a little over 12 years. We work alongside one another, and I get paid more than them because I have a higher job bid, and it tears me up inside. But it's become a sort of fait accompli now. There is understandably a lot of bitterness against the union by these guys now, who formerly had been union activists in some cases. So the company got what they wanted. But we're in the bigger sense just reflecting the weakness of the labor movement as a whole.

The UMW has not been a leader in the social struggles in the coalfields for a decade now. I think the UMW has become too much a part of the demise of the labor movement. The UMW stood up more than the rest of the labor movement at the beginning of the whole Reagan onslaught, and I was really proud of our union. We have continued, off and on, to do some very important things, like the Pittston strike. But the whole outlook that is being reflected is that the only way that we can save unions in this country is to close our eyes to everything that's going on in other countries and hold tight to the few jobs we have left, by any means necessary, even if this means changing our stance on issues like health care.

There's a deep sense of pessimism and defeat in the leadership of the labor movement in the United States, some of which is understandable and some of which comes from an incredible lack of vision and lack of willingness to tell the truth to the membership—that we're in a tough time, that we may have to take concessions, that the only way we're going to move forward is to see ourselves as part of something bigger than just those who happen to be dues-paying union members. Unless we see ourselves as part of a fight for a whole class of people in the United States and around the world, we're doomed to this irreversible snowball. I'm very pessimistic now about the state of our union as it is presently led. I think there is a real need for a militant coal miners' union that has the audacity to go out and organize, and I hope against hope that the UMW can shake itself back to being that kind of union. But I get more pessimistic about it every day.

Right now we have a real divide between the leadership and membership of the union. There is no trust in the union that the leadership is telling the membership the truth about what's going on. And that lack of trust, unfortunately, is based on repetitive experience. You can't build an organization like that. I'm still committed to the fight for a strong union. Unfortunately, it may turn into a situation where we have to replace the union that we have now and where a lot of people have to suffer between now and then. If it could happen inside the UMW and save the union from its own suicide, then that would be the best thing.

We need a new union movement that has some social consciousness and enthusiasm to it. For example, think about the Anita Hill hearings. I used this as an example when I spoke at the City University of New York. I said that I'd heard that after Anita Hill testified, women's organizations were flooded with calls from women wanting to know how to file sexual harassment charges. And I asked the question to the labor people there: Were your unions called? Did anyone look to their unions as a place to go to answer these kinds of questions? Were any unions present on the grounds raising issues on behalf of the labor movement? [Anita Hill] may not be a union member, but she certainly represented the issue of people having the right to work with dignity on their job. The unions did not present themselves as champions of that.

Nor are they representing themselves as champions of any other cause that affects working people. The homeless don't look to the unions the way the unemployed did in the thirties. If a Roosevelt were to come along now, he'd have to look with a magnifying glass to see what hit Roosevelt in the face in the thirties: a labor movement that was starting to rise to its feet to champion poor people and the unemployed, as well as people with jobs.

I think everybody is realizing that the economy has less room to give than it did in the thirties. It's so fragile—the bank crisis, and the deeper structural weakness. I don't know how much room that there is for new programs like the New Deal that were developed in the thirties.

Also, the atmosphere in the world as a whole is really frightening right now, because in those places where people strove to make changes to benefit themselves—and I'm thinking mostly right now about Nicaragua and Cuba—much has been undone to isolate them. The breakup of the Soviet Union and eastern Europe and the lack of clear direction there is also very frightening. The idea of the United States as invincible—that Operation Desert Storm [the 1991 Gulf War between the United States and Iraq] helped to perpetuate and deepen, following on the heels on the United States' invasion of Panama—all of these seem like bigger obstacles than those facing people in the Depression years.

It's a very frightening world when you have a child and think about what kind of world they're going to grow up in. But if there is to be a movement to represent the needs and aspirations of working-class people, it's got to be something very different from what has become extremely limited and myopic, which calls itself the labor movement in the United States today.

The labor movement has been marginalized not only through the actions of our enemies but also through our own actions. I can understand why a black person who has been struggling to make ends meet, who has gotten no recognition from a union, sees a job opportunity come up by crossing a picket line. What interest have unions shown in that person? Why should that person think that by crossing the picket line he is hurting anyone? I'm not condoning scabbing by any means, but it's not enough for us to stand on the side and call people scabs. We have to recognize how we got to where we are today, where that kind of activity is much more prevalent.

There is another thing in which I think the unions are culpable: the growth of selfishness and the lack of connection with any other part of the human race. Unfortunately, it's the kind of idea that a lot of unions have helped to reinforce. The whole anti-imports campaign was designed to get people to protect their own jobs and to hell with workers who happen to work in Japan or in El Salvador. You can drive—or you could a few years ago—and you would see big billboards put up by the AFL that said, for example, "Buy Indiana," "Buy Western Pennsylvania," "Buy Pittsburgh." As if our role as unions is to scramble for the crumbs that we can get for our own individual people, whatever "our" means. This is very dangerous. The unions didn't start that whole process, but they really fed it.

On the other hand, we have waged some very big battles that have permanently made a difference in people's consciousness, a fact which the media would like us to forget. Since the thirties the civil rights movement has happened. Black people have a completely different idea of themselves, as do whites about blacks, than they did at that time. That's an empowering thing. Women have been empowered by the women's liberation movement and what followed from it. Affirmative action programs, despite their being undermined now, have empowered a generation of blacks and women and Hispanics, who have in their memories, if not in their daily lives, the knowledge that they can be more than society tells them they can be. In the environmental movement people who call for recycling are now seen to be on the right track instead of being called crazy hippies. It's become social policy.

There is still very much alive today a cultural flowering that didn't exist before. We also have ties with people around the world that were unthinkable before. Many people who have been involved in these movements have traveled around the world and have links with, for example, the South African freedom movement. Those things may pale in importance when [we're] looking at the obstacles that we're up against, but they are there to build on.

# 23

# THE GREAT BRITISH STRIKE OF 1984–1985

## Sylvia Pye

*"I know I'm not a special person, but I know that if I stopped doing what I'm doing now, it's one more person who would be out of sight, and there are very few of us left fighting."*

*In Great Britain women and girls who worked as "pit brow lasses" [surface workers at coal "pits" or mines] in the nineteenth and early twentieth centuries ignited a violent debate on women's work. Political and public pressure on the pit brow lasses followed passage of a law in 1842 excluding women's underground employment at coal pits. A small number of women continued to work on the surface past the turn of the twentieth century, especially in Lancashire, where Sylvia Pye now lives.*

*Pye never worked as a miner, but she and other pit canteen workers come closer than any other women in the British coal industry to being women miners. More important, Pye rose to a position of leadership in Women Against Pit Closures (WAPC), a group that resisted the government's program of pit closures. In the early 1980s women miners in the United States developed close ties with WAPC, and mutual support flowed between the two countries during the bitter 1984–85 British strike.*

*Jolted out of a comfortable life by the death of her husband, Pye went to work at the pit canteen and emerged as one of Britain's most visible coalfield activists.*

I was born in Newton-Le-Willows, in Lancashire. It's not specifically a coal village, because we had cotton mills and other industries as well as a lot of coal mines. This area is part of northern England's industrial center. Years and years ago my mother's people came over from Ireland and were in mining

Sylvia Pye (second from left) and other activists at the Parkside Pit Camp.
Photo courtesy of Sylvia Pye

disasters, but for me there was no connection between that time and when I married into a mining family.

My mother came over from Ireland to visit an aunt in the early 1920s, just as this community was being developed. She met my father, who worked on the houses as a plasterer. She went back to Ireland, and he pursued her by letter and any way he could. They got married in Ireland. My eldest sister was born over there, and then they came back and lived here. I'm the youngest of three children and live now across the road from the house where I was born and where my mum and dad lived all their married lives.

My dad was a very silent man, a good man. He always said, be sure to join a union, and be in the Labour Party. But he never went into detail about why we should do that. I've been told by other people that when there was trouble, he fought to improve conditions.

My parents were strict disciplinarians. They taught me to be respectful of other people. Up until my mother died I would never argue with her, even when she was very wrong, because I respected her. Mother always worked at part-time jobs. She worked in factories up until me being in my teens. Mother was very Irish, a real gentle person. My mum was a housewife, first and last. My dad expected that of her. He accepted that she had to work, but he didn't like it very much. He was one of these men who expected a woman to stay at home and be exactly what she should be. He was a bit staid in that respect.

I might be being a bit prejudiced, but I think Irish people have a way all of their own. They're so welcoming, so giving. I think they're beautiful people, and my mum was one of those. I have an auntie, who she came over to stay with, who is 94 years old. They were both brought up amongst the real horrors in Ireland. They were very, very poor. My mum's time was the Black and Tan time, when the British soldiers were there in the 1920s to squash the rebellion and occupy the country.

My family was in Dublin, where most of the rebels were. One of the stories is that my mother's brother was shot by the Black and Tans as he was walking on the street. He wasn't involved in anything, but because the troops had made that mistake, they said my family was involved and immediately went to the house. My mother was pretty young. Her mother was bedridden. They tipped her mother out of the bed and searched the bedclothes and ransacked the house. They were supposedly looking for weapons. They found nothing, of course. I only found out about this in recent years, because my mother never spoke of it. I am proud of my Irish connections. It made me very aware of the feelings that I have, why they are so strong.

*After England entered World War II, Pye's village was frequently bombed because of industrial sites in nearby Liverpool.*

We were in a very dangerous position here. We lived near Liverpool, and there were air raids at the docks. A lot of bombs were dropped on factories I lived very close to. We had air-raid shelters in every garden, made of metal. People who were well off had stone shelters. If my mum didn't have time to get us into the shelter, she'd shove us under this concrete slab in the pantry that held up a lot of the house. I was very young, but I remember the house shaking. Everything could have fallen down around our ears, and we would have been perfectly safe. But she couldn't fit in with us, so she got under a table.

My dad wasn't in the army, but he was in one of these trades where he had to work away at the docks. We didn't see him for a long time. Then it got to where he couldn't get work, and he had to travel different places to find it. When I was young naturally I thought I would go to work, because that's what everybody did.

*Because she was artistically talented, Pye decided to try to become a draftswoman.*

I was very good at drawing and wanted to be an artist. As I grew older, I wanted to be a draftswoman. It was very male-dominated at that time. There were draftsmen, but I was going to change things. I was a tomboy. I didn't like dresses and preferred to play with boys. When I was into my teens, that's what I was aiming to be. But I needed a scholarship, and I was never considered for one. It was only kids with moneyed backgrounds that were ever chosen. It never really bothered me at the time, but looking back on it, I realize I had no chance. It didn't matter that I was clever. I was always near the top when we had exams. But it was your family connections that mattered. I accepted it, and everybody like me accepted it. But when I left school, I didn't have the qualifications to do what I wanted to do, so I just started work.

The first job I took was in a photographer's studio, but my ambition still was to be a draftswoman. I went to an engineering works where they made locomotives. I went into the draft office as a junior, hoping one day I would get there. I worked and went to night school and took exams. You had to take the exam at the place you worked. If you passed, you would do the draft work. I did quite well in my regular exams but never passed their exam. I didn't understand it, but I do now. I was never intended to be given that position.

After that, I went into the factory and loved it. I was making axle pads for locomotives, and I was earning really good money, which I'd never earned before. I worked there for quite a while and then decided I wanted to go back to an office. I worked at a foundry office and worked as a time clerk, doing the men's wages in the computer room.

*In 1962 she married her longtime sweetheart, who came from a mining family.*

My husband came from a long line of miners: his father, his grandfather, his grandfather before him. His dad decided his son wasn't going to be a miner, because there were better choices. He worked at the viaduct, and I met him at the factory. I had just turned 15.

We courted for a very long time and got married in our mid-twenties. Ford Motor Company opened up a factory in Liverpool, and he became a supervisor there. Our first years of marriage were full of strikes. There were three or four a year. It was very difficult with three babies, because I didn't work then, and I was sitting at home wondering where my next shilling was coming from. We had a small television, and we'd see Ford executives meeting in London. I said to Bill, "That's not fair. They're not living in the same world as we are." I said to him, "When are you going to go back to work?" And he'd say, "I can't do that. We'll get by." And we did.

*In 1972 and 1974 bitter coal strikes broke out across the country.*

The miners were locked out. There was a huge mass picket at the Saltley Gate coking plant, with the police unable to control the thousands of trade unionists who went on strike and turned up at Saltley to support the miners. It's a thing that has stayed in everyone's mind. The miners are thought of as having brought the government down. And the Tories never forgave them for that. They still haven't. That's where a lot of our problems in 1984–85 began.

I was with Bill when the 1974 strike was going on. We spent time with his father, who was on strike. There were always people around talking about the strike. Women were involved in that strike, but I don't think groups of them got together and opened soup kitchens. It still had nothing to do with me, and they didn't discuss any problems. Bill and I were cushioned, because we weren't directly affected by the strike.

Around 1983 there was a printers' strike just a few miles from here. I think that's where a lot of the police learned their tactics they used on us in '84–'85—like what we call the "wedge." The riot police in riot gear with shields formed a point, like an arrowhead formation. They would move in and split a crowd, and then they would make two circles and surround them and batter the hell out of them. The printers' strike was the first time the wedge was used.

*In 1980 her husband died in his forties.*

He was very young. My kids were quite young. I'd always worked in an office, and he'd worked at Ford's. We worked shifts and managed to stay with the kids. We would fit around one another, and they were never too much on their own. We were really, really close. We had a good marriage. He was a good husband and father. We would always talk things out. His death knocked me completely out for quite a while.

For a long time I was very bitter. People kept saying, "You've got three children, you're got to pull yourself together." But I didn't care. It took me quite a long time before I realized they were right. So I gave up working to devote myself to my kids. Well, that was the wrong move completely! And they realized it. My oldest son Billy was only 15, but he started taking his father's place. That's probably what made me sit up and say, I can't do this any longer. I took hold of myself and decided to go back to work, not just for financial reasons but because I needed to get something together with my life.

I had a friend who worked in the pit canteen who said they wanted some temporary workers. When I went for the interview the personnel manager was a man who had been a good friend of my husband. He said, "You don't really want to work in a canteen, do you? I can put you on in the office." I thought, what the hell do they think I am? I said, "No, I'm quite able to do the job, I've been a housewife with three kids." I went temporary for two weeks and never left.

The canteen was like a cafeteria, run by the [British] Coal Board. There was a canteen at every pit head. Miners could get breakfast, or a snack, and a meal at dinnertime and a tea at teatime. It was open from five o'clock in the morning until ten o'clock at night.

*As a canteen worker at the pit, she was on the mine payroll and enjoyed the camaraderie of the miners and her coworkers.*

I worked the morning shift. It was unbelievably hard work. There were massive floors, and it was always filthy because a lot of the men came up black and dirty. But honestly, I loved it. When we clocked out, our work was done. When you work in an office, you take it home with you. I was always tired, and I always had a bad head. But there you clocked on and did your job, and the atmosphere was brilliant. The women were so friendly, and the men were more gentlemanly than office bosses. I heard bad language, but it was never directed at me. They always apologized if I came into a conversation where I'd overheard it.

Meanwhile, Billy had left school, and he wanted to be a miner. I still thought mining was very bad and very dangerous. I was also aware that my father-in-law was alive, and he hadn't wanted his son to be in the mines. He was a pretty big figure at that time, because Billy had no dad and we needed him. When Billy said he wanted to be a miner, I told him he'd better go talk to his granddad. He came back and said his granddad thought it was great, because it was totally different now. So Billy went down the pit.

Then Andrew, my second son, wanted to go down the pit. He was very clever. He should have gone to university. We had a really big battle. But he went in as well. He would laugh and tell Billy, "I'm going to be your boss one day." He was working at the pit and going to college and doing quite well. He'd only been at Parkside a few months when the strike came. He was at the top of his classes. But when the strike was over, the [Coal Board management] told the lads they couldn't finish their education. It was punishment for being involved in the strike.

*Strikes authorized by the National Union of Mineworkers (NUM) in April 1984 had broken out. Strikers were protesting the pit closures proposed by the British Coal Board, which operated the nationalized industry.*

I started in the canteen in late 1981, so I'd been there three years when the strike started. Even then I really didn't know what was going on. I'd got used to strikes during my marriage, but I wouldn't say I knew why they were happening. All I knew was that all the men were out. My sister helped educate me. She was a lot older than me and a lot more militant in those days. She had lived union all her married life.

The union said they didn't want us to go out at the canteen. There were only three of us out of 13 who were willing to go out. At first it was probably better that we were there, because we learned about how angry the other women were at the miners. We'd been friends and gone out socially. If we hadn't stayed, we wouldn't have known the women had those feelings.

We stuck it out for a couple of weeks, and then we told the union, we're going out. I didn't want to stay at home. I wanted to be active in the strike. So they arranged to be on the picket line to stop our bus, and we three gladly got off.

We went off to a meeting, and the young lads were talking about setting up a kitchen. When somebody mentioned "kitchen," everybody looked at me and said, "Well, she's a canteen worker, so she should know what to do."

Soup kitchens had been set up by men in other places, but not here. There were 1,600 men at the Parkside pit alone. They were traveling everywhere to get a meal. We found two premises, one for an actual soup kitchen and another one where women and children could go. Men went there as well, but there were times when they were just off a picket and they needed to let off steam. We decided to keep the two places a little bit separate, and it worked very well. We had a family center and a soup kitchen, which I ran with the help of my youngest son. By that time he had been arrested a couple of times, and the condition of his bail was that he wouldn't go anywhere near picketing at Parkside. So he came and helped me in the kitchen. My sister helped me as well.

*Through her strike activities she met women veterans of earlier battles.*

I found a couple of women who had been around in 1926 during the big miners' strike. By this time they were quite old. Their stories were brilliant. One lady lived just around the corner from me, and we had never known she had been through that strike. She told me she could still remember the names of scabs and scabs' families, and she could still tell me where they lived after nearly 60 years. I said, "That's amazing," and she said, "No, it's not." She said, "There are just things you never forget. In another 50 years you will be able to tell who did what to you in 1984." And it's a fact.

She said the police were animals in 1926. Her father had his head split open by a police baton. I was still trying to come to terms with what I'd been brought up to believe about the police. I thought the police were the best things on two legs. We would tell our children, if you ever get lost, go find a policeman. I had great respect for them.

Before I went out on the picket lines, Billy and Andrew and Kenny would come and tell me stories about the police, and I would say, "But what were you doing to them? Weren't you provoking them?" Then I started to go on picket lines, and I saw the police were not acting like human beings. They weren't like sons and fathers and brothers. They did most of the provoking and most of the attacking themselves. The Coal Board brought in people from the army and the special forces. They would show up, batter people, and vanish.

The police started watching our house very early in the strike. The lads didn't have a meeting place at first, and they came here. The phone was tapped. Even when they found other places to meet, this place was still a target for the police. They used to stay around here all night, waiting for people to go out. When I started going out to the soup kitchen, I got up at half-past four. When I left the house, the police would follow me to the soup kitchen, where I went and turned all the boilers on and started the morning breakfast.

We had three or four hundred miners eating at the kitchen every day. I only had a canteen stove. How we managed I don't know, but we fed them all. I used to buy 25 loaves of bread a day, and there were 24 slices of bread in each loaf. Andrew helped me quite a lot with the bookkeeping, to tell me what food we needed. All the food was donated except for bread and milk.

We used to make leaflets up and leaflet an area. Then we would ask for a parcel of food and stand there and see if they would give us anything. All the families with children got an extra parcel for the kids. We really did an amazing job. About 20 people worked around the clock at a central place, separating the food and putting it into parcels. We took care of hundreds of families. Each one would get a treat now and again, like a tin of salmon. We did it fairly. Vans would take the food out to different areas. Here we were lucky because here we had other industries. In the pure mining villages people were desperate because everybody was out. We filled a huge van with extra food and took it to a village in Yorkshire. There they had 150 families, and they were only getting a food parcel every month.

Donated money was pouring in. Single men didn't get a food parcel, and they didn't get any dole money. If you had a wife and kids, they got a little something. A single lad got nothing except his picket money, which was about a pound a day, or two dollars.

*With the strike call, support groups began forming throughout the British coalfields.*

When the strike started in '84 there were groups springing up all over the place. They were dominated by women: miners' wives, mothers, aunties, everybody. Women Against Pit Closures had a very large rally in London in the first three months of the strike. I've never seen a rally as good as that one. We actually stopped London on that day. Quite a few women were on the platform and spoke, which was unusual. I don't know where the idea first came from. I do know that Ann Scargill [wife of the NUM president Arthur Scargill] and Betty Heathfield and others rallied around and sent for other women. All the women met in Chesterfield, and the national Women Against Pit Closures was formed on the 7th of July, 1984.

Here in Lancashire there was a support group for every pit. We had 11 or 12 groups, plus the central food bank and our soup kitchen. When the national group met, a Lancashire woman came back and said we needed a group in central Lancashire and in other areas. We were so busy we hadn't been able to send even two women around to meet other women. So we all sent one woman to our central meeting. Then we decided to have a central area meeting every week, and we moved it around in Lancashire so people wouldn't always have to travel such long distances.

Lancashire as a group picked one delegate to go to national meetings every month. We got to know what was going on and to take what was happening here to the national group. We needed that central link because before that we didn't have a clue.

We set up our own bank accounts. At first we were in with the union, but then the union's funds started to get sequestered. We decided to pull our money out at every area level of the women's groups. The Coal Board didn't come after our money, because we had nothing to do with the National Union of Mineworkers. We were completely separate as an organization, although we worked with them. We had autonomy. We made our own decisions.

*She was elected as a representative to the national committee of Women Against Pit Closures.*

When the national group [was] set up, we said we would be 75 percent mining-related and 25 percent other groups, mostly from other unions. We sat around the table, with delegates from each of 24 different coalfields and representatives from other support groups from the Labour Party and from London. Each area would give a report, which I found very valuable, to know exactly what was going on. The area groups tried to solve their own problems, and only if they were unsolvable would they come back to [the] national [group]. Being part of that committee to me was something really brilliant. Then we would come back and report to our local meeting.

The gays and lesbians formed a miners' support group. Before that I wouldn't say there was any connection between us. They had been under a lot of pressure, which we didn't know anything about. We hadn't wanted to know anything about it. Then we became aware of their fight, which was similar to what we were facing. They sent a group to meet the Welsh miners, and they were nervous, because Welsh miners are traditionally macho. But they formed a really good bond. And now they invite them back every so often. I was invited to speak at some of their meetings and went to some of their clubs for the first time in my life. I've always been one of those people who said, "Oh my God, I don't really want to know about anything like that." But I've got some really good [gay] friends now.

*Some strikers opposed the arrival of women on picket lines.*

At first a lot of the men didn't like it, especially the younger men with younger wives. They didn't like the women hearing the abuse. They didn't like the fact that we were there watching their reactions to things. We went through some battles to get women to stay there, to get the sanction of their husbands. They were very protective. They used to make sure you were at the back and away from the police.

Usually they didn't arrest as many women, but on some picket lines the police really made a beeline for the women. One picket I was on, they brought in horses. There were students there, and they went straight for them. The miners couldn't do anything about it because they were being split up by the horses. They ripped one girl's T-shirt straight off her back and threw her down. She ended up in court in a miner's jacket.

The biggest problem for the men was when we started to appear on the picket lines as an organized body. It didn't go down well at first. We had to make the men feel that we were going to work with them and not push them out and not become macho women. That lasted a couple of months, and then they settled down and accepted us.

A turning point for the men came when they saw the effect we had on the picket lines. When women started to appear on picket lines, the police started to be intimidated. It was really strange, because you could turn up on a picket line that was mostly men and there would be a few policemen. As soon as a few more women came, police would arrive from all over the place. I don't think they knew what to expect. We gave the police the run-around. If we decided to walk down the road at a peaceful picket, suddenly hundreds of police would appear because of a few women. And if we could gather a few hundred police in a place where women had gone, it took them away from someplace else, you know.

I still find it strange, because we're still experiencing the same thing today [at the pit camp]. It's as if they think we have some sort of supernatural powers. We were at that time beginning to organize a lot of things, and we found ways to organize things without them finding out.

At first it was difficult for me to believe how violent the police were. My son Andrew was only 17, and he was savagely beaten at a picket line. His chin was out of line, his nose was out of line. He was mangled. By the time they took the X rays, the bones in his face and head were so swollen they couldn't check him for underlying fractures.

Other lads were held back and couldn't tell who was being beaten, except Billy, who went in to help Andrew. They beat him, too, but not as badly. Then they threw them both in jail. When I got them both out, I took them both to a hospital and told the hospital where they'd been, so it went on their records there that all their injuries had been sustained while under police custody.

None of the men in my family ever said they had to go back to work. But I know plenty that did. I know plenty of good women who couldn't stop their husbands from going back. My heart broke for the women, because then they couldn't work with us anymore. They felt betrayed and felt they had betrayed us. A lot of the families were kicking the miners out because they weren't going to work. I was lucky because we were all on strike.

There were also women who never came to the picket line. These are the ones whose men mostly went back to work. They stayed home, and they didn't know what was going on. I used to sit in groups and say, "I understand how these women feel. I've been there." All they knew is what they read in the papers or saw on the television, about how bad the miners were. But they never told the real story.

At our social center we brought people in from social services who could help the families. We used to have these sessions where people could come in and ask about anything. It worked very well. People who were buying their houses and had mortgages could come in and find out what resources were available. Other people couldn't afford their life insurance premiums. In rented property, electric bills had to be covered. For 12 months we took care of almost every social need. People educated themselves, and women who had completely relied on their husbands got involved and found they had a lot of talent.

*Miners' wives traveled widely to organize support for the increasingly bitter strike.*

We started getting invitations for Women Against Pit Closures to go all over the world. Every area got the chance to send one or two delegates on trips. In this country we found out where people from different areas were going to be, and you'd go to a different area. I went to East Germany with a party of children and some women, I went to Russia with the peace movement, and I came to America twice. Other women went to Australia and America and Russia.

It was just before school holidays that the strike started, around Easter break. Really small children got together, because there was nothing else to do with them. And they would draw pictures and write poems about what was happening at that moment in time. That generation, I think, got a very deep sense of what was happening.

There were times when children were on the picket lines. Both parents were there, and they would have to be there. They were always in the soup kitchens and the social centers. They became a big part of what was going on. They had to be included a lot of the time because both parents were involved.

When the men got over the idea of their wives being on the picket lines, I think it made them a lot more aware of what it meant that their families were standing with them. The picket lines especially were better conducted. If a picket line got violent, the men would stand in front of or circle anyone who they felt was particularly vulnerable. The kids stood up for themselves with other kids during the strike, even to the youngest ones. They would not be intimidated. I think it taught them a lot.

Christmas of 1984 was going to be particularly bad, because we'd been out on strike for a long time. We expected to have nothing. But those kids have probably never had such a good Christmas before or since. Support came from all over the world. They didn't just get one toy or two, they got loads. There was plenty of food. There were so many parties that we were actually turning down children's parties because we couldn't get to all of them. That was a really fantastic time, the unity between the miners and the wives. You lost some friends during the strike, but you found new ones. And these are friends you've still got.

*In March 1985, a year after the strike began, the National Union of Mineworkers called an end to the strike, and miners returned to the pits. The women's organization remained active.*

Margaret Thatcher and the Tories wanted to get rid of the unions, and particularly the NUM, because of the miners' strikes in the 1970s that had brought down the government. That was their goal, and they did it. They wanted to smash the unions and get rid of people's ability to have a say in their working conditions.

Most of the men who hadn't been jailed or arrested or anything went back to work. But some of the men who were sacked never worked again. Eleven years later we've still got 49 of those men who have been blacklisted. These are the most militant people.

Some of the men wanted the women to just disintegrate, and it didn't happen. Organizing ourselves into a national body made us strong. Our efforts blossomed into a women's organization that has continued. Some of the union committee members were amazed when I said to them, "We're still carrying on to handle any problems that come up." They said, "But it's all over!" We said, "No, we're fine, we've now got a good organization, and we're carrying on." We had our own funds. Some men would have loved for us to go away, but it's a good thing we didn't. We were absolutely necessary after the strike, because that's when the problems did start.

Families were facing having to pay the bills that had been frozen during the strike. People were threatened with losing their homes, being evicted, and being taken to court for debt. Electricity boards in some areas were saying the electricity that they'd let people have during the strike was on loan and they wanted to be paid back.

We were able to put people in touch with lawyers. We got court cases squashed. We won a lot of cases that we hadn't had time to work on during the strike because we were too busy. By the time the men went back to work, we were in touch with people who could help. In fact, we took a lot of that work off the union, because we had the time to do it.

Some women did go back home, but they changed their ways. A lot of women stayed active. Some went into the Labour Party. A couple of MPs [members of Parliament] started out as ordinary housewives during the '84 strike. We came up with quite a few good poets and songwriters. People got together and did tapes. They probably never realized they had that talent before. We found a lot of people who could sketch and draw. Other people went on to adult education.

After the strike Women Against Pit Closures approached the NUM about women joining the union as associate members. We wanted to find out things for ourselves instead of being told them, and sometimes not being told at all. The area executives had to take it to their meetings. Other areas voted for it, but in Lancashire we didn't get it. I said, "After all we've done, we've been voted against?" Billy said it was presented that the women wanted to come in and take over the union. It scared the men to death. Eventually we did get associate membership through the national union.

After the strike we were able to arrange public meetings and do nitty-gritty things the NUM couldn't do. Women Against Pit Closures was really on the ball. The NUM has a rule that if Arthur Scargill comes to speak in an area, he has to be invited by the union's area executive. But if he comes to a women's meeting, it doesn't apply. So we would invite him, and then the miners would come out to the meeting.

*All of the men who stayed on strike until the end marched with Pye when she and two other canteen workers returned to the pit to work.*

The miners all marched into the canteen with me, the 249 who were still out on the last day. They all came into the canteen, even the ones who weren't on that shift. It was really very difficult. There was a strain among the canteen workers. We had been very friendly before. The first few days were absolutely horrendous, because we had miners coming in saying, "I'm not having her serving me." Normally the three of us who had joined the strike had worked together, but they split us up. A couple of weeks later there was a cutoff at work. They found work for the other ten people in the canteen and sent the three of us home.

I worked more than two years after the strike ended. I developed arthritis all through me, and it was becoming quite dangerous for me to lift the huge pans. Also, my mother was very ill at the time, and I needed to look after her. So I left the pit and came out.

*She says the strike changed her personally and politically.*

Well, I changed drastically, didn't I? I was a real mouse before. When I was married, I was cushioned and happy. The only thing I joined was the PTA. I found another complete side of myself. Although I won't take anything away from my previous years—because I was happy, because I did have a good life and I certainly had good parents and a fantastic husband—I don't feel that I lived. Do you know what I mean? I lived, but not really. I didn't live in the real world.

I think I'm living more fully now, because I'm aware of what's going on and other people's problems, not just my own. There are a lot of things to be done. I know I'm not a special person, but I know that if I stopped doing what I'm doing now, it's one more person who would be out of sight, and there are very few of us left fighting.

I think everyone is needed, because there are so few people today who are taking a stand. The years have changed, not just my life. Everything has changed. There is less employment, and people haven't got a choice anymore about whether to work.

All this has gotten worse since the strike ended ten years ago. We have no apprenticeships anymore, and we have no people going into the crafts. We've gone back hundreds of years. Our forefathers started fighting for what we had, and now we've lost it. Our national health care, which was very good, is now crumbling and is going to be privatized. Our national rail service is in danger. They've taken the nationalized coal industry. They've taken the electricity and the gas and the water. Everything is privatized.

After the strike we quickly lost seven of eight of the pits in Lancashire. There wasn't a fight around most of them. The first one was a very militant pit, and they intended for it to close immediately. We did fight long and hard with the union, and in the end the men gave in. They told us, "Look, don't come back, we're giving in." After that there wasn't a great deal of fight around the pits, because they were moving around from pit to pit, finding workers who wanted retirement and were ready to go.

Parkside was one of the newer pits, with a very young workforce. When they told them they were out of work, there was nowhere else to go to. Used to be there were always other pits. There was always a choice. A favorable report had come out showing that Parkside pit had high standards and plenty of coal. The government report came out shortly after that, and it was completely different. They said the working relationships were really bad and there was no coal.

*In October 1992 the British government announced that it planned to close 31 pits nationwide. Ten were slated to close immediately, including Parkside. The announcement brought widespread public protest.*

When the announcement was made, the better part of the country was up in arms. There were huge demonstrations in London, even what we call the blue-rent pearl [wealthy] people. They were coming out of the Savoy Hotel and cheering us on.

It became obvious that, although it was a good campaign, something was needed to highlight the problem. The marches and rallies were so huge that the government started making promises. They said they would study the problem, but we knew it was a ploy to placate people. We started a new campaign to keep the pits viable so they could be reopened.

*Leaders of WAPC discussed setting up "pit camps," based on the women's peace encampment at Greenham Common military base.*

We knew we needed some sort of a campaign. Some of our women had held discussions with the Greenham Common women, and that's where the idea came from. The Greenham Common camp was going on during the miners' strike. We used to go up to their peace camp and help out whenever we could. They were women who didn't really want men to be involved. But we didn't feel our fight was completely a women's struggle. Although there were women's camps and women were leaders in a lot of activities, we were actually fighting for the men's jobs. We didn't feel that they shouldn't have a part in the running of the camp, although I will admit they have always taken our orders. It's been mostly our camp. We've led the demonstrations. But the men were very involved. They were invited to do night-duty shifts at the camp. There was a bit of contention about whether that should be happening, but we stuck to that, and I think eventually most of the camps did include and welcome the miners.

In fighting for the mining industry, the fight has never been just for women. We started out in '84–'85 supporting the men, and it wouldn't make sense to exclude them now. It would have just alienated them even more, when we needed to get their confidence. I think they feel better about women's groups working as long as they feel they have some sort of input into it.

Ten pits were threatened with closure immediately, and we decided to set up camps at those pits. When we were talking about it at the national meeting, the idea was just being thrown around. We had asked a lawyer to come into our meeting to explain what our legal position would be. Then somebody said, "Let's just do it." That's exactly what happened. This meeting was in December, before the Christmas holidays, and we set up eight camps within a week.

The government planned to close Parkside in October 1992. The men were laid off then, but they had to go to work every day at their shift times to pick up their checks, even late at night. This was another form of intimidation.

Meanwhile, the Coal Board was offering them so much of their redundancy [severance] payments to give up. They told the men if they waited they wouldn't get it. So men began worrying. By June we'd lost about 100 men from the original 800, but the rest hung in there.

In June 1993 all the workers were gone. They wanted us to leave the pit camp, but we said, "No, we're staying." We couldn't stop them from laying off the men, but we've been able to actually stop them from filling in the shaft, and then maybe, when they realize they need this coal, they'll go back in. They're realizing it now, but they're not about to go back and open it because that would mean admitting a mistake. So they're willing to lose one of our best assets just to get rid of the NUM.

At first the government said they only needed 12 pits in the country. Now they're saying they might need 32. We have 14 pits left open and working, and we've got others that have been closed but haven't been filled in. So they could still reopen Parkside. But they won't.

When they privatize, they'll make all the workers redundant, close the pits for a few weeks, and reopen and contract the work. If the workers don't do what they're told, they'll be sacked. And unfortunately, so many miners now are demoralized and have nowhere else to go, and they'll get their workforce.

*On 18 January 1993 the Lancashire Women Against Pit Closures opened the Parkside Pit Camp to keep vigil at the pit entrance.*

In the first 12 months of the pit camp, we had five occupations. We tried to have them at crucial times for the campaign, to get our point across. On International Women's Day in March we held a day to honor pit brow lasses. We borrowed clothes from museums, even the little coal trucks that they used years ago, and clog shoes. The women went down on the pit yard, and the management had a fit. They didn't know what the women were doing. The men thought it was brilliant.

There was an underground occupation at Easter by women from Yorkshire. It generated national media attention. We can't always get that, because there is a block against NUM news or Women Against Pit Closures.

We sent women up for the tower occupation to force the government to reopen negotiations with the men. Four of us climbed 260 feet to the top of a pit tower. Stone steps went part of the way up, and two steel ladders went up to the roof. The men at the pit said they would never go up there. We carried a tent up and all our food and about 16 liters of water. We loaded up everything in backpacks.

When we were out there, there was an area in the middle of the roof that threw off slight radiation. We had to stay away from that. We had 40-mile-an-hour winds while we were up there, and there was only a low wall. We were warned by the office to lie flat against the wall. It was a horrendous weekend, really bad weather. But it didn't deter us. They would have had to blast us down.

We had an occupation in what we called the "liner train cabin" on the far side of the pit, which could be seen from the motorway. It's where the trains used to go beneath and the coal used to drop through a chute into the coal wagons. We wanted to observe what was going on at the other side of the pit, so we had an occupation there for 11 days. Management blocked off the windows—so the women were in total darkness—and put guards and guard dogs around it.

On our occupations we always take in handheld radios. On the tower we had brilliant contact because we were in the open. When they were inside they would have to hold them out of a window. In the liner train occupation we broke in on the seventh day and took them more supplies even though [management] had a massive guard presence.

*After several quiet months the pit camp rose to action when trucks of crushed limestone arrived to fill the mine shafts.*

We got up one morning and saw the stone lorries coming in, and we knew they were ready to start filling the pits in. Our goal had always been to stop them. We went out and bought chains and locks, along with a steel rope.

At half-past four in the morning we chained ourselves to the steel rope to form a physical human barrier. There were about 14 in the "chain gang," and supporters stood at the side. We were on that chain for 24 hours a day, and it lasted for two and a half weeks. People were only coming off for two and a half hours at a time to have some sleep and something to eat. It was very cold, and there were thunderstorms.

We built up a small fire, because we were quite a way from the camp. We built a makeshift shelter and moved one of our porta-loos up so people didn't have to physically go away from the line. We wrapped blankets around us and wore waterproof clothing. After about three or four days the lorries stopped, but we stayed until we were absolutely sure.

About a week into our chain gang, a management group drove in one day. We had an occupation going on in the pit pumphouse. They were put there to watch the compound where the stone was coming from. We let management through, and later that evening all the floodlights were suddenly shut off. The pit at night was [normally] quite busy and very bright, with the lights and the sound of the pumps. But suddenly at midnight it was plunged into complete darkness.

I was on the line at the time. There were seven women and two men on the line. Management started driving out of the pit with all their headlights on, and we stopped them. My son and a girl went over and asked them two questions: why had all the lights gone off, and is it safe? They said, "No comment," and backed off into the pit.

By this time we had called the local police to find out if they knew what was going on. We had a fairly good relationship with them by now. We hadn't caused any violence. They actually watched out for us. The local police had come up and left with us two constables, a man and a woman. A police car came up, and an inspector came up with three other constables and marched up to the line. He said we were imprisoning management against their will and that we would be arrested. We said, "That's not true. We're not stopping them. We just asked them why the lights went off and if it was safe. We need an answer, because we've got people inside in occupation." He said, "I don't have to give you answers to that." Then he said, "You're in breach of the peace," and off he marched.

Half an hour later the local police were still there. I had gone back down to the camp. We also had another picket at the back gate. I had just finished a live radio interview. At about 12:30 A.M. it was really dark. I heard a noise and went to the door and was absolutely amazed to see these two riot police standing in front of me with their truncheons and shields. I thought I was seeing things, so I shut the door.

I opened the door again, and they were still there. I looked past them to see massive numbers of policemen pouring out of vans with dogs. Going up the drive there were three huge blue vans with the massive headlights, with these riot police running up by the back of them. It was like I was watching a film. I was quite slow at first to realize what was going on. I shut the door, ran to the back of the porta-cabin where the control radio was, and got on the radio and told the people at the top of the drive that the riot police were coming. I didn't know what else to say because it was such a shock.

I went back to the door, opened the door, tried to go out, and the police said, "You're under cabin arrest." And they shut the door. So I was left inside. I didn't really know at that time what had gone on at the top. I just heard a crackling on their handset radio. Afterwards they said that they had gotten my message that the riot police were coming, but all they saw in the darkness were the headlights and the visors glistening. The riot police were all dressed in black.

Several people were chained onto the steel rope. The riot police cut off the chains of some of the women and pushed some of them aside. There was a very tiny woman. They threw her in a van. They made a beeline for my son Billy. One of the policemen hit him twice in the face and knocked him to the floor. Fortunately he didn't retaliate. Then the police withdrew and went as quickly as they came. It was an act of intimidation.

We were quite shocked for a while. We went and had a cup of tea, and went back and took care of Billy, and then went back on the line. A few hours later the local police came. They were absolutely gob-smacked that we were still sitting there, because they thought we wouldn't stay.

We had two witnesses from the local police force who had witnessed this. The riot police came from a different district altogether. They didn't know the local police were there. Those two went back to the local police station, and they put in an internal complaint, which the police have to investigate.

The head police inspector said it never happened. Then he went off and checked and came back and said it did happen, because there was an internal complaint. He wrote us a letter that there were only 12 riot police there, but we know there were at least 75.

The riot police are special government police. They don't really exist, if you know what I mean. They're there for special occasions. They are the police of the state. They have truncheons and riot shields. They wear helmets and visors and like a black boiler suit and massive big boots and bloody black truncheons. It's quite obvious that the Coal Board must have sent them in. Maybe for them it was just a night out and an exercise. But for us it was an intimidation.

One funny thing happened that night. We had put up a set of dummies and called them the Pit Family. We had a man, a woman, and a daughter, and the daughter was pregnant with a bag of coal. When the riot police came, one of them went up to her with his truncheon and said, "Move!"

We had completely stopped the stone from coming in. We came back down to the camp and took down the chains and relaxed into a situation where we were going to stay there in case they tried to do it again.

Then the Coal Board's strategy changed, and they just walked away. They thought we would get fed up and leave, but we knew they would come back. In early spring 1994 we heard they were going to start again to bring stone in to fill the pit in. We knew they would have to remove us physically to keep us from doing again what we had done before. The queen's bailiff from the high court served us with an eviction order in April to remove the camp. They said they would forcibly remove us if we didn't leave peaceably, and we refused.

The court order was made out to Women Against Pit Closures, "persons unknown," and Sylvia Pye. I ended up in court, and they held a closed hearing. It was just me and one woman from Women Against Pit Closures. We didn't have a case against the eviction, but he [Pye's solicitor] thought we could fight to get the names off the order of the group and myself.

The judge said at one point that, though the campaign had been laudable, it was now time to give back the land to the people who owned it. She gave them the injunction to remove us "forthwith," and then their solicitors asked for license to sue me personally for all costs. My solicitor was so shocked he was speechless. Then he strongly objected, but she granted the request.

The eviction took place a few weeks later. There were six of us at camp at the time. We knew something was happening when suddenly the road went quiet. One cab that got through stopped and said, "There is a massive police force at the end of the road." Then he drove off. Three of us went into the porta-cabin and barricaded ourselves in. Three others stayed outside, and we had a runner who was going to leave the camp with the important documents and a complete phone list. She was to phone the press, because they would not allow anyone in with a camera. We had the BBC on 24-hour standby.

Then the police came. They closed off two main motorways, and they couldn't divert traffic. There were hundreds of police in vans and motorbikes, even parked miles away. There were two police helicopters. There were about 30 bailiffs and 120 police to arrest six of us. The three people we had outside did not resist and were carried across the road.

Then they came to the cabin. They were on top of it and around the sides and they were at the door. They completely battered the cabin down. But it took them a lot longer than what they thought. It took them about 25 minutes to actually batter it down. They took the man first and dragged him out. Then they took a woman, who had bad arthritis and didn't want to be manhandled. Then they came to me. There were two huge men and two women. They dragged me, and I resisted as much as I could.

When they got me to the door, they had a bit of a difficulty, because I'm not that small in stature. The two women had to go down the steps first, and they dropped me onto the floor before they got me to the other side of the road.

The police had brought trucks with three huge rolls of steel fencing and wire and cement. Before the police came to the pit, two people had laid across the road in front of the three steel wagons with the fencing. The girl was pushed to the side, and the man was arrested. They arrested and charged two men that day but none of the women. Then the police put up that fencing faster than anything I've ever seen in my life.

They told us we could come back and take our belongings, and unfortunately we did that, because they let the cameras in and that's what showed on television. It looked like we had walked peaceably off with our belongings and nobody touched us. It was so wrong. They planned that. The sheriff and the police inspector said on TV that it was all dignified and peaceable. We'd given interviews, but ours weren't shown.

*Pye's legal defense dragged on as she waited to hear the total amount she was being ordered to pay.*

It started off that I was going to pay all the court costs, and then I was charged for the removal of all the camps by the bailiffs. Believe me, that operation couldn't have been bigger if we had been the IRA or the Nazis. We heard later the market square about half a mile away was completely cordoned off, and there was more police in there. There were hundreds of them in vans, and the traffic police were there for the roadblocks.

We're waiting on the cumulative figure. It will go to court, and they'll ask me if I'm going to pay, and I'll say no, and then they can order bailiffs into my home and take anything that belongs to me. So I've given all my stuff away.

I'm now being told I can be arrested at any time. If I'm seen down there, they're telling me I can be arrested, even though they haven't restricted my activity as part of the court action. We've put a picket on this morning. I stayed across the road. It's very difficult for me, but it's better for the campaign, because we don't want people to get removed yet. Everybody is important to the campaign.

For a few days after the eviction we just stayed out in the open. Across the road there is a school that has been closed down and is now derelict. This is where we continue to be, excuse my language, a pain in the ass to British Coal, because we're still stopping their operation. We're there 24 hours a day, seven days a week. We're like a family. Two babies have been born to women involved in the camp. We call them Pit Camp Baby Number One and Number Two. We had Number One's first birthday party not long ago. We're there to stay as long as we can hold out.

I think inevitably they are going to win unless we can really get a lot more support from the public and the labor movement. To be successful we need a lot more people, and to keep it in the public eye. We are trying desperately to make people realize that what we're doing can be done in other places and that you don't have to be frightened of the law.

*Women Against Pit Closures has created a pit-camp-on-wheels that travels in a caravan through the country to build support for their efforts.*

We took the caravan up to Yorkshire and we've also been outside town halls where they're making cuts on the public sector. We're showing that we're not just fighting against pit closures. So we go protesting wherever.

We'll go on fighting at Parkside because that's where we are, that's where we live. Other camps have folded for lack of support, but we intend to stay. If at the end we lose, at least we've delayed them now for over two years, and we're a real pain in the ass, and we've cost them an awful lot of money. Maybe they will try to get it back from me personally. If I go behind bars—if that's what it takes—I am prepared for that, and others are as well.

# 24

# SANCTUARY
## Bonnie Boyer

*"When I went in the mines I was single, depending solely on myself for the first time in my entire life, and that mine became my sanctuary. That's where I became a woman full of self-esteem and in control of what was going to happen in her life. So the rewards have been immense for me."*

*Bonnie Boyer was 12 years old when she decided to reject the norms that enforce social injustice, and this consciousness has guided her life ever since. She takes pride in her role as a caretaker in her local union and in her family and sees that as a vital part of an activist's life. Boyer has emerged as a leader in the UMWA, CEP, and the Coalition of Labor Union Women (CLUW).*

I grew up at a time when every kid should have grown up—in the 1960s, when television was making people conscious of what was going on. Shortly after Martin Luther King was killed and the Watts riots happened, I became aware that things weren't right socially. That was when I made a conscious decision that I wasn't going to be part of any establishment that would discriminate against a person because of their color or sex. I was 12 years old.

There were mixed feelings, and controversy, in my family because there was racism in my family. My mother was raised as a Jehovah's Witness and believed that all races were equal. So she combated that, and my grandmother did, too. She was an American Indian. But I remember my uncles not understanding. So it was the women who were saying racism was wrong, and the men saying, "Oh, they're black, and we don't care about them."

It didn't really matter to me what anyone said, because I could see with my own eyes that it was wrong. If kids are taught hatred, they hate. And if they're taught equality, they believe in equality. I was raised Lutheran and

Bonnie Boyer and her son, Mitchell. Photo by James W. Harris

was taught equality by my grandmother, my mother, and my church. I believed it because it's what the Bible said.

*She grew up in a working-class neighborhood in Kitanning, Pennsylvania. The family was close-knit and stable despite a divorce.*

I grew up in a poor white neighborhood. There was only one black family, and they were interracial. We lived near them. Most of the people there were the working poor. Everybody looked down their nose at us, but it was a

Bonnie Boyer and section crew. Photo by James W. Harris

good childhood. My mother was overly protective and strict, but it was very family-oriented, and I had Grandma there most of the time.

My grandmother came from Pennsylvania. My grandmother never remarried after my grandfather died. They had farmed. After he died, she raised six children on poverty level by cleaning and being on public assistance.

My father and mother had four children. I have twin sisters, who are older, and a brother younger than me. My parents were divorced when I was very young, and my dad moved to California when I was two, and my mother remarried when I was six, so most of my family structure was my mother and my stepfather.

My stepdad is this big burly guy, but he's gentle and real generous about helping people out. He was a township supervisor. He raised four kids that weren't his and never showed partiality. He is the one who really made me aware of unions. I don't remember him ever sitting at the table and saying, "This is what the union is doing for me." I think it was more unconscious. It wasn't what he said. It was what he did. He would go to meetings, he would talk about beating the bosses. I didn't realize the impact until many years later.

Even if I didn't see it while he was working, I saw it afterwards when he lost his job as a steelworker and finances were hard to handle. When we were younger there were a lot of layoffs, so the older kids in the family were accustomed. But the younger kids weren't used to it. All of a sudden they couldn't have everything they wanted. It was hard, because my parents were broke, and he was in his mid-forties, with high blood pressure. The only job he could get was a paper route.

After the Watts riots, I decided I wasn't going to do anything that would make me part of a system I hated. I became very conscious of the whole world around me. I remember having bouts in school with teachers. The sociology teacher once made a very sexist remark. He said women shouldn't be working, and that they should leave things alone. Then I saw in the paper that his wife had divorced him, and I clipped it out of the paper and brought it to class. Needless to say, it caused a controversy. Then I remember an art teacher asking me if I was going to settle down and have a family. And I said, no, that I was going to have my own life.

*As a teenager, she longed for independence but also wanted a family.*

I was young and stupid. I thought I could work in stores, and that I would be successful. They were lollipop dreams and fantasies. I was poor, so I never thought about the possibility of an education.

I got married at 18 or 19. I thought I could be married and have my own life, and that was a misconception on my part. He was an educated man, and he had a good job working on computers with the federal government. And it wasn't a necessity for me to support myself, but I kept working as a store clerk, or administrative clerk. I always thought it would work out, but it didn't. We were too young.

I think women need to make a conscious decision very young to say it's okay to have a husband and children, but you have to have other commitments in life—whether it's to your church, a women's group, or a political group. You need to stay active in society. I think one of the biggest problems women have is giving up a part of themselves for the family, or the house. Later on they feel cheated. Instead, they should try to have more, whether or not it means conflicts with the husband. I think that's one of the biggest reasons for divorces after about nine years of marriage. Somebody's not having their own life. It's not always the woman, but I tend to believe that most of the time it is.

*A separation from her husband prompted her to seek work in the mines.*

I didn't think of the coal mines until I had left my first husband. We separated, and I read in the paper that the coal mines were hiring women.

My sister's ex-husband was a coal miner and told me about a company that was hiring. I needed a job bad, but I really didn't want to go into the mines. But then when I heard they made ten dollars an hour, I said "That's really what I want to do!" I was back living with Mom and Dad, and it wasn't a good situation, because I wanted to be more independent. I was 23.

I got called in and interviewed, and they told me they'd had problems with women being on comp [workers' compensation]. My stepdad called some guys he knew at the mine where I wanted to apply. It supposedly had the longest expected life, and I wanted job security.

I am a first-generation coal miner. My stepdad didn't understand my decision, and I think my mother still hates to tell people that her daughter is a coal miner. My grandmother, who had been widowed with six children, was the one who was really supportive. She said she wished she had had that kind of opportunity.

*She later met her second husband, Gary, who operates a cogeneration plant.*

My husband said, "You don't have to do this. I'll take care of you." He didn't want me to enter the mines. He doesn't like me in there, and it's not because of the men. It's because of the dangers and the stress related to work.

He's hardworking, he's dedicated, he's a good family man. He has very few faults. Even my girlfriends say, "Where did you get him? He's perfect." He doesn't have any big causes. His big cause is letting me have my big cause, I think [laughs]. Sometimes we argue about politics because I'm so liberal and he's more of a conservative liberal.

At the time I got hired I didn't know any women miners. There was this awful stigma that the only women who worked in the mines were whores. Some bosses' girlfriends were hired. They didn't last long. The men would go tell their wives about acts of sex in and around the mine, and the bosses' wives would find out, and the woman would end up quitting. It was a vicious cycle of people telling people stories. The wives felt protective. They didn't want the women going after their husbands.

I was the first woman in my local who would attend functions where the wives would be. After they checked me out as far as my intentions and they saw I was strictly business, they were all right. I'd say now, after nearly 11 years, [their suspicions are] gone. I've been lucky, I guess, because I kept going and hanging in there, and things have gotten turned around. I went to the Catholic pie social with two miners' wives. Some of them are my best friends.

*At the mine, she faced resistance in learning to operate equipment.*

Everything I've been trained on, I had to fight for. The company didn't want to train me. Our company thought women made good buggy runners. I've heard that so much, I'd like to scratch their eyes out. It became a struggle to get trained on the miner and other equipment. It still is. After ten years, I'm still in the training process.

I didn't start filing grievances until about the seventh year. I think it takes that long to learn their whole system of discrimination. I've talked to other women miners. It's a turning point. It's like, you've taken all you can tolerate, and you start filing grievances, and you start pushing things to the limit. You turn your whole attitude around, you don't care anymore, and you're going to fight the company.

With the union, I knew that I would have to be committed to gain the support of the guys and learn how to protect myself with the contract. Getting active in the union was a decision I had made before I entered the mines. Labor rights was another tie to women's rights and civil rights. I was in Harrisburg during Three Mile Island, so I became involved during the "no nukes" campaign. I was in my early twenties. That was another decision I made about going into the mines. I thought, I'd rather have coal than nuclear energy.

At the mine I worked with three women and about 350 men. The filthy talk made me cringe. When I get mad I can filthy-talk with the best of them. But it's the vulgarness, not just the curse words, that bothered me. There are just a couple of men who get really filthy and vulgar. Some of the men don't like it either.

At first I ignored it. But about three months after I started one guy came on my shift and said he was going to do filthy things to me. He was walking the air courses, and he was telling me what he was going to do to me. I had a pick hammer, and I told him I was going to stick it through his head if he didn't get out of my way. We had a big battle. It was my first one. The next day he stood up on the bench in front of about 60 of the men and said, "I want you to know that I made a grave mistake, and I owe Bonnie an apology." And I'm sitting there thinking, he is just trying to humiliate me further. That's all I could think about. We didn't speak for about a year. I detested him. I would not forgive him.

He was one of the wildest and craziest of the bunch, and the other men knew if I wouldn't back down with him, that I wouldn't back down with any of them. And that broke the ice for me. He actually did me a favor. At a very early stage I had to take a stand.

After that they knew I wouldn't tolerate being harassed or abused. And I haven't had to deal with a lot of that. Behind my back I have heard really filthy things said about me, but never to my face. I've overheard things, but it doesn't bother me now.

It was one of the older miners who first became a friend and confidant. He said, "I'm not gonna let them mistreat you." He was the man I told I was pregnant when I was carrying Mitchell in the mine. I knew he wouldn't go blabbing to the rest of the guys.

*She continued to work through her pregnancy but saw a doctor often to monitor her prenatal health.*

I didn't want to cause fear in the men, because they had never worked around a pregnant woman, and I didn't want to cause controversy. It was my first baby, and I was really worried that I was doing the wrong thing, that I was gonna hurt the baby. I wanted to make sure if anything did happen, they could get me out. I told the superintendent, the president of the local, and this older man who was on my crew. He was great about it. He didn't act like I was going to kill myself.

I'd known other women who'd carried babies at other mines, but I didn't know if they did the same kind of work I did, if they were doing as much strenuous work. We were lifting crossbars.

At the time it was really scary. It was traumatic, because you're going through your first trimester, and you don't know what you're doing anyway. I had heard so many horror stories at the women's conferences about deformities, about women miners who had had bad pregnancies, and birth defects, and things of that nature. It scared the hell out of me, and my husband wanted me to quit.

I went to the doctor's very regularly—every two weeks for six months. I gave him some research from Kentucky about women working heavy jobs during pregnancy, and the study showed that if you'd done it longer than six months and your body was adjusted, it would be okay. He didn't take me off from working until I was six months and three weeks pregnant. My blood pressure started to fluctuate, and I had to stop working until I had the kid. I decided that I'd be tested, and as long as everything was okay, I'd work through it. I felt that I was young and healthy. The only thing I worried about was the fact that there is less oxygen underground. The baby's brains have a lot to do with the oxygen level in the mother's body. That's what scared me the most, that I would have a child who wasn't intellectual! [laughs]. But that didn't happen.

I was very tired during the first trimester. I'd work and I'd sleep, and that was it. I didn't have any morning sickness really. I had a muscle spasm in my back one time at work. I thought, here goes, I'm having a miscarriage. I didn't even leave the mine. It was during lunch, and I laid down on a bench. I worked through it, and nobody knew about it.

*Her coworkers were unaware of her pregnancy until a few weeks before she left work.*

I didn't start showing until I was about six months. So the men knew about it only about three weeks. One guy said he didn't want to work with me on the miner, that he was afraid and didn't want to deal with it. I said, "Fine." I just stayed on my shuttle car. I had a good job. For the most part, I don't remember it being a big deal. My pregnancy showed them all that I was a pretty normal person, that I was going to have a family. They teased me and said that I'm one of these people who want everything, that I want it all. And it's true. I am kind of spoiled, and I do want it all [laughs].

Since I only had three years in the mine, I only got 21 weeks of union benefits. I stayed off eight weeks afterward because I had a C-section. I didn't want to go back then, but I had to. I would like to have had more time off after he was born, but it didn't work out that way. So Mitchell's lived his whole life with a coal-mining mom.

On my first day back I bawled the whole way to work. I was already worried that I was on second shift and wouldn't see my son by the time he went to school. And now it's a reality, but I don't cry about it. It's not like I planned to separate the two parts of my life. I always planned to keep working.

When you have a new baby you're so tired, your whole life revolves around it and the need for sleep. You don't have time to think about anything but taking care of the baby and getting to work.

A coal mine is one place where you have to concentrate on what you're doing, no matter what else is happening in your life. If you have family dying at home, you have to put it out of your mind and do your job, because you have all these other people that you're responsible for.

One time I told the company, "I didn't come here to die." And it's true. I think the mines are fairly safe. It's the management that makes it unsafe, and the distrust, and the things that have been going on for generations and generations in the mines. No one should die in the mines. No one should be maimed in the mines. But it happens. It happens to everybody that spends any amount of time in there.

*In 1989 she suffered a serious burn accident underground.*

I was working at a battery-charging station, changing a battery on a coal hauler, when the power cable blew up. I had 5 percent of my body burned. It was mostly all second-degree burns, with spots of third-degree.

It's horrifying to be on fire. It's undescribable. The only thing going through my mind was, I have to get this fire off me. It was on my leg, but the flames were coming up. I thought, oh my God, we're going to have an explosion, and everybody's going to get killed. I was thinking while all this was going on, what is the most imminent danger here? All I could think about was whether or not the torches, with all the voltage burning, would ignite. I started feeling responsible for the whole mine even while it was happening. I didn't know I would be that much in control of what was going on.

I rolled in the mud to put out the fire. I had a mechanic there, and he grabbed a fire extinguisher and took care of that. I remember saying, "Don't touch me, don't touch me," when he was trying to help. They had teased me that, if I ever got hurt, they were going to take my clothes off [laughs]. And he kept saying, "Oh Bonnie, I'd never do that!" And I said, "No, just don't touch my leg!" One of the guys helped me walk to the fresh-air entry. The crew came down to see. I wanted to be carried out. I remember crying and thinking, this is going to be the most ungodly scar. And it is. They kept asking me if I was all right. And I kept saying, "I'm going to be scarred to hell and back" [laughs]. I remember looking up at the faces of my crew. They were devastated. They were really grieved that this had happened. And I thought, these guys really care.

The accident was hardest on Mitchell, I think. He was four. I stayed in the Kitanning hospital instead of going to a burn unit in Pittsburgh so he could come back and forth to make sure I was still there. He was afraid I wasn't coming home. He'd come to the hospital, and I'd have tubes in me. It was just morphine and sugar solutions to keep me from dehydrating, but it's rough on a little kid to see your mother with tubes up her arms.

It was hard on my husband, too. I've had crushed feet and neck injuries, and it hurts him. It's like seeing a battered woman, but it's your wife. But it was good in a sense, because when I went back, with all the fear I was dealing with, he knew how real my commitment was.

*She battled recurring nightmares of the accident. When she returned to work, her crew showed its support.*

I had to force myself to go back. It was the first time I thought, I could die there, and I could die so quick, and my family would be gone. I had dreams that it was going to happen again, and this time the electrical part would get me and kill me. If the cable would have hit me, it would have been 1,200 volts. I would have been electrocuted. Now I realize that the dreams are part of the process of psychological healing, but at the time I had a lot of fear. Going back, I had to do the same job on a daily basis. I'd have anxiety attacks and I'd have chest pains when it was time to change batteries.

I think the mechanic had the same anxieties, because he saw what happened.

There was some talk while I was off that I wouldn't change batteries again. I think they respected the fact that I didn't ask anyone to do it for me. I wouldn't ask someone to do something I was afraid to do, because what if something happened to them? Then you'd really have something on your mind.

The crew changed toward me when I came back. When I came back I noticed the guys were buddying with me more. Before, they'd be reserved about patting me on the back, even touching me, thinking, she'll have me in the office if I lay a hand on her. But when I came back there was a show of affection that had never been expressed to me before. They actually said they missed me, and that was something I never thought I'd hear! [laughs]. I never expected to hear that if I stayed there 20 or 30 years, because men aren't like that.

On a crew you get close. Everybody has a role. Every section needs a clown, and I'm the crew goof-off. I like to carry on and have fun. We even have someone who does impersonations. We're a good combination. If they want to fight with someone about the union, they always come to me. I took a lot of slack over the Senate confirmation [hearings] of [Clarence] Thomas [for the Supreme Court]. It was good, because they'd ask me what my opinion was. And they'd razz me and call me "Anita" and I'd call them "Thomas" [laughs].

Guys will talk to me about things they won't talk to their wives about, or to other guys about. If they're expecting their first baby, they will ask me questions. It's like a brother-sister thing. Or if they've had a fight with their girlfriend, they'll ask me, "Do you think I did the wrong thing?" Sometimes they'll ask me things on a personal level that I don't think they'd ask each other. And they know if they tell me something, it doesn't go anywhere else.

In the mining industry you have to depend on each other, regardless of your color or your sex. So I think that does make you closer, and I don't know if that's even true in all nontraditional jobs, that type of dependency.

*The death of her father in California triggered deep feelings of grief, and her crew's compassion helped her cope with the loss.*

When my dad died, it came as a shock to them—and to me, too—how they reacted. I cried down there. I was really bawling. They had a very compassionate reaction. It was like, "Are you going to be okay?" and, "Let's get her out of here as quick as we can," and, "We're sorry that you're going through a rough time." It was an understanding that was real helpful when I went back to work. I mean, they're an integral part of your life, and it's always better if it becomes a caring relationship. It's good, and it's healthy. I feel good about these men, even though we can fight and they can pick at me and we can carry on. It's an open thing, and it's not because a lot of them love the person I am, or because I love the person they are, but it's an understanding that we are who we are.

With a couple of the men, I feel it will be a lifelong friendship. We may get laid off and not see each other often, but when we do it will be like seeing my brother in California. Everything just clicks back into place, and it's like we haven't been apart since birth. But it takes respect before it becomes friendship. Something makes them have a lot of respect for you, and they think, I want to get to know this person a little bit better. It's not something that I've intentionally done to spark the relationship. I forgot about if anybody was going to like me and decided that I was gonna be who I was, and that I was going to be outspoken about what I believed in.

I think my honesty and my openness in the predominantly male workforce is what drew them to me, because a lot of women wouldn't go in and say, "I believe in women's rights, and I believe in civil rights," especially because the workforce is predominantly white men. And they think, jeez, why is she saying that? She knows she's going to start a controversy every time she opens her mouth.

A psychologist once told me that my problems stemmed from the fact that sociologically I was a man. He said, "You think like a man, you work like a man, you act like a man." I said, "So I'm a man just because I had to fend for myself in this life?" It blew my mind. I thought, something wrong with me socially? I thought, no, there's something wrong with you, buddy [laughs].

There are always going to be certain stigmas until we start teaching our children at birth not to let society dictate what is correct, so they will not see gender as an issue.

*In raising her son, Boyer has tried to teach him about the value of fairness and equality.*

I try to raise my son nongendered. There's times that I have to shake my head, because he comes out with things that are sexist. I was going to pick up beer for my husband, and Mitchell said, "Mom, women can't go get beer." That was something he and his dad did in the truck. It was a male thing to do. So I said, "Well, get in the car." I just put him in the car, and we went and got a case of beer so he could see this was not a male-female thing. Children can pick up subtle things like that, and you don't realize it. They start dividing what males should do and females should do. You really do have to watch your children closely. I was more apt to wrestle around with my son than my husband was, because I didn't want him to think that was something that Mom couldn't do, too. When he was very little I tried to implant certain memories and certain things, like bringing him to women's conferences.

I also try to take Mitchell to places where he is in interracial situations. Since we live in a predominantly white neighborhood and his school is mostly white, I've put him in situations where there were more black children than white.

I would like nothing more than that Mitchell be a lot like his father. My husband has had a major impact on my life. He has always been a caretaker. He is the kind of man I want, a family-oriented man who doesn't have qualms about taking care of his children and doesn't have qualms about his wife leaving home and having her life, too. From birth until they're grown up, it's important for boys to take care of themselves and learn to be a primary caretaker in the household.

*She devotes long hours to working with her union and the Coal Employment Project.*

By the time I combine the Coal Employment Project, the Coalition of Labor Union Women, and the United Mine Workers, it probably takes two or three days a week out of my life. But then, that is my life, and my family has got to understand that. It's like Nelson Mandela said: he spent half of his life in jail, and he said, "Well, the struggle is my life." And I guess that's maybe how women have to think if they really want change. They're going to have to put forth some struggle. And I don't consider myself nothing like Nelson Mandela, but this is my life.

When we came down for the pro-choice rally in Washington, D.C., I had acute bronchitis. I'd been sick for two months steady, and I couldn't get over it. That morning my husband didn't feel that I was in good enough health to be going. He was afraid my lung would collapse. I was really sick, but I started getting pumped up for the trip.

I was all charged up, running around the house, grabbing sleeping bags. He came out of the kitchen, and I was on the phone talking, and he just looked at me. And he has this look like—you can see that he's glad that I'm who I am, I'm bubbling over, and he makes me bubble over that much more because he ends up getting excited that I'm so excited about it. That type of understanding is the greatest thing there is, I think, in a marriage. And that made my day better. It always makes my day better when I get to see that look on his face. He looks like, "Okay, you win, and I'm glad for you" [laughs]. I think it makes for a healthy marriage.

*When she is underground and away from coal production, the mine soothes her spirit and calms her mind. She says the mine's darkness helps her think more deeply about her life.*

One time when I was little a friend said that, if someone was out to hurt you, they could see you in the daylight but in the dark you could hide. I feel like I can go anywhere in that mine and lay down and turn my light out and not be afraid of anything. If you're working out in society, or working in stores, or in the hospital, you never know what you're coming in contact with.

In the mine you're working with nature, and you can usually trust her. Everybody who even talks about a mine talks about it as being female. She might get wild, but I just feel like that's my sanctuary, in the darkness of that mine. That's where I can think completely clearly if I'm alone and it's completely black. I like the blackness. I like the darkness. It's like self-hypnosis.

I don't know how to explain it. My day can be so full and that half-hour lunch break is the only time I have to regroup and relax. A lot of times, when I was having trouble in there or having a bad day at home, I would just turn out the light and go through my mind and say, what could I have done to prevent an argument, or to make Mitchell go to gym class [laughs]. Just crazy things.

Of course, there's always danger. I feel the sanctuary in a spiritual sense but not in a physical sense. I don't feel like it's going to be safe for me all the time. But in a spiritual sense I feel that it's a place for soul-searching.

In the mine I never feel closed in. Our mine has four or five miles of entries. It's a house with many rooms. Your mine becomes like your life, and your childhood and your adulthood and everything in between. It's a house with many rooms because there are places where it's bad and there's danger, and in your life there are places where it's bad and there's been danger, too. I've even thought of writing a short story on this, because it lays so heavy with me.

When I have to crawl through an air course on a safety inspection, because of fear I feel a kind of imprisonment there until I get to the other side. I know I'll get through it, but when I get to the other side, I'm relieved, like I have been relieved to get through other things in my life.

And when I go and see good roof and good conditions, I think about sitting in my living room and glancing at my cupboard full of china doll figurines with the little light shining on it at night. So all the time it's a reflection on my life. It will soothe my mind, the reflections of the mine and the reflections of home. I paint pictures in my mind all the time.

Music does the same thing. I remember the first time I walked into the mine. One of the old albums I had was *Emerson, Lake and Palmer*. I entered the mandoor and thought of "Welcome back my friends, to the show that never ends. Come inside, come inside." Do you remember that? And that's exactly the thing I thought about when I walked through the mandoor. Like, oh God, here's the circus [laughs].

My mind has a way of soothing anything for me, which is good. Some people meditate. I let thoughts flow through my brain. They say that's wrong, that your mind should be blank. But jeez, I'd hate it to be that way [laughs]. It relaxes me.

You love the mine that you work in, you love the sanctuary of it. I don't know why. I think men feel it, too. I used to hear these pensioners say coal mining was in their blood, that their daddy did it, and their granddaddy did it, and their granddaddy before them did it. Well, I have no ties to coal mining before me, but it's true that it gets in your blood.

I think that even the men feel that type of unity with the mine, because they always refer to it as a woman. And I think it's like their relationship with a woman. They're usually either very close or very far apart. It's a natural relationship. It's like nature is saying, "Encompass me, I'm here to provide for you." And I think that's why they give the female gender to the mine—because it's there to take care of all these people. It's our livelihood, and for the most part in modern history it's a generous livelihood.

If I feel that way about it, and I'm in sync with it, I feel like I'm not going to be bitter. My grandmother had a hell of a life, but she wasn't about to die bitter. And I guess I've always thought in my mind, I'm going to be just like Grandma. I'm going to be happy, no matter what happens. If I'm poor, I'm going to be happy anyway.

You have choices in life, and I think one of the major choices in life is whether you're going to harbor hate and be bitter about things or whether you're going to love and enjoy and forgive and be happy about things. I think that's what life's test is. It's not that I don't get mad and could choke some people. But for the most part I'm very forgiving, and I mend fences, and I feel bad if I don't.

I think bitterness always comes from repeated battering and how you deal with it. You can let it eat you alive, and some people let it do that to them. Other people shrug it off, ignore it, don't let it get to them, or have other ways to fight it. I think I'm lucky that way. I'm a fighter. And I can understand someone getting bitter because they can't scream and they can't talk and they can't fight back. And I think that's the type of people who harbor a lot of bitterness, and I feel sorry for them.

*Working in a traditionally male occupation, she believes, has provided her with greater freedom and equality than would be available in other jobs.*

When I went in the mines I was single, depending solely on myself for the first time in my entire life, and that mine became my sanctuary. That's where I became a woman full of self-esteem and in control of what was going to happen in her life. So the rewards have been immense for me.

Traditional women's jobs wouldn't have done that for me, because in most of them women work in a system that's male-dominant, where males are the high breadwinners. In nontraditional jobs women are at an equal pay level and an equal work level. In there, men will say, "You're getting paid the same as me. You'd better do your share." And in traditional women's work I don't think that's the case. I think women do their share there, but they're not getting paid for it, and they're not getting the respect that you do in nontraditional work.

And when you're a married woman with one child, economically you're at a better status than a lot of the men. So it has evened things out a lot. It makes a difference in how they feel about you. And the men I work with know I'll do anything. There's nothing that I won't do. I think that's what they like about me. If it means I have to get down on my hands and knees and scrub a floor at the unemployment assistance fund office, I'm not too good to do it. But I'm not too good to stand on the front line of a picket line in a dangerous situation, and they've seen me there, and they know I'll do it. I'll do whatever it takes. I think that it took them about five or six years to realize that.

*During the Pittston strike in 1989–90, Boyer helped organize strike support and coordinate activities in her area.*

The Pittston strike was the first time that I was on the front line of a picket line. I was really pumped up about the strike. I was pumped up because they were pulling in Jesse Jackson, they were pulling in all the social movements. I thought, it is time, it is time we are doing this. It was great. It was fabulous. I believed that everything we were doing was right. I loved it because we were fighting for the pensioners' health care as well as for a contract.

I was pretty highly involved in the structure of the strike in this area. I was what they called a general, with authority over maybe 60 people. It really scared me, because I thought, oh my God, these guys aren't going to come because they're going to hate this authority! And I even raised the question to people who put me in this position about what would happen if the men didn't like the idea of my being here.

One time, we were on a picket line at a rally in Indiana, and I had to stand up and sing "Solidarity Forever" in front of a crowd, because I was the only one who knew all the words, Later, at a union meeting, the guys got these little songbooks, and they were singing "Solidarity Forever" like a choir, while a guy on one of my teams came walking in [with] a present. It was like a trophy, and we laughed. I was the only general who got one. It was like, you've done a fine job, and this is a souvenir.

At Pittston the mineworkers tried to build the solidarity that Martin Luther King talked about. He said during a garbage workers' strike that civil rights and labor rights cannot be separated. Somewhere along the line we lost that. We tried to regain it at Pittston, but shortly after we won it was forgotten again.

That's something that women in particular need to remember. That's what's going to beat the government's attack on civil rights, women's rights, union rights. I don't care what other people say about something I see as wrong. I don't care what the government says, or anyone else. I have the right to say it's wrong. That's why I'm here, that's why I'm an American. Don't get me wrong—I think we have the best system there is, and I'm not anti-American. But I'm not always pro-government.

We're going to have to unite and take charge to get what we need. In ten years I've gone from [being] a person who was shy and who was scared to death to travel by herself to someone who goes anywhere alone and who is very open and very verbal. I guess the mine has helped me there, too, along with the failure of my first marriage. That combination made me a better person. It was rough. But for me that was the route to a better life.

My first marriage was difficult because there was a love for my ex-husband, a lot of broken dreams, and I had to overcome the feeling I had failed. By the same token, it took me into the mines. It was the first time in my life that I felt very confident that I didn't need anybody, that I was going to survive. It was a relief.

When I think back on it, everything tied me to the mines. I believe that I'm there for a reason. I'm there because that is what allows me to tie everything together. To be who I am. That's what allowed me to get involved with the women's movement and civil rights issues, to become involved with political issues dealing with poor people, old people, things I believe are right. I haven't gotten all the answers as to why that's my place in life. There's not a doubt in my mind that it's why I'm here.

*In 1993 Boyer's local union was affected by a widespread selective strike called by the UMWA after negotiations stalled on a new contract with the Bituminous Coal Operators Association. She was on strike for nearly a year and views the strike as the beginning of a new phase in the union and in her own life.*

It was the beginning of a growth in awareness in our local, and it was an awakening for me. We had to wake up to reality. We were out on strike for seven months, and then we came back to work and had a layoff. We saw that we were losing jobs rapidly and the industry was fading away. The mining workforce is becoming extinct, and women have taken the brunt of it. We were pioneers, and 20 years later we're dinosaurs.

The strike made me more decisive about what I want to do within the union, and what I want to do within the Coal Employment Project. The rank and file need to take charge of their destiny, and I think they will. We're working longer hours now, and we've lost some jobs. But if you back people up against a wall, they're going to fight back.

I am tired of the good-old-boy system, which seems to be getting worse. In the 1960s people made progress, but now there is more tokenism in the labor movement. They throw women and blacks and Hispanics a couple of bones. That's not progress. The women have had a good working relationship with the UMWA, even if we didn't always agree.

The issue in the strike was double-breasting, where a union company splits off and runs partly nonunion. I was on the outreach committee and talked to other unions and women's groups and the Coalition of Labor Union Women. They were great. There are some brilliant women in that network who have served as mentors for women in unions. The small group of nontraditional women in CLUW have tried to educate the organization about the fact that you can still be a gentle-hearted woman and do men's work.

Those of us who work nontraditional jobs don't want to be behind a desk. Most of us thrive on physical work and love heavy equipment. I feel as comfortable in my muddy suit and work clothes as I do in a designer dress and shoes. I spend most of my waking hours underground, but my house reflects the feminine side of me. I go to work in the dark, but I come home to pastels.

As far as my job, I'd like to get 20 years in and be out at 43. That would be great. I'd have a good pension, and then I'd have enough links to still be involved in everything. I don't want to change. I want to be a coal miner. But I've come to the realization that I'll probably have to change. I'm scared, because I've been a miner so long. It's who I am. It's provided me with the money to do the things I want to do. It's provided me the self-esteem and a way to be real involved in the union and in these groups.

If I get laid off, I would like to go back to school in pre-law and politics. I have worked for years in the political arena in the union. I've lived it, and I've worked in it. In law, because it's a private-industry thing, I can do it at 45. Actually, if you have a few gray streaks, they'll probably say, "That's the one we want—a lot of experience there" [laughs]. And I am experienced, because I'm streetwise and socially adept. I would love to be a lawyer, but I would love to be a 20-year coal miner, too.

*In 1994 the coal industry was deunionizing and laying off workers, while production continued to increase. If a new generation of women enters the industry, Boyer wants to be the first to help them.*

There may not be any new women right now, but if there is any one of us left alive when the coal industry starts booming again, I know we'll be right there for them. I don't think any of us would let a woman go into the mines without the support system we have created for ourselves and other women.

Other women broke the ice for me. I didn't know it at the time, but they made a huge difference. Look at the women who worked in World War II. We are the daughters of that generation. That's what we have to try to be to young girls, to show them they have the ability to do these jobs.

Another important thing for the future is that all minorities come together under one umbrella and stop letting this mighty white government cause dissension among us. Abortion splits women right down the middle. But the real issues in the United States are economic, like the fight for pay equity for all people. I'd like to see a slimmer margin in that white male dollar to a woman's 70 cents. I'd like to see it move up to at least 80 cents during my lifetime.

I've lived in an environment where my family's women have always died in poverty. I know a lot of people at 20 don't think about getting a job with a pension. But at 22 years old that's all I wanted, because I saw what a hard time my grandmother had.

I want to be able to take care of myself. I want to pick and choose. Even if it is in a nursing home, I want to have my own little paid-for corner so nobody can push me around. If I can have my own corner and still go out and vote, it would be okay. A pension is really important to me to prepare for my old age. I want to live a long time, and I think I'm going to. I want to live to be 100, so I can reflect on my life.

# Afterword by Kipp Dawson

There are two ways in which even the most attentive reader might miss the full meaning of the histories in this powerful book. Each narrative can be read on its own for its drama and for the picture it gives of a strong, persistent woman. Or, read from the vantage point of the mid-1990s—when few women still work in coal mines in the United States and unions and the labor movement, which were critical to most of their experiences, have dwindled to mere shadows of their previous selves—these stories can seem like personal remembrances of times so different as to have little relevance today.

The women whose stories make up this book are representative of, and only a small part of, an important phenomenon in recent American history that is particularly significant in these hard times. While each of these women is unique and has a special story to tell, what is even more important is their collective ordinariness and the qualities they shared—and share today—with their working-class sisters everywhere.

During the 1980s, and continuing into the 1990s, working-class people in the United States (and most of the "West," including Sylvia Pye's Britain) have found themselves being dramatically pushed backward economically and losing much of their collective strength as unions have shrunk in size and power.[1] The women we have met through this book were among the approximately 4,000 women miners who were more determined, willing, and able than most other workers in the United States to stand up to, and resist, the attacks on workers, minorities, and women during the "Reagan years." As a whole, coal miners in the United States, organized in the United Mine Workers of America, and in Britain, through the National Union of Mineworkers, were among the most tenacious and more successful examples of dynamic, nationally organized resistance. Miners managed to hold onto more of their union than most workers did during those years, and in the United States they were able to stave off, if not stop, the worst of the attacks. In both countries women were key to these battles.[2]

In the United States women miners played a unique role, not only in their

Judy Simons, Ohio miner, 1983

own union but in the labor movement and the working class as a whole. From 1977 on, through their organization the Coal Employment Project (CEP)/Coalmining Women's Support Team (CWST), women miners organized annual conferences, workshops, national and international tours, a rich monthly newsletter, and a network of communication and solidarity that arguably led them to play a more consistent and longer-term educational and solidarity-organizing role than any other labor organization in the United States. The CEP began as an advocacy group for women seeking, and working to hold onto, jobs in the mines. Quickly it became a network of women miners working to win acceptance in and build their union, fight problems specific to women miners, and campaign for improvements in their lives and those of all miners. From there it was an easy and natural step for the CEP to become a champion of, and gathering place for, union fighters from many unions and several countries. During the mid-1980s, when most union gatherings were dominated by discussions of how much to give in, CEP conferences were spirited rallies of women miners and their supporters on behalf of the UMWA, women's rights in the UMWA and the mines, other workers on strike in the United States, and unionists from other countries,

Longwall crew in Peabody's Harris No. 2 coal mine in West Virginia, 1982

with a special emphasis on, and relationship with, British miners and their wives.[3] As several women in this collection testify, CEP experience helped many women miners gain self-confidence, leadership training, connections with other women miners and other workers, insight, and a perspective that few American workers had available to them during those years.

While these women miners were not able to stem the antilabor trend or leadership backsliding that still cripple unionism in the United States, their work gives us both a rich heritage and a source of hope for the future. The common thread in their stories is a mixture of tenacity and need that characterizes the lives of far more women than just those lucky enough to have found their way into the coal mines during the brief openings of the 1970s and 1980s. Millions of women share the need to find decent-paying jobs through which they can support themselves and their families and gain dignity and self-respect. Millions of women need the kind of support these

Vira Rose, mine foreman, McDowell County, West Virginia, 1983

women miners forced from their union and organized for themselves through the CEP.

These women miners were unique in that they found jobs that many women would have jumped for—often only after long fights to force the operators to hire them, and sometimes sliding in on the coattails of other women. Either necessity or nature, or in most cases a combination of both, gave these women the extra strength to "make it" in an especially difficult work world. These same qualities led many of the women to become activists in, builders of, and leaders of the Coal Employment Project and gave the CEP its unique character, its tenacity and independence. Women miners also brought these characteristics to their union and strengthened it in the process. But it is critical to appreciate that these qualities are not limited to the few women who found their way into the mines. The women who became miners, and those who speak through this book, are much more like than unlike their working-class sisters across the country and, indeed, around the world.

It is equally important to understand, as many women miners grew to, that no woman could have been hired in any mine in the United States in the 1970s and 1980s had it not been for the atmosphere and victories created by the civil rights movement, followed by the women's rights movement. Whether they knew it or not at the time, each of the women telling her story in this book, and every other woman miner, was able to burst through barrier after barrier because others had successfully challenged similar barriers in *society as a whole.* In fact, coal mines owned by major steel companies were among the first to hire women as the result of a successful affirmative action court battle initiated by black steelworkers in 1964.[4] The women who broke into the male world of coal mining certainly contributed to social change and were contemporaneously hailed as symbols of women's progress in particular. But even the strongest among them could not successfully have challenged those stone-hard closed doors alone.

In the United States of the mid-1990s, it is easy to forget—or to grow up never knowing—that Americans have forged powerful movements that mowed down inequities and injustice and that working-class people, and women in particular, have taken on and beaten back forces seemingly much larger than themselves. Should such amnesia, or ignorance, become pervasive in this country, our children will be doomed to atomized, dehumanized, impotent lives. The stories these coal-mining women tell, and the collective story of their energies and achievements and growth, are important weapons in the fight against that ignorance.

Never before have we, and our children, so needed examples of people ready and willing to reach out to others in need to forge a common fight for humanity, for justice, for dignity, for a future for our planet. We have other bigger and more significant sources of such examples in U.S. history, including especially the civil rights movement. Still, this small group of unusual women provides us with one particularly important example in that they were females in a male world, working-class women fighting in a time that was bad for working-class battles, female "rednecks" and African Americans and Navajos and Anglos from Alabama to Kentucky to Utah to Wyoming who forged iron-hard links with one another and with people like themselves on every peopled continent. Working underground, they reached for the stars. If these women could do what they did, surely many more can jump similar hurdles in the future, when collectively the people of this country have remembered their commonality and their need to work together to build a future worthy of our children. Those who do so will have the struggles, trials, defeats, and above all the collective growth of their female coal-mining predecessors to build upon.

# Appendix 1: The Interviews

*Part 1: The Women Before Us*

Cash, Arna. Madison, West Virginia, 15 September 1980.
Crawford, Elizabeth Stull. Dillonvale, Ohio, 31 July 1982.
Dolin, Irene Adkins. Julian, West Virginia, 23 May 1992.
Fulford, Alice. Columbus, Ohio, 30 July 1982.
Hawkins, Audrey Smith. Barnesville, Ohio, 30 July 1982.
Hepler, Billy. Lowber, Pennsylvania, 22 November 1982.
James, Alice. Helper, Utah, 24 June 1991.
Kelly, Madge. Rock Springs, Wyoming, 20 June 1985; telephone interview, 18 August 1994.
Krmpotich, Helen Korich. Rock Springs, Wyoming, 19 June 1985; telephone interview, 18 August 1994.
Lunsford, Minnie. Dartmount, Kentucky, 12 August 1983.
McCuiston, Ethel Dixon McNight. Cumberland, Kentucky, 6 May 1980.
Netchel, Martha. Pottsville, Pennsylvania, 13 March 1983.
Rigas, Liberty. Martins Ferry, Ohio, 12 April 1983.
Rogers, Delpha. Corbin, Kentucky, 4 September 1980.
Smith, Ethel Day. Evarts, Kentucky, 25 June 1980.
Steele, Susie. Woodbine, Kentucky, 29 June 1980.
Stevens, Elizabeth Zofchak. Woodsfield, Ohio, 1 October 1982; Poolesville, Maryland, 24 March 1992.

*Part 2: The New Miners*

Anderson, Ruby. Pound, Virginia, 4 March 1981.
Angle, Barbara. Keyser, West Virginia, 26 February 1981.
Askew, Eska. Madisonville, Kentucky, 30 October and 21 December 1991.

Baker, Pam. Jenkins, Kentucky, 6 May 1980.
Baldwin, Pat. Madisonville, Kentucky, 31 October 1991.
Barber, Sandra Bailey. Mayking, Kentucky, 2 March 1981; Abingdon, Virginia, 26 February 1994.
Beverly, Kathy. Dorothy, West Virginia. 6 July 1980.
Boone, Shirley. Crichton, West Virginia, 22 September 1983.
Bowen, Nancy. Williamson, West Virginia, 1 March 1981.
Boyer, Bonnie. Shelocta, Pennsylvania, 19 November 1991; Washington, D.C., 26 March 1993; telephone interview, 26 August 1994.
Brock, Brenda (with Pat Estrada). High Splint, Kentucky, 9 May 1980; Harlan, Kentucky, 20 May 1992.
Brock, Tammy. Beckley, West Virginia. 26 June 1980.
Brown, Patricia. Bessemer, Alabama, 25 August 1983 and 6 March 1994.
Burden, Lorene "Granny." Anton, Kentucky, 1 November 1991.
Davis, Carol. Waynesburg, Pennsylvania, 14 October 1992 and 25 May 1994; telephone interview, 5 August 1994.
Dawson, Kipp. Pittsburgh, Pennsylvania, 20 November 1991.
Dorsey, Sandy. Martins Ferry, Ohio, 12 April 1983.
Estrada, Pat. Knoxville, Tennessee, 23 January 1981.
Fraenzl, Clare. Bentleyville, Pennsylvania, 20 November 1982.
Fraley, Patsy. East Point, Kentucky, 22–23 May 1992.
Fraley, Tom. East Point, Kentucky, 23 May 1992.
Griggs, Charlene. Carbon Hill, Alabama, 21 August 1983 and 6 March 1994.
Harman, Joyce, and Barbara Angle. Petersburg, West Virginia. 7 August 1983.
Horner, Martha. St. Louis, Missouri, 24 and 26 June 1994.
Howard, Bonnie. Ermine, Kentucky, 6 May 1980.
Huitt, Joy. Arlington, Virginia, 12 September 1986; Helper, Utah, 23 June 1991; telephone interview, 22 July 1994.
Hyche, Shirley. Tuscaloosa, Alabama, 7 March 1994.
Keene, Dorothy and Mitchell. Oceana, West Virginia, 16 June 1983.
Kifer, Ruby Carey. Chandler, Indiana, 4 September 1986.
Laird, Elizabeth. Cordova, Alabama, 4 February 1981, 23 August 1983; telephone interview, 16 August 1994.
Lattimore, Geraldine. Birmingham, Alabama, 4 February 1981.
Lewis, Barbara. Lynch, Kentucky, 29 October 1991.
Lindsay, Libby. Crawley Creek, West Virginia, 24 May 1992.
Luna, Evelyn (Evie Tsosie). Kayenta, Arizona, 11 November 1986; Washington, D.C., 24 March 1992; telephone interview, 19 August 1994.
Magan, Doris. Institute, West Virginia, 30 June 1981.
Mayernik, Margi. Bentleyville, Pennsylvania, 20 November 1982.

McGill, Cheryl McMackin. Arlington, Virginia, 17 January and 1 February 1992.

Miller, Michelle. Bentleyville, Pennsylvania, 20 November 1982.

Miller, Rita. White Plains, Kentucky, 31 October and 21 December 1991.

Molineaux, Janice. Washington, D.C., 19 January 1992; St. Louis, Missouri, 25 June 1994.

Nance, Pandora. Madisonville, Kentucky, 30 October and 19 December 1991.

Parsons, Pamela. Penrod, Kentucky, 20 December 1991.

Pitts, Rosa. Marianna, Pennsylvania. 27 April 1983.

Pye, Sylvia. St. Louis, Missouri, 26 June 1994; telephone interviews, 26 and 31 July and 3 August 1994.

Raisovich-Parsons, Linda. Welch, West Virginia, 12 August 1982.

Robichaud, Julie. Anton, Kentucky, 1 November 1991.

Rose, Vira. Skygusty, West Virginia, 1 February 1983; Gary, West Virginia, 24 May 1992.

Shine, Paulette. Morgantown, West Virginia, 11 April 1983.

Simons, Judy. Martins Ferry, Ohio, 12 April 1983.

Thomas, Andrea "Ting." Marianna, Pennsylvania, 24 May 1994.

Tipton, Janice. Bessemer, Alabama, 25 August 1983.

Totten, Cosby. Tazewell, Virginia, 7 February 1981; Thompson's Valley, Virginia, 29–30 August 1983 and 25 May 1992; telephone interviews, 16 and 30 August 1994.

Williams, Mavis. Cumberland, Kentucky, 29 October 1991.

# Appendix 2: The Old Works

*The "old works" are mined-out, abandoned areas of underground mines. They often are the site of ghostly tales, and they offer clues about earlier mining methods and the workers who employed them.*

*One story was passed on to me by Linda Raisovich-Parsons, a UMWA safety specialist who worked underground in McDowell County, West Virginia. She recalled working with a mason in the 1970s to build a "stopping" or block wall after a crew mined into an old works of a older, neighboring mine. Walking into the old section, they discovered old tools and antique rails laid for small coal cars. At the dusty dinner hole, Linda and her coworker leafed through a stack of old newspapers from a quarter-century earlier. The chalked date of the last fireboss, who checked the section, was still marked on a timber prop.*

*Like the old works of an underground mine, this section contains pieces of the past, in dusty, outdated language, a sampling of documents that may shed light on the women who have gone before, and public reactions to their presence in the mines.*

## From *Black Diamond* 15, September 14, 1895

It used to be a very common custom for girls to work in the coal mines of Great Britain, and it is only of comparatively recent years that they have discontinued to do so. The last that was heard of this particular custom was the employment of girls on the "Pit Brown," known so well as the "Pit Brow Lassies." In some places, however, young girls of tender years are still hard at work in the depth of the coal mines, and with very disastrous results.

In this connection it has been learned that four athletic young girls find daily employment at a small coal mine in the Mahanoy Valley, several miles from Shamokin, Pennsylvania. The colliery is owned and operated by Joseph Mans, a hard-working German, who says he has simply introduced the customs of the fatherland in having his four daughters assist him in preparing the fuel for market. The girls are six-footers, good looking and well-formed, each tipping the scale in the neighborhood of 200 pounds. Katie, aged 20, has charge of the breaker; Annie, aged 16, runs the mine pumps and breaker engine like a veteran engineer; Lizzie, aged 18, drives a mule attached to a gin for the purpose of hoisting the coal from the slope; and Mary, aged 19, sees that the slate is picked from the coal by her little brothers, whom she helps in the work. Mans formerly worked in the mines at Shamokin, but during the past twelve years, with the assistance of his wife, who runs the farm, and their daughters at the mine, he has managed to buy this coal mine and a large amount of timberland besides.

## *Those Black Diamond Men*: A Tale of the Anthrax Valley

*This 1902 novel, by William Futhey Gibbons, was based on social conditions in Pennsylvania's anthracite coalfields when child labor was common. The "breaker boys" bent over conveyor belts sorting coal in dark, windowless tipples that served as factories at the mine mouth. Gibbon's story features a "breaker girl" who tries to hide her gender. This passage concerns the discovery by the breaker boys that one of their coworkers is a girl.*

### Chapter 22: Strike of 'Malgamated Terrors

It was toward the last of March, a year after the fire, that the new breaker was finished and ready for work. The strike had slowly worn out the endurance of the men during the long winter, and now that work was offered again, most of the men were ready to take their old places. But if the men had been starved into submission the boys had not, and they were only waiting for an opportunity to show their power. The opportunity came during the first day's work in the new breaker. It was Jim Owens who called Mick Phelan's attention to it as soon as work was over for the day.

"Say, Mick, have y' seen the new slate picker?"

"Aw, there's no new boy come to the works. Don't I stand where I c'n see the office? There's been nobody there to-day but a dago woman and a gir-rl."

"Sure, that's just what I'm tellin' y'. She got the job near me that young 'Spike' Dolan used to hold."

"What's his name?"

"Haven't I told y'! It ain't a he, it's a she."

"It's whata-a-t?" Mick Phelan's jaw fell in sheer amazement. Mick was the bully of the breaker, but the new girl could have vanquished him, if she had been present to take advantage of his collapsed condition. "A gur-rl, is it? a gur-rl—and a dago at that! The howly saints! Aw, it's a lie y're tellin' me, Jim Owens."

"It's no lie. Bat McCarty told me he was [near] the office door when the woman and the gur-rl come in. She's to take Dolan's job in the mornin'. I've heard me father say that there was places in the old country where the gur-rls worked around the mines the same as the boys—"

"Let sich foreigners as thim work as they likes in their own country and let them stay there. But if they comes to Ameriky, they must do as we does. It ain't dacent to be sendin' a gur-rl into the breaker, an' I won't have it. We'll hold a meetin' after supper. Sound a call as y' go down the lane."

Mick Phelan's word was law, for was he not the king of the slate pickers and chief of the 'Malgamated Terrors? But like some other leaders he chose to seem to defer to the will of his gang of Terrors from the Mudluck breaker.

. . . Taken one by one and away from the Mudluck breaker the Terrors were not half so bad as they wished to appear. But when they had congregated, unwashed, in the lee of the culm dump, each urchin with a reputation for noise and mischief to sustain, it was apt to go rather hard with any unsuspecting stranger who might chance to pass.

The Amalgamated Terrors owed their organization to a former strike which had involved all the railroads of the region. While their fathers were organizing a strike in sympathy with the railroad men, the boys, being thrown out of work, organized out of sympathy too. Their first effort in the line of sympathy was to stone the trains manned by non-union men who had been taken on to fill the places of the union strikers.

. . . Mick and Jim took their way to the top of the culm heap which lay next the village of Mudtown and paused a moment, before descending to the meeting place of the Terrors, to give the final signal by which a meeting of the club was called, three short blasts followed by one long one blown upon Mick's grimy fingers.

Most of the members of the 'Malgamated Terrors had come up by this time. Many smoked and all of them swaggered, each one copying some trick of manner or speech admired in his father or some stage hero. Mick Phelan was beginning to take more interest in pugilists than in mere actors. Mick had begun to outgrow the small suburban ambition of being "the toughest kid in Reagan's Patch" and was beginning to send out challenges for pugilistic encounters with the "breaker bullies" or the "feather-weight" champions of other communities.

As for the sending of these documents, that was really managed by the faithful satellite Jim Owens, who signed as well as wrote them. Mick could not have signed his name to anything, for he did not know how to write, his entire education consisting of the first three lessons in the primer, which had been thumped into him by various teachers during the intervals of playing hookey which had occupied the two years when the state had his name upon its schoolroll.

Among the last to arrive was an under-sized, unwashed imp known as Bat McCarty, the clown of the breaker and hence a privileged person. Looking about the ring of assembled Terrors, he asked in a high-pitched voice, "Who'll lend me a match?"

Several were proffered. Accepting the first one offered, the gamin prepared to strike it on his trousers' leg. Then suddenly arresting his hand, he made a motion towards his mouth as if he had forgotten his pipe, tapping his pockets one after another in succession.

"Sure, now, if I on'y jist had the loan of a pipe an' tobacky, Mick Phelan, I'd be havin' a shmoke!"

A shout of laughter greeted Bat's sally, and Mick, with the instinct of a budding political boss, produced both pipe and tobacco.

Jim Owens stated the question before the meeting: "What are the Terrors a-goin' to do about the dago gur-rl?" Mick Phelan, as befitted his dignity, said nothing. But with the foresight of a true politician he had prompted several of his henchmen to express his opinion. This soon set the passions of the Terrors ablaze.

The meeting was a stormy one and the angry Terrors clamoured for an opportunity to show their strength. Here at last was something of importance which called for united action on the part of the society. At the end of an excited outburst, Mick kicked against the door to enforce silence and then announced the deliberate action of the club, delivered in his best high-tragedy voice,: "This is more'n the 'Malgamated Terrors is a-goin' to stand. Brother Terrors, I order a str-r-rike!"

Mick did not rise the next morning when his mother called him. When his father was on strike he was accustomed to lie abed late. It was not until his mother appeared by his bedside that he rose. Even then he did not tell her why he slept so late. When he had reinforced his courage by his breakfast he ventured to tell her that the breaker boys had agreed to go on strike. For a moment she stood with her arm outraised holding the knife with which she had been cutting bread for the children. Then dropping the knife she towered over Mick threateningly.

"So you will go on strike?" she said, reaching down for his coat collar. "So you will go on strike, will you?"

"Hold on!—Mother!—Howly snakes—Don't!—Wait a minute!—Do you mind,—we've got to strike.—We can't have—a gur-rl dago—pickin' slate—"

Mrs. Phelan had wound her fingers into Mick's collar in order to steady her son while she punctuated his protest. The last vigorous spank, a well-directed and uplifting stroke, fairly landed him in the yard.

"Now gwan out o' this. To wor-rk wid y'. It's strikes enough we've had in this house," she added grimly as she resumed the task of cutting bread for the frightened younger Phelans, while Mick took his disconsolate way to the breaker, being devoutly thankful that his father's work made it necessary for him to leave the house earlier than the breaker boys. His father's temper was not so quickly quenched as his mother's.

## Chapter 23: A Scab Slate Picker

It was highly unfortunate for the ambitions of little Pippinella Jindy, who aspired to become a slate picker, that such an experience as that of Mick Phelan, when he announced to his mother that the 'Malgamated Terrors had gone on strike, had also befallen nearly every other member of the organization. The backbone of the strike had been effectually broken by the blows of the respective mothers of the Terrors, but the boys had been stung by the process into a kind of unreasoning resentment against poor little Italian Pippinella, as though she were directly responsible for their sufferings.

By some sort of intuition Pippinella seemed to understand this. For this reason she kept away from the breaker, although the family sadly needed her wages.

. . . If the miners had suffered during the strike, the unskilled Italian and Slavic labourers, whose wages even in good times seldom rise to a point where they can save anything, suffered still more keenly. When the men in the union had given up and gone back to work again, these foreigners who had helped to break the strike came in for such a share of hatred and persecution that they could hardly find work of any sort. There was some slight excuse, therefore, for their sending their sons to work in the breaker even when they had not reached the legal age, since their families were almost starving.

So it came about that beside the 'Malgamated Terrors there were in the breaker a number of Italian boys, bullet-headed, stocky little fellows, mostly under age, who had been set to work by their parents to help support their families in their dire need. These boys fell under ban with Pippinella and life became miserable for them.

When the mirth of the community over the strike of the 'Malgamated Terrors had somewhat subsided, Pippinella prepared herself to go back into the breaker again. The members of the organization had been on the watch for her daily, and every morning the two factions lined up on opposite sides of the valley between the culm dumps, the Terrors for offensive warfare and the Italian boys for mutual protection.

On the morning when Pippinella returned to work, Mick Phelan and his crowd appeared on the culm dump which overlooked the staircase leading to the new breaker, threatening the Italian boys with violence if they attempted to man the screens. The crowd of younger boys, being without a leader, stood huddled sullenly together out of range of the party on the dump, uncertain whether to fight or run.

As the whistle blew there came up a strange figure wearing a man's coat, with the sleeves cut off at the elbows, and with old blue overalls covering the legs. The shoes were pointed at the toes and had evidently once belonged to a lady. The hair was cropped short, revealing a bullet-shaped head and a fighting jaw.

"Hi! Here comes the king of the dagoes!" shrieked Mick.

When the ragamuffin attempted to mount the stairs he was greeted with a fusillade of stones, flung against the side of the breaker so as to make a terrifying noise.

In spite of this, the boy seemed about to pass up into the breaker. The crowd huddled below wavered for a moment and would have followed. But at that instant Mick Phelan raised his voice to its most malignant note, crying, "Scabs! Scabs! Kill the scabs!"

The ragamuffin on the steps came back, waving his arms. "Scab!" he shrieked. "He call-a you scab!" Then in Italian he screamed some challenge which stiffened up the wavering line and sent it charging up the steep side of the culm bank. The leader did not wait to see whether the crowd followed, but with splendid courage raced toward the top some two paces ahead of the others.

It was but a brief battle. Mick was an ancient enemy and straight for him the column rushed. But before they came into personal conflict a heavy lump of slate shattered the leader's ankle and he went over the edge of the bank in a lump. A hoarse shout from the breaker boss who had come out to see what was the cause of the tumult scattered the combatants and in a few minutes the boys were in their places while the limp figure of the leader was carried into the office.

Half an hour later as Dr. Creigan was examining the unconscious child he looked up quickly into the faces of the men gathered outside the door. "Why this child is a girl!" he cried. "Who sends a girl to the breaker?"

Presently Pippinella opened her eyes and shivered. She looked wonderingly into the faces of the men for an instant. Then she tried to rise. "I must get-a to the breaker," she said.

"My child, how does it come that you work in the breaker?" asked the doctor.

"I no mak-a blame," she said sullenly. "Mick Phelan, he mak-a fight."

"Why do you work in the breaker? Girls oughtn't to work in the breaker."

"My mudder she say we must-a have job. We hungry. I mak-a work only li'l time. My brother Domenico he com'a work soon. He too li'l mak-a fight.

I com-a first day, mak-a fight with Mick Phelan. Domenico, he com-a next day. No fight; strike all gone; all nice."

"Well, you won't make much more fight for a while."

"Will Domenico los-a job, if I no work-a today?" she asked anxiously, half rising only to sink back again in pain. "You tak-a me in breaker; I sit on-a bench, pick-a slate all day."

Just then Pippinella's mother came in. Nobody knows how the news of such trouble travels, but it had reached Reagan's Patch almost as soon as the telephone message had reached Mr. Hatton.

"We'll send her to the hospital in my carriage," Hatton said.

But Mrs Jindy did not agree to this. She gathered Pippinella into her arms as if she had been a baby, and in spite of the injured foot, would have carried her off to her home. When at length she was persuaded that there would be no effort to take them both away to the hospital, she allowed herself and the child to be led by Mr. Hatton to the carriage, still holding Pippinella in her arms and crooning soft Italian endearments over her.

When they reached the wretched little tenement, Pippinella, who had not cried out while the doctor set the bones of the foot, burst into loud wails at the sight of the pointed-toe shoe slit from top to sole by the doctor's lancet.

"Never mind," Mr. Hatton said, "I'll send you a pair of shoes with all the buttons on them, which won't need to be tied on with rags."

"Will breaker boss keep-a job for me till I get-a well?"

"No, Pippinella, we can't have girls working in the breaker; but I have been talking to your father and I'll give him a job as watchman of the breaker, where he can make a great deal more than you could picking slate. You must hurry up and get well and go and Domenico must go to school. I think Mrs. Hatton has some sort of plan for what she calls a 'settlement house,' where there will be all sorts of games and pictures and fun for you and the rest of the children of the Patch."

To which Pippinella only replied, "Mick Phelan, he no let-a me play games in the Patch. But Mrs. Hatton mak-a all right." Then looking at her foot, as if it were through it that the good times were to come, she added, "Won't Mick Phelan be mad-a 'cause I no break-a his foot?"

## Superstitions

*Although most taboos against women miners were transmitted through oral traditions, some were later recorded. The following passage comes from* Coal Dust on the Fiddle: Songs and Stories of the Bituminous Industry, *by George Korson (Hatboro, Pa., Folklore Associates Inc., 1965), in the chapter titled "Craft Superstitions and Legends."*

One of the most curious superstitions is that a woman's visit into a mine is disastrous. It is probably the most common superstition in the coal fields, anthracite as well as bituminous. Miners will cite instances of disasters following in the wake of a woman's visit. To outsiders they may appear purely coincidental, but the mine workers insist upon a cause-and-effect rationalization.

One of the places where I ran into the superstition was at St. Charles, Virginia, on March 13, 1940. I had recorded several miners' ballads in a shack there, and while my recording machine was cooling off I questioned the mine workers regarding occupational superstitions and other types of folklore. The very first thing they told me was that they stayed away from a coal mine that had been visited by a woman. I reminded them of the widely heralded visit to the Willow Grove mine at St. Clairsville, Ohio, by Mrs. Eleanor Roosevelt, wife of the President. Would that make any difference? No, they were sorry, but not even a woman of Mrs. Roosevelt's prominence and good deeds could be excepted from the general belief in the evil spell which a woman cast over a coal mine. I smiled politely and made a mental note of the fact that they were being a little too hard on Mrs. Roosevelt, if not all womanhood. So the Willow Grove mine was hexed! Well, several days later I was in Birmingham, Alabama, when I read the dire news—the Willow Grove mine, the identical one which Mrs. Roosevelt had visited, had blown up with a toll of 72 lives. The disaster had occurred on March 16, just three days after my talk with the St. Charles miners. I still think it was simply a coincidence, but I have a hunch that when next I set foot in St. Charles I shall be cornered and told, "We told you so!" . . .

The vision of a maiden in white appeared before miners in the No. 6 mine at Buck, Oklahoma, over thirty years ago. Most of them believed it to be a bad omen, but not Charley Withers, a young shot-firer who lived with his mother. His mother shared the fear of the other miners and begged him to quit his job, but Charley only laughed at her. Each evening she met him halfway down the hill as he came out of the mine and walked home with him. Finally he yielded to her pleas and quit his job. They were to meet halfway down the hill from the mine for the last time. This, however, happened to be the day when the maiden wraith, her face more troubled than ever, reappeared before the miners working on an early shift. They dropped their tools and hurried out of the mine. Charley Withers met them as he was going to work and they warned him not to enter. Charley, however, laughed at their fears and continued walking. Twenty minutes after he had started working, an explosion tore up the mine, hurling tons of debris high up in the air and completely destroying the superstructure. Charley and another shot-firer named Sam McKinney were killed instantly, on the very spot where the vision of the maiden in white had appeared. . .

The origin of this occupational superstition is lost in antiquity. Russian miners believed that it came out of the distant past, when in the northern twilight beyond the Volga the Russian peasants observed the vampire souls of women leaving their river caves for their flight through the night air in search of prey. The more literal-minded Italian mine workers declared simply that it was woman's delight to poke her nose into places where she had no business that led her into inspecting coal mines; the fates, angered by her temerity, showed their displeasure by unloosing a calamity such as an explosion, fire, or a cave-in.

*In another volume of his work, however, Korson noted beliefs that the luck brought by women need not all be bad. The following passage is drawn from Korson's* Minstrels of the Mine Patch: Songs and Stories of the Anthracite Industry.

There are . . . numerous women operators of mines through the state who enjoy local fame. "Mountain Annie, " "The Mulligan Queen," and "Mother" Phelan are names familiar to miners in the northern camps. Josie Bishop is a familar name in the Mojave desert country. Some miners believe that a woman's entering a mine will change the luck, for good or for bad. . . . One man came upon a $600 pocket after taking a woman into a mine near Ophir, and after that frequently took women down.

## From *Coal Age*, Vol. 13, No. 6, Feb. 9, 1918

### "Women Learn First Aid on Bureau of Mines Rescue Cars"

"With the wives of miners and other women members of mining committees as pupils, the United States Bureau of Mines has entered upon an active campaign to reduce casualties. Eight "mine-rescue cars," each with a crew of men highly trained in modern rescue metholds, have been equipped and are making the rounds of the mining communities. Five hundred women have taken the course given and the bureau has been overwhelmed, since the declaration of war, with applications from women who wish to replace men on the cars."

## "Employment of Women in Underground Work of All Kinds"

*In the late 1920s, the International Labour Office studied women working in underground mines, with a view toward banning women miners in every member nation. After prodding from Japan, where tens of thousands of women worked underground, the ILO issued a series of reports with summaries of conditions in each nation. The reports were published in the early 1930s, when economic depression had gripped much of the world. This excerpt is from a 1935 report, "Employment of Women in Underground Work of All Kinds."*

The question thus brought before the Eighteenth Session of the Conference is one which may be considered as ripe for international action, since the prohibition of the employment of women on underground work in mines is one of the oldest and most widespread provisions of labour legislation. In some of the more important mining countries of Europe such measures have, in fact, been in force for 75 years or more. In underground work in mines, and in particular in deep-level mines, the danger attendant on the extraction of heavy substances is complicated by a number of other risks. It is essentially an exhausting form of work. Hence the employment of women on such work was one of the earliest abuses of the industrial system to call forth the protests of humanitarian minds. A century ago, when protective legislation was still non-existent, the conditions of underground employment for women—who were allotted the worst paid and most unpleasant kinds of work, such as carrying loads without any mechanical aid, and were obliged, owing to the high temperature of the ill-ventilated underground galleries, to work half naked, being herded together in a promiscuity offending against all the laws of human dignity—gave rise to a movement of opinion in favour of legislation to put a stop to such abuses. The British Act of 1842 to prohibit this kind of work for women was one of the earliest measures of labour legislation, and indeed, the very first to deal with adult workers. Other European countries gradually followed suit. In Austria-Hungary the Mines Act of 1854 empowered the mining authorities to take any measures they considered necessary in this respect. In 1874 underground work for women was prohibited by law in France, in 1876 in Luxembourg, in 1878 in Germany, etc.

The custom of considering this form of work as unsuitable for women gradually became so well established, at least in European countries, that even countries in which women had never been employed on underground work frequently inserted a provision prohibiting this practice in their general labour legislation.

In certain Oriental countries, however, the practice of employing women below ground in mines has persisted, although on an increasingly restricted scale, probably owing to the fact that work is there performed by family groups. With the introduction of technical improvements and the mechanisation of mining operations, the conditions under which women are now or were until recently employed are certainly less deplorable than those of the women workers in the English mines of 1840; while such underground employment is also now subject to regulations which to some extent mitigate its disadvantages. But even under these improved conditions employment on underground work in mines is still a form of work arduous enough to make its total and early abolition for women extremely desirable.

*The ILO's strategy worked; in 1929, Japan reported having 36,759 women miners; by 1931, the number had dropped to 8,147. This passage from Japan's entry in a 1933 report refers to regulations affecting pregnant miners.*

Under Section 15 of the Mining Regulations hitherto in operation, a pregnant woman may cease work four weeks before her confinement and may not be employed for the six weeks following it, provided that at her request she may be employed four weeks after her confinement on work which the doctor has pronounced harmless.

Section 16 provides that a woman who is nursing a baby may demand time for nursing it up to the limit of twice a day for thirty minutes, provided that the holder of a mining right need not grant time for nursing to a woman engaged in underground work if he has received permission to this effect from the Chief of the Mines Inspection Bureau, and has made the necessary provision for the feeding of the babies of the women concerned."

## The Case of Ida Mae Stull

*In the 1930s, Ida Mae Stull challenged a state eviction after she was ordered to stop working in an Ohio mine of which she was part owner. The following excerpts are drawn from the local newspaper, the Cadiz* Republican.

### From "Stops Woman As Miner" (Feb. 1, 1943)

A Cadiz woman, Ida May [sic] Stull, was until last week probably the only woman coal miner on the records. By order of James Berry, state Superintendent of Mines, she was ordered to quit her work, under the Ohio code which forbids employment of women for such a purpose. Mrs. Stull came to Cadiz about four years ago as a housekeeper but kept up her outdoor work, which she had done all her life. When only 12 years old, she carried a lantern for her father as he wheeled coal from a mine. Manual labor, such as digging wells or cisterns, or helping put up buildings, is what she likes better than housekeeping.

So when Harry Wolfe, for whom she keeps house, started to open up a

Ida Mae Stull.
Photo from an Appalshop film, "Coalmining Women"

mine a mile out on the Jewett Road, last summer, she was given a partnership in it and worked alongside of him cutting props and making the entry. Underground she would run out six or seven cars of coal a day. Mr. Wolfe called deputy inspector Andy Mullen from Adena to pass on his mine and Mullen reported that she was working there. The order from [Superintendent] Berry followed.

But this may not end the matter, for it is contended that, as a partner in the mine, she is entitled to work, for she is not an employee working for a regular wage, but one of the operators. Taking this action seems to be stretching the law a long, long ways, when there are so many more places to apply it. Mrs. Stull was making her bread and butter in the way she preferred and without injustice to any one else, as a miner.

## From "Woman's Story in Movies" (Feb. 22, 1934)

When Universal Newsreel photographers offered last Saturday, to tell her story, she gladly put on denims, boots and cap, so that a nation might see how a working woman helped pay off a mortgage on the house and kept off of charity by making her two dollars a day, digging coal. . . .

"No real miner wants to fool around outside, shoveling and loading coal, when there's real work to do inside," she says. "Just when the hard work was done and we were ready to make some money, they come along and say I've got to work in the kitchen. Any woman can wash dishes. I want to go in there underground with my pick and shovel, for Harry is going to have a hard time holding the house just working by himself, in the cold season.

"Maybe some of those women who stay pretty in beauty parlors won't like pictures of me shoveling coal and sawing wood, but it won't hurt them to see how some other women in the world have to work to buy bread and butter."

## From "Woman Back in Mine" (Jan. 24, 1935)

Ida Mae Stuhl [sic] is back at her job of digging coal near Cadiz, away from the pots and pans, by a ruling of Attorney General John W. Bricker given James Berry, Chief of the Ohio Division of Mines. Because she helped cut the props for the mine and make the entry, and is not under contract of hire, her title of the "only woman miner" has been restored after a year of nationwide publicity. . . .

The Attorney General, in his opinion, says that "the legislature did not intend to prohibit females from working at these lines of work when they were not hired but were just working for themselves . . . We think that notice may be taken of the fact that a housewife, especially one residing on a farm, is frequently engaged in work in the performance of which she is often required to lift articles weighing in excess of 25 pounds. The Legislature did not intend that she should be prohibited from performing such work. Nor did it intend that a woman who owned a coal mine, and operated the same herself, should be fined for employing herself at such work, since it would be impossible for the operator or owner to employ herself."

## Minutes of the United Mine Workers of America, International Executive Board, April 17, 1944

*UMWA President John L. Lewis took action to "adjust" the employment of women at union mines during World War II.*

PRESIDENT LEWIS: Last year during one of our New York meetings, during a Policy meeting in 1943, a question came before the Board, and the Policy Committee, as to the policy in the employment of women on the outside of mines. The question came up from District 29 and District 22. The Union Pacific had undertaken to employ about half a dozen women on the pit head in some Wyoming mines, and a certain coal company or two in District No. 22 had employed some women. The matter was discussed in the meeting and a position taken by the organization against the policy permitting the women to work in and around these mines. The officers of District 22 were here, and the officers of District 29 were here. In 29, to the best of my knowledge, the matter was promptly adjusted, and I had assumed it was adjusted out in Wyoming. A short time ago the Chair began to hear reports, emanating from government circles, that the War Manpower Commission and Military Department and the draft authorities were examining the policy of augmenting manpower of the country through the employment of women around coal mines, and it was pointed out women were being employed in coal mines, and I asked Board Member Mates to check with the officers of District 22 as to what degree the policy of the organization had been executed in Wyoming, and found it had not been enforced, that from the original six or a dozen women that were employed at the time the Policy Committee took action, that we have now some either 74 or 87 women employed by the Union Pacific Coal Company on its tipples. What do they do?

BOARD MEMBER MATES: Picking [sic] the boney out of the coal, and also working in machine shops in various sections of District 22.

PRESIDENT LEWIS: The position of some of the government departments is that that establishes a precedent and is and opens a new way to augment manpower.

It is my opinion we ought to appoint a Commission of two members of the Board to visit District 22 and see what is going to be done there and what is being done there to enforce the policy of this organization. I don't know why a District does a thing like that in contravention to the precise and meticulous policy of the Union. They attend these meetings here and sometimes go back and seemingly beguiled by the siren word of some coal operator. We know Mr. Eugene McAuliffe is a persistent man, but I hope he is not so persistent that our organization will undertake to set aside its policy to keep him in a good humor. In any event, I think the matter should be investigated.

BOARD MEMBER MATES: There is also a State law in Wyoming against it.

BOARD MEMBER GASAWAY: I move the Chair appoint a Commission to look into the entire matter in District 22, or any other District for that matter.

The motion was seconded and carried.

The Chair appointed Board Member Henry Allai, District 14, and Board Member O.F. Nigro, District 15, as a committee to visit District 22 on the question of the employment of women in mines.

### From "Speak Your Piece," in the Whitesburg, Kentucky, *Mountain Eagle*, March 3, 1983

*This write-in column, with unsigned opinions, resembles a radio talk show in print and was very popular in this weekly newspaper in eastern Kentucky.*

To all the coal mining wives and daughters speaking their piece. I am a lady coal miner. I do respect myself and so do your husbands I work with, except for a few. And the reason they don't respect us women who work in the mines is because they can't trust and respect their own wives. The other reason is because none of us will have anything to do with them. There are a lot of good men working where I do. I may be greasy and nasty with coal dust like you say, but it does wash off. Some things just never wash off of nasty-mouthed women. I don't want any of your husbands and I never went to work to get one. I went to work to support my children. I don't need to go in the mines to be with your husbands or boyfriends. Most of them would see me on the outside as well as inside the mines if I wanted to be with one of them. But I don't. I am a woman coal miner of seven years and proud of it. Thank God for speaking your piece and telling the truth, lady coal miners.

* * *

To the man who said he was going to marry a dirty, greasy coal mining woman. You just admitted the only reason you married her was for her paycheck and not for her looks. Like you said, the only reason a woman has a man is for his payday. You admit the only reason you want her is for her payday.

* * *

To the man who said it's okay for women to work in the mines. If it's okay, why won't my husband let me go get a job in the mines? I'd like to make as much money as the next woman does. You know what I get a week? I get $20 to spend. That's not much money, you know?

* * *

To all who speak their piece. I am a coal miner's daughter. Have I got a word for all the women who say that coal-mining women are just out to get their kicks when they go in the mines. My mom is a coal miner and she goes underground to work, not to play. She did not take a job away from your man. Maybe they are just too sorry to get out and look for a job the way she had to. When she comes out dirty and nasty they would probably look at her just the same as they would inside. So, to all you foul-mouthed women, try the mines. You'll find out it's not as easy as you let on.

* * *

To the man who said he was going to marry the dirty, greasy woman he worked with. Well it's my husband. The next time you look over your shoulder and your buddy has a sneaky grin on his face you'll have to wonder how he knows more about your wife than you do. Remember, what's good for the goose is good for the gander.

* * *

Concerning jealousy. Marriage is great, peace is greater. Married women should stop being so jealous of women who don't have husbands, and may be without one because of some misfortune. If you handle your own affairs you don't have time to judge. Women coal miners and women without husbands, keep up the good work. Remember there is more to a man or a woman than good looks. Quality is great.

* * *

To the man who said he is a truck driver and his wife is a coal miner. If your wife was doing anything do you think she'd be dumb enough to tell you? Never. If you find it out, you'll have to find it out on your own. Don't sit around waiting for someone to knock on the door and tell you. They won't.

* * *

For all you ladies who think your coal mining husbands would take the coal mining ladies. Believe me, if he wants a woman, she doesn't have to be a coal miner who he workes with underground. Remember to look higher up. My husband took a secretary and this is what she's getting: Saturday night, rooster fight; Sunday, rooster fight. When he isn't at a rooster fight he's fishing. Then there's squirrel and deer hunting. I wish her luck because this is all she's getting.

* * *

If the women can work in the coal mines more power to them. Some of them have children and they have to make a living.

* * *

To the widowed woman coal miner with four children. My hard hat is off to you. Your comments were sincere and well worth reading, which is a lot more than I can say about most of what goes in this column.

* * *

Women of Letcher County, I think it is time we took a united stand to have these coal companies start hiring and working us women who need to work and take care of our families in a decent way, instead of having to depend on welfare and food stamps. The American Civil Liberties Union will help us and I, for one, intend to contact them for help and show these companies we will stand up and not be denied our rights because we are women.

* * *

To the man who said he was leaving his wife for the woman coal miner he worked with. How would you have felt if your wife had broadcast all over Letcher County she was leaving you for your best buddy who works beside you? It wouldn't have felt very good, would it? You should have thought of that before you put it in the paper.

# Appendix 3

**Hiring of Women Coal Miners, 1973–1989[a]**

| *Year* | *% Women* | *No. of Women* | *No. of Men* | *Total* |
|---|---|---|---|---|
| 1973 | 0.00015 | 1 | 6,713 | 6,714 |
| 1974 | 0.0015 | 71 | 45,501 | 45,572 |
| 1975 | 1.0 | 274 | 27,983 | 28,257 |
| 1976 | 2.0 | 445 | 22,064 | 22,509 |
| 1977 | 3.4 | 764 | 21,493 | 22,257 |
| 1978 | 4.2 | 830 | 19,036 | 19,866 |
| 1979 | 11.4 | 555 | 4,301 | 4,856 |
| 1980 | 8.7 | 355 | 3,730 | 4,085 |
| 1981 | 9.2 | 261 | 2,567 | 2,828 |
| 1982 | 8.6 | 174 | 1,845 | 2,019 |
| 1983 | 7.9 | 42 | 490 | 532 |
| 1984 | 7.9 | 53 | 615 | 668 |
| 1985 | 6.8 | 27 | 373 | 400 |
| 1986 | 7.2 | 19 | 246 | 265 |
| 1987 | 6.3 | 29 | 433 | 462 |
| 1988 | 6.2 | 25 | 378 | 403 |
| 1989 | 8.1 | 40 | 453 | 49 |
| Total[b] | | 3,965 | 158,221 | 162,186 |

a. Statistics are based on records of chest X rays routinely give to newly hired coal miners; these records are compiled by the National Institute for Occupational Safety and Health, Chest X Ray Division, in Morgantown, West Virginia.
b. Does not reflect layoffs.

# Notes and References

## *Introduction*

1. Statistics on women's hiring in the coal industry are based on records of chest X rays routinely given to newly hired miners by the National Institute for Occupational Safety and Health (NIOSH), Chest X-Ray Division, Morgantown, West Virginia.

2. Libby Lindsay, "Twenty Years in the Mines: A Tribute to Women Miners," *CEP News* (January 1994): 3.

3. Evelyn Kroker and Werner Kroker, *Frauen und Bergbau: Zeugnisse aus 5 Jahrunderten [Women in Mining: The Evidence of 500 Years]* (Bochum, Germany: Selbstverlag des Deutschen Bergbau-Museums, 1989), an anthology of articles published in conjunction with an exhibit held 29 August–December 1989 at the Mining Museum in Bochum, Germany.

4. Angela V. John, *By the Sweat of Their Brow: Women Workers at Victorian Coal Mines* (London: Routledge & Kegan Paul, 1984), 11–20.

5. Patricia J. Hilden, "The Rhetoric and Iconography of Reform: Women Coal Miners in Belgium 1840–1914," *The Historical Journal* 34, no. 2 (June 1991): 422.

6. Regine Mathias, "Women as Workers in Japanese Mines," *Frauen und Bergbau*, translated for the author by Rebecca Ripple.

7. International Labour Office, proceedings of the International Labour Conference, Geneva, 1933, 1934, 1935.

8. See Christina Vanja, "Women in Mining: An Overview," *Frauen und Bergbau*, 11–29.

9. Ibid; see also George Korson, *Coal Dust on the Fiddle: Songs and Stories of the Bituminous Industry* (Hatboro, Pa.: Folklore Associates, 1965); and June Nash, *We Eat the Mines, the Mines Eat Us* (New York: Columbia University Press, 1979).

10. Christine Vanja, "Mining Women in Early Modern European Society," *The Workplace Before the Factory: Artisans and Proletarians 1500–1800*. Ithaca, N.Y.: Cornell University Press, 1993, 103.

11. Ronald L. Lewis, *Black Coal Miners in America: Race, Class and Community Conflict 1780–1980* (Lexington: University Press of Kentucky, 1987), 5. Also see Robert S. Starobin, *Industrial Slavery in the Old South* (New York: Oxford University Press, 1970), 166.

12. Korson, *Coal Dust on the Fiddle*, 203–4.

13. *Black Diamond* 15 (14 September 1895): 335.

14. Florence P. Smith, "Chronological Development of Labor Legislation for Women in the United States," United States Department of Labor Women's Bureau bulletin no. 66-II, Washington, D.C.: U.S. Government Printing Office, 1932.

15. *Coal Age* 13, no. 18 (4 May 1918): 841.

16. R. Dawson Hall, "The Labor Situation," *Coal Age* 13, no. 19 (11 May 1918): 886.

17. Floyd W. Parsons, "Employment of Women in Mining," *Coal Age* 13, no. 19 (11 May 1918): 1.

18. Benjamin F. Davis, "No Place for Girls at Mine," *United Mine Workers Journal [UMWJ]* 29, no. 1 (9 May 1918): 11.

19. "Girls 'Fired' from Mines," *UMWJ* 29, no. 2 (16 May 1918): 11; see also *Coal Age* 13, no. 19 (11 May 1918): 886–87.

20. *Pottsville [Pa.] Republican*, 3 May 1918.

21. Benjamin F. Davis, "No Place for Girls at Mine," 11.

22. "Girls 'Fired' from Mines," *UMWJ*, 29, no. 2 (11 May 1918): 11.

23. Parsons, "Employment of Women in Mining," 1.

24. Dave Corbin, "Ida Mae Stull, Intent, Fought for Her Rights," *UMWJ* 88, no. 17 (1–15 February 1982): 12.

25. Florence P. Smith, "Labor Laws for Women in the States and Territories," U.S. Department of Labor Women's Bureau bulletin no. 98, 1932.

26. Corbin, "Ida Mae Stull, Intent, Fought for Her Rights."

27. Elizabeth Stull Crawford, interview with author, Dillonvale, Ohio, 31 July 1982.

28. Irene Adkins Dolin, interview with author, Julian, West Virginia, 23 May 1992.

29. "Necessity for Women in Coal Mines Does Not Exist," *UMWJ* 53, no. 23 (1 December 1942): 9.

30. Ibid.

31. UMWA International Executive Board minutes, 17 April 1944 (see also appendix 2).

32. Kelly interview, 2 June 1985.

33. From author's correspondence with Olga Smodey Majnik, 19 July 1991.

34. Kelly interview, 2 June 1985.

35. Interview with Ethel McCuiston, Cumberland, Kentucky, 6 May 1980.

36. "1969 Handbook on Women Workers," U.S. Department of Labor Women's Bureau bulletin 194, Washington, D.C.: U.S. Government Printing Office, 1969.

37. Statistics on hiring in the mining industry were compiled by the NIOSH Chest X-Ray Division (see n. 1). Annual hiring statistics were published each year in *Coalmining Women's Support Team [CWST] News!*

38. Kipp Dawson, "Women Miners and the UMWA, 1973–1983," occasional paper no. 11 (City University of New York, Center for Labor-Management Policy Studies, July 1992).

39. George Vecsey, "4 Women Seek Jobs as Miners, and Man's World Is in Conflict, *New York Times*, 9 September 1972, 25.

40. Matt Witt, "Textile Workers Seek Jobs in Mines," *UMWJ* 84, no. 8 (15 May 1973): 12. See also "The Rank and File Speaks: Should Women Be Coal Miners?" *UMWJ* 84, no. 8 (15 May 1973): 13.

41. Wanda Harless, "Women Miners," letter to editor, *UMWJ* 85, no. 12 (1–15 June 1974): 21.

42. Pam Schuble, "Woman Member," letter to editor, *UMWJ* 84, no. 15 (16–31 July 1974): 12.

43. George Getchow, "Women in the Pits: Kentucky Mother, 29, Endures Dirt, Danger to Work as Coal Miner," *Wall Street Journal*, 29 November 1976, 1.

44. Shirley Boone interview with author, Crichton, West Virginia, 22 September 1983.

45. Roger B. Trent and Nancy Stout-Wiegand, "Attitudes toward Women Coal Miners in an Appalachian Coal Community," *Journal of Community Development Society* 18, no. 1 (1987): 1–14.

46. Cosby Totten interview with author, 16 August 1994.

47. Sandra Bailey (Barber) interview with author, Mayking, Kentucky, 2 March 1981.

48. Linda Raisovich (Parsons) interview with author, Welch, West Virginia, 12 August 1982.

49. Brenda Brock interview with author, High Splint, Kentucky, 9 May 1980.

50. Elizabeth Laird interview with author, Cordova, Alabama, 23 August 1983.

51. Sue Thrasher, "Coal Employment Project," *Southern Exposure* 9, no. 4 (winter 1981): 48.

52. *CWST News!* 1, no. 1 (June 1978): 2.

53. Dawson, "Women Miners and the UMWA," 15.

54. *CWST News!* 2, no. 6, 2. CEP newsletters carried extensive coverage of legal settlements in the Consol case and others.

55. *CWST News!* 1, no. 2 (July 1978): 2.

56. *CWST News!* 1, no. 5 (November 1978): 1.

57. *CWST News!* 1, no. 10: 1.

58. *CWST News!* 2, no. 5: 3.

59. This issue was raised at almost every CEP conference from 1979 to 1993, as well as in a meeting between 15 CEP members and UMWA Vice President Cecil E. Roberts in Washington, D.C., on 15 April 1994.

60. *CWST News!* 5, no. 5: 3.

61. Patricia Brown interview with author, Bessemer, Alabama, 25 August 1983.

62. Martha Horner interview with author, St. Louis, Missouri, 24 June 1994.

63. See the NBCWA of 1988, for example. The antidiscrimination language remained intact in two subsequent agreements.

64. Bonnie Boyer interview with author, Shelocta, Pennsylvania, 19 November 1991.

65. Carol Davis interview with author, Waynesburg, Pennsylvania, 14 October 1992.

66. "Women in the Mines," *Newsweek*, 7 December 1979, 74–75.

67. Data compiled by Linda Raisovich-Parsons of the UMWA Department of Occupational Safety and Health. From 1973 through 1993 the victims included: Marilyn McCusker (1979), Eleanor Bown (1980), Jackie Herran (1980), Mary "Cat" Counts (1983), Linda Thompson (1984), Nannette Wheeler (1984), and Gloria L. Smith (1990).

68. Ben A. Franklin, "Women Who Work in the Mines Assail Harassment and Unsafe Conditions," *New York Times*, 11 November 1979, 30.

69. A. P. Watson and C. L. White, "Workplace Injury Experience of Women Coal Miner in the United States" (Oak Ridge, Tenn.: Oak Ridge National Laboratory, 1985).

70. "Angie" interview with author, Morgantown, West Virginia, 22 February 1981.

71. *CWST News!* 3, no. 1: 6.

72. Jan L. Handke and June Rostan, "Survey of Pregnant Women Miners: Report of a Preliminary Study." The researchers interviewed 26 self-selected women who had been pregnant while in a mining job. Reproductive history interviews were administered to gain information on reproductive outcome and occupational exposures.

73. Brenda Brock, interview with author, High Splint, Kentucky, 9 May 1980.

74. "Angie" interview with author, 22 February 1981 (see n. 65).

75. A list of resolutions passed by conference delegates was published annually in *CWST News!*

76. *CWST News!* 5, no. 11: 1.

77. *CWST News!* 8, no. 4: 3.

78. Cosby Totten, "For Our Children: The Coal Miners Fight for Parental Leave," *CWST News!* 5, no. 11: 3.

79. The role of the SWP is an important part of CEP history that has not yet

been published. This information comes from the author's conversations from 1979 until 1994 with CEP members and SWP members.

80. In 1988, at the beginning of the corporate campaign against Pittston, I raised the issue of broad auxiliary organizing with Joe Jurczak, coordinator of the corporate campaign. He agreed to hire on lost-time wages two UMWA women from southwest Virginia: Cosby Totten and Catherine Tompa. Totten, Tompa, and I helped organize a regional network of approximately 12 local union auxiliaries and participated in the occupation and other acts of nonviolent protest.

81. Marat Moore, "Ten Months That Shook the Coalfields: Women's Stories from the Pittston Strike," *Now and Then* 7, no. 3 (fall 1990): 35.

82. From discussions in September 1989 with male UMWA organizers, who informed me that the UMWA strike chief ordered in all-male meetings that no women would be informed of or involved in the occupation planning.

83. Interview with Cosby Totten, Thompson's Valley, Virginia, 25 May 1992.

84. From my conversations with auxiliary members during the 1993 strike. In another example, one woman who was a UMWA member and miner's wife was elected president of her local union auxiliary in Illinois.

85. From my notes of the CEP conference, Pittsburgh, Pennsylvania, 24–26 June 1988.

86. *CWST News!* 8, no. 7: 1.

87. From conversations with Joy Huitt during and after her reelection bid in 1990.

88. The union's official history, *United We Stand: The United Mine Workers of America 1890–1990*, states that Trumka filled international staff positions with "more female miners" during his administration, a clear overstatement. When Trumka took office in 1983, UMWA member Linda Raisovich (Parsons) was already a member of the international union's safety staff. The UMWA hired the author, a former member, in 1983, but added no rank-and-file women to the staff.

89. *Wall Street Journal*, Feb. 6, 1996, p. 1. See also Shail J. Butani and Ann M. Bartholemew, "Characterization of the 1986 Coal Mining Workforce," Bureau of Mines Information Circular 9192, Washington, D.C.: U.S. Department of the Interior, Bureau of Mines, 1988.

90. Davis interview, 14 October 1992.

91. Totten interview, 16 August 1994.

92. Lindsay, "Twenty Years in the Mines," 3.

## *Afterword*

1. For one comprehensive assessment of the impact of the 1980s on the American working class, see Kevin P. Phillips, *The Politics of Rich and Poor: Wealth and the American Electorate in the Reagan Aftermath* (New York: Random House, 1990). See also Tinsley E. Yarbrough, ed., *The Reagan Administration and Human Rights* (New York: Praeger, 1985).

2. For accounts of the role miners' wives played in British miners' struggles in the 1980s, see *The Cutting Edge: Women and the Pit Strike* (London: Lawrence and Wishart, 1986); Norma Dolby, *Norma Dolby's Diary: An Account of the Great Miners' Strike* (London: Vano, 1987); Jill Miller, *You Can't Kill the Spirit: Women in a Welsh Mining Village* (London: Women's Press, 1986); National Union of Miners, *A Century of Struggle: Britain's Miners in Pictures 1889–1989* (Sheffield: National Union of Miners, 1989); Jean Stead, *Never the Same Again: Women and the Miners' Strike, 1984–1985* (London: Women's Press, 1987); and Joan Witham, *Hearts and Minds: The Story of the Women of Nottinghamshire in the Miners' Strike, 1984–1985* (London: Canay Press, 1986).

For reports on American women miners and their relationship to union battles and their union, see *Coal Mining Women's Support Team News!*, the monthly newsletter of the CEP. See also Kipp Dawson, "Keeping Hope Alive: Women Miners, the United Mine Workers of America, and the Union Movement of the United States, 1984–1986" (tutorial thesis, Chatham College, Pittsburgh, 1994), in particular ch. 3, "Women Miners' Solidarity History."

3. The work to document the history of the Coal Employment Project has just begun. For initial efforts, see Dawson, "Keeping Hope Alive," and Dawson, "Women Miners and the UMWA."

4. Beginning in the 1960s, black steelworkers organized protests and brought court actions that led to rulings in 1974 that both the steel companies and the United Steel Workers of America (USWA) were in violation of title 7 of the Civil Rights Act of 1964 (*U.S. v. Bethlehem Steel Corp.*) and to two consent decrees, signed by the federal government, nine major steel companies, and the USWA, outlining affirmative action programs in the mills. By this time the massive women's liberation movement had emerged, and black steelworkers incorporated some of the problems and needs of women workers into their demands; the consent decrees set affirmative action goals in hiring, training, and promoting black, Hispanic, and women workers. It was under the impact of this development that the largest doors began to open to women seeking jobs in the coal mines. For an account of this development, see Herbert Hill, "The NAACP and the Struggle for Full Equality," an address to a workshop at the NAACP convention, July 1975, Washington, D.C., reprinted in *Last Hired, First Fired; Affirmative Action vs. Seniority* (New York: Pathfinder Press, 1975); and Philip Foner, *Organized Labor and the Black Worker, 1619–1981* (New York: International Publishers, 1981), 351.

# Selected Bibliography

## Women Coal Miners in the United States

"The America Coal Miner: A Report on Community and Living Conditions in the Coal Fields." The President's Commission on Coal. Washington, D.C.: U.S. Government Printing Office, 1980.

Aumiller, Grace. "In the Mines." *Mountain Life & Work* 51 (February 1975): 36–37.

Axelrod, Jennifer. "Appalachia, Women and Work." *Magazine of Appalachian Women* 3 (January/February 1978): 7–11.

Bales, J., and C. White, "Protective Equipment in the Mining Industry: A Survey of Women Miners." Oak Ridge, Tenn.: Coal Employment Project, 1984.

Barnard, Thomas H., and Brenda J. Clark. "Clementine in the 1980s (EEO and the Woman Miner)." *West Virginia Law Review* 82 (summer 1980): 899–936.

Bernard, Jaqueline, and Elmer Rasnick. "Dark as a Dungeon Way Down in the Mines." *Ms.* 3 (April 1975): 21.

Bowman, Lois. "Coal Mining and Women." *Mountain Life & Work* 54 (July 1978): 10.

Burns, Barbara. "Why I Decided to 'Go Underground.' " *McCall's*, September 1977.

Butani, Shail J., and Ann M. Bartholemew. "Characterization of the 1986 Coal Mining Workforce." Bureau of Mines Information Circular 9192. Washington, D.C.: U.S. Department of the Interior, Bureau of Mines, 1988.

Church, Ruth. "Women Miners—In the 40s and Today." *Mountain Life & Work* 50 (November 1974): 14–15.

"Coal Mining and Women: A Special Issue." Jennifer Axelrod, ed. *Mountain Life & Work* 54, July 1978.

Dawson, Kipp. "Your Sisters Underground—Part One." *Sojourner* 15, no. 9 (May 1990): 20–21.

———. "Your Sisters Underground—Part Two." *Sojourner* 15, no. 10 (June 1990): 13–15.

Franklin, Ben A. "Women Who Work in Mines Assail Harassment and Unsafe Conditions." *New York Times*, 11 November 1979, 30.

Getchow, George. "Women in the Pits: Kentucky Mother, 29, Endures Dirt, Danger to Work as Coal Miner." *Wall Street Journal*, 29 November 1976, 1.

"Girls at Mines Quit Work." *Pottsville [Pa.] Republican*, 3 May 1918, 7.

"Girls in Coal Breakers." *Pottsville [Pa.] Republican*, 1 May 1918, 3.

Hammond, Judith A., and Constance W. Mahoney. "Reward Cost-Balancing among Women Coalminers." *Sex Roles* 9, no. 1 (January 1983): 17–29.

Kernan, Michael. "Sorrow at the Rushton Mine." *Washington Post*, 24 June 1981, B1.

Klemesrud, Judy. "In Coal Mine No. 29, Two Women Work Alongside the Men." *New York Times*, 18 May 1974, 16.

"Lady Miner Digs Her Job." *Ebony*, October 1974, 116.

"The Militant Women Mining the Coalfields." *Business Week*, 25 June 1979.

"Miners Threaten Strike Unless Women Are Fired: They Go." *Coal Age* 47, no. 12 (December 1942): 113.

Moore, Marat. "Hard Labor: Voices of Women from the Appalachian Coalfields." *Yale Journal of Law and Feminism* 2, no. 2 (spring 1990): 199–238.

———. "Women Go Underground." *A Model of Industrial Solidarity? The United Mine Workers of America, 1890–1990*, ed. John Laslett. State College, Pa.: The Pennsylvania State University Press, forthcoming in spring 1996.

———, and Connie White. "Sexual Harassment in the Mines." Coal Employment Project booklet. Knoxville, Tenn.: Allied Printing, 1989.

"No Women in Mines Now." *Pottsville [Pa.] Republican*, 4 May 1918, 6.

Osman, Samira. "Women Underground." *Marie Claire*, November 1988, 38–44.

Serrin, William. "One Fight for Women's Rights: A Coal Miner's Life and Death." *New York Times*, 8 November 1979, A1.

Stallsmith, Pamela. "Digging for a Living." Richmond, Va., *Times-Dispatch*, 2 October 1994, A1.

———. "Standing Firm." Richmond, Va. *Times-Dispatch*, 3 October 1994, A1, A3.

Sullivan, Allanna M. "Women Say No to Sexual Harassment." *Coal Age*, August 1979.

Tallichet, Suzanne E. "Gendered Relations in the Mines and the Division of Labor Underground." *Gender & Society* 9, no. 6 (December 1995): 691–705.

———. "Moving Up Down in the Mine: The Preservation of Male Privilege." *More Than Class: Studying Power in U.S. Workplaces*, ed. Ann E. Kingsolver. Albany, N.Y.: State University of New York Press, forthcoming in 1996.

———. "Moving Up Down in the Mine: Sex Segregation in Underground Coal Mining." Ph.D. dissertation, The Pennsylvania State University, December 1991.

Trent, Roger B., and Nancy Stout-Wiegand. "Attitudes toward Women Coal Miners in an Appalachian Coal Community." *Journal of the Community Development Society* 18, no. 1, 1987.

Trillin, Calvin. "U.S. Journal: Central Pa.: Called at Rushton." *New Yorker*, 12 November 1979, 182–89.

Vecsey, George. "4 Women Seek Jobs as Miners, and Man's World Is in Conflict." *New York Times*, 9 September 1972, 25.

Watson, A. P., and C. L. White. "Workplace Injury Experience of Female Coal Miners in the United States." Oak Ridge, Tenn.: Oak Ridge National Laboratory, n.d.

Wilkinson, Carroll Wetzel. "A Critical Guide to the Literature of Women Coal Miners." *Labor Studies Journal*, spring 1985, 25–45.

"Women Coal Miners: Special Issue." *Mountain Life & Work* 55, no. 7 (July/August 1979): 3–29.

"Women in the Mines." *Newsweek*, 17 December 1979, 74.

"Women Work at Coal Mine" [Wyoming]. *Coal Age* 47, no. 11 (November 1942): 90.

Wyoming Advisory Committee. "Workplace Conditions in Wyoming: Women and Minorities in the Mineral Extraction Industries." Report to the U.S. Commission on Civil Rights, March 1982.

Yount, Kristen. "Women and Men Coal Miners: Coping with Gender Integration Underground." Ph.D. thesis, University of Colorado, Boulder, 1986.

## Women and Coal Mining in Other Countries

Foster, Kathleen. "Vigil in Britain: Women Activists Fight for Miners' Right." *New Directions for Women* 22, no. 4 (July 1993).

Hilden, Patricia J. "The Rhetoric and Iconography of Reform: Women Coal Miners in Belgium 1840–1914." *The Historical Journal* 34, no. 2 (June 1991).

India [government of], Ministry of Labour, Labour Bureau. "Socio-Economic Conditions of Women Workers in Mines." Chandigarh: Labour Bureau, 1979.

International Labour Office. "Employment of Women on Underground Work in Mines of All Kinds." Reports of the International Labour Conference, Geneva, Switzerland, 1933, 1934, 1935.

John, Angela V. *By The Sweat of Their Brow: Women Workers at Victorian Coal Mines*. London: Routledge & Kegan Paul, 1984.

Kroker, Evelyn, and Werner Kroker. *Frauen und Bergbau: Zeugnisse aus 5 Jahrunderten [Women in Mining: Evidence of 500 Years]*. Bochum, Germany. An anthology of articles published in conjunction with a 1989 exhibit at the Mining Museum in Bochum. In German.

Ray, [Shrimati] Renuka. "Women in Mines." Tract no. 2, All–Indian Women's Conference. Aundh: Aundh Publishing Trust, 1945.

Vanja, Christina. "Mining Women in Early Modern European Society." *The Workplace Before the Factory: Artisans and Proletarians 1500–1800*. Thomas Max Safley and Leonard N. Rosenband, eds. From October 1990 colloquium at the University of Pennsylvania. Ithaca, N.Y.: Cornell University Press, 1993, 100–117.

## Women Miners and the United Mine Workers of America

Brittain, Carol. "First Women Delegates." *Mountain Life & Work* 52 (October 1976): 36–37.

Dawson, Kipp. "Women Miners and the UMWA, 1973–1983." Occasional paper no. 11, City University of New York, Center for Labor-Management Policy Studies, July 1992.

Fox, Maier B. *United We Stand: The United Mine Workers of America, 1890–1990*. Washington, D.C.: United Mine Workers of America, 1990.

Grandstaff, Herald. "Couple Meets, Weds in UMWA Coal Mine." *UMWJ* 92, no. 11 (16–31 October 1981): 6.

Hall, Mike. " 'We Dig Coal . . .' Is Compelling Story." *UMWJ* 93, no. 11, (16–30 June 1982): 8.

"Health, Safety Tops Women Miners' List." *UMWJ* 93, no. 3 (16–28 February 1982): 7.

Moore, Marat. "Navajo Women Host CEP Conference." *UMWJ* 104, no. 7 (August–September 1993): 18.

———. "300 Attend Women Miners' Conference." *UMWJ* 97, no. 9 (September 1986): 8–9.

———. "Women Miners Hold First Western Conference." *UMWJ* 96, no. 7 (August 1985): 16–17.

———. "Women Miners Take Charge, Reaffirm Roots." *UMWJ* 103, no. 7 (August–September 1992): 14–15.

"150 Attend Women Miners Conference." *UMWJ* 100, no. 7 (July–August 1989): 17.

"The Rank and File Speaks: Should Women Be Coal Miners?" *UMWJ* 84, no. 8 (15 May 1973): 13.

"The Rank and File Speaks: What We Want Is Equal Treatment." *UMWJ* 94, no. 6 (July–August 1983): 14–16.

"Reverend Tells Companies: Get Right with God." *UMWJ* 94, no. 6 (July–August

1983): 7. Concerns the death of Mary "Cat" Counts in the McClure mine explosion in southwestern Virginia.

Saltz, David. "Women Miners Urge Solidarity." *UMWJ* 93, no. 14 (1–5 August 1982): 7.

"Senate Committee Rejects Strunk's Nomination." *UMWJ* 98, no. 9 (September 1987): 7.

"Today's UMWA Miners." *UMWJ* 90, no. 1 (January–February 1979): 23–24.

"UMWA Joins Affirmative Action Case." *UMWJ* 90, no. 3 (April 1979): 4.

"Union Sponsors Women's Conference, First in History of UMWA." *UMWJ* 90, no. 10 (November 1979): 8–9.

Witt, Matt. "Textile Workers Seek Jobs in Mines." *UMWJ* 84, no. 8 (15 May 1973): 12.

"Women Coal Miners Spread the Word." *UMWJ* 105, no. 6 (August–September 1994): 16.

"Women Miners Build Solidarity at Conference." *UMWJ* 106, no. 4 (July–August 1995): 13.

"Women Miners Call for Unity and Activism." *UMWJ* 102, no. 7 (August–September 1991): 10–11.

"Women Miners Celebrate Solidarity." *UMWJ* 101, no. 6 (July 1990): 17.

"Women Miners Conference Focuses on Negotiations, Parental Leave," *UMWJ* 98, no. 9 (September 1987): 6.

"Women Miners' Conference: 'We're Working to Build the Union.'" *UMWJ* 95, no. 7 (July 1984): 16–17.

"Women Miners Hold 10th Conference." *UMWJ* 99, no. 9 (September 1988): 20.

"Women Miners in West Virginia File Suit against Consolidation." August 1981, 19.

## Creative Work by and about Women Miners

Angle, Barbara. "Any Man: Kate's Story." *UMWJ* 95, no. 12 (December 1984): 7–11.

———. *Rinker*. Washington, D.C.: Crossroads Press, 1979.

———. *Those That Mattered*. New York: Crown Publishers, 1994.

Barret, Elizabeth, producer. *Coalmining Women* (film). Whitesburg, Kentucky, 1983.

Cizmar, Paula. *Death of a Miner*. New York: Samuel French, Inc., 1982.

*Coal Mining Women's Support Team News!* The Coal Employment Project newsletter contains poems and other writing by women coal miners and supporters in the United States and abroad.

Dickens, Hazel. "Coal Mining Woman" (song). © Happy Valley Music, 1981.

Gibbons, William Futhey. *Those Black Diamond Men: A Tale of the Anthrax Valley*. New York: Fleming H. Revell Company, 1902.

Kahn, Si. "Coal Mining Woman." Song written about Sandra Bailey Barber.

Lyon, George Ella, and Peter Catalanotto. *Mama Is a Miner*. New York: Orchard Books, 1994.

Massek, Sue. "What She Aims to Be." Song recorded with the Reel World String Band.

Moore, Marat. "Because the Earth Is Dark and Deep." *The American Voice*, no. 22 (spring 1991): 73–88.

"They'll Never Keep Us Down: Women's Coal Mining Songs." Cambridge, Mass.: Rounder Records.

"We Dig Coal: A Portrait of Three Women." Franklin Lakes, N.J.: State of the Arts films, 1981.

Wheeler, Billy Ed. "Mama's Going Down in the Mines" (song).

Zola, Emile. *Germinal*. New York: Penguin Books, 1983.

## Women's History and Labor History

Balser, Diane. *Sisterhood and Solidarity: Feminism and Labor in Modern Times*. Boston: South End Press, 1987.

Beyer, Clara M. "History of Labor Legislation for Women in Three States." United States Department of Labor Women's Bureau bulletin no. 66-I. Washington, D.C.: U.S. Government Printing Office, 1929.

Clark, Paul F. *Miners Fight for Democracy*. Ithaca, N.Y.: New York State School of Industrial and Labor Relations, Cornell University, 1981.

Corbin, David Alan. *Life, Work and Rebellion in the Coal Fields: The Southern West Virginia Miners 1880–1922*. Urbana: University of Illinois Press, 1981.

Deaux, Kay, and Joseph Ullman. *Women of Steel: Female Blue-Collar Workers in the Basic Steel Industry*. New York: Praeger, 1983.

Farr, Sidney Saylor. *Appalachian Women: An Annotated Bibliography*. Lexington: University Press of Kentucky, 1981.

Foner, Philip. *Women and the American Labor Movement: From the First Trade Unions to the Present*. New York: The Free Press, 1982.

Fox, Maier B. *United We Stand: The United Mine Workers of America 1890–1990*. Washington, D.C.: United Mine Workers of America, 1990.

Gabin, Nancy F. *Feminism in the Labor Movement: Women and the United Auto Workers 1935–1975*. Ithaca, N.Y.: Cornell University Press, 1990.

Gluck, Sherna Berger, and Daphne Patai, eds. *Women's Words: The Feminist Practice of Oral History*. New York: Routledge, 1991.

Green, Archie. *Only a Miner: Studies in Recorded Coalmining Songs*. Urbana: University of Illinois Press, 1972.

Greenwald, Maurine Weiner. *Women, War and Work: The Impact of World War I on Women Workers in the United States*. Ithaca, N.Y.: Cornell University Press, 1990.

Hall, Jacquelyn Dowd, James Leloudis, Robert Korstad, Mary Murphy, Lu Ann Jones, and Christopher B. Daly. *Like a Family: The Making of a Southern Cotton Mill World*. New York: W. W. Norton & Co., 1987.

John, Angela V. *By The Sweat of Their Brow: Women Workers at Victorian Coal Mines*. London: Routledge & Kegan Paul, 1984.

Kingsolver, Barbara. *Holding the Line: Women in the Great Arizona Mine Strike of 1983*. Ithaca, N.Y.: ILR Press/Cornell University, 1989.

Korson, George. *Coal Dust on the Fiddle: Songs and Stories of the Bituminous Industry*. Hatboro, Pa.: Folklore Associates Inc., 1965.

Lehrer, Susan. *Origins of Protective Legislation for Women 1905–1925*. Albany, N.Y.: State University of New York Press, 1987.

Lewis, Ronald. *Black Coal Miners in America: Race, Class, and Community Conflict, 1780–1980*. Lexington: University Press of Kentucky, 1987.

Long, Priscilla. *Where the Sun Never Shines: A History of America's Bloody Coal Industry*. New York: Paragon House, 1989.

Milkman, Ruth. *Women, Work and Protest: A Century of Women's Labor History*. Boston: Routledge & Kegan Paul, 1985.

Moore, Marat. "Ten Months That Shook the Coalfields: Women's Stories from the Pittston Strike." *Now & Then* 7, no. 3 (fall 1990): 6–12, 32–35.

"Mrs. Roosevelt Visits Mine Pit in Illinois." *New York Times*, 17 June 1936, 28.

Nash, June. *We Eat The Mines, The Mines Eat Us*. New York: Columbia University Press, 1979.

"1969 Handbook on Women Workers." United States Department of Labor, Women's Bureau bulletin no. 294. Washington, D.C.: U.S. Government Printing Office, 1969.

Orwell, George. *The Road to Wigan Pier*. London: Victor Gollancz Ltd., 1937.

Pidgeon, Mary Elizabeth. *Women in the Economy of the United States of America: Employed Women under NRA Codes*. A revision of United States Department of Labor Women's Bureau bulletin no. 155, 1937. New York: DeCapo Press, 1975.

Smith, Florence P. "Labor Laws for Women in the States and Territories." United States Department of Labor Women's Bureau bulletin no. 98, 1932.

———. "Chronological Development of Labor Legislation for Women in the United

States." United States Department of Labor Women's Bureau Bulletin no. 66-II. Washington, D.C.: U.S. Government Printing Office, 1932.

Stinson, Helen. *Fatalities in West Virginia Coal Mines 1883–1925*. Dallas, Texas: self-published, 1985.

Trotter, Joe. *Coal, Class and Color: Blacks in Southern West Virginia 1915–1932*. Urbana and Chicago: University of Illinois Press, 1990.

Waldman, E., and B. J. McEaddy. "Where Women Work: An Analysis by Industry and Occupation." *Monthly Labor Review* 97, no. 5 (May 1974): 3.

Walshok, Mary Lindenstein. *Blue-Collar Women: Pioneers on the Male Frontier*. Garden City, N.Y.: Anchor Books, 1981.

Wilkinson, Carroll Wetzel. *Women Working in Nontraditional Fields: References and Resources 1963–1988*. Boston: G. K. Hall & Co., 1991.

## Papers and Collections

The Coal Employment Project Records. Archives of Appalachia, East Tennessee State University, Johnson City, Tennessee. 100 linear feet of chronological files, collected periodicals, videotaped news and feature stories, *Coal Mining Women Support Teams News!*, photographs, videotapes of conferences, meetings. Other records are kept at the CEP office, P.O. Box 682, Tazewell, VA 24651.

Marat Moore collection. Archives of Appalachia, East Tennessee State University, Johnson City, Tennessee. 57 audiotapes and 1 linear foot of transcripts of women coal miners. Clippings and UMWA materials on the 1989–90 Pittston coal strike.

# Index

*Note:* Entries in which the last name of the narrator is capitalized (as in FRALEY, Patsy) refer to testimony.

# The Author

Marat Moore is an award-winning writer who worked as an underground miner in West Virginia. She is the former associate editor of the *United Mine Workers Journal* and worked as an organizer during the 1989–1990 Pittston strike. She lives with her husband in Washington, D.C.

Photo from Eart Detter.